About Island Press

Since 1984, the nonprofit Island Press has been stimulating, shaping, and communicating the ideas that are essential for solving environmental problems worldwide. With more than 800 titles in print and some 40 new releases each year, we are the nation's leading publisher on environmental issues. We identify innovative thinkers and emerging trends in the environmental field. We work with world-renowned experts and authors to develop cross-disciplinary solutions to environmental challenges.

Island Press designs and implements coordinated book publication campaigns in order to communicate our critical messages in print, in person, and online using the latest technologies, programs, and the media. Our goal: to reach targeted audiences—scientists, policymakers, environmental advocates, the media, and concerned citizens—who can and will take action to protect the plants and animals that enrich our world, the ecosystems we need to survive, the water we drink, and the air we breathe.

Island Press gratefully acknowledges the support of its work by the Agua Fund, Inc., The Margaret A. Cargill Foundation, Betsy and Jesse Fink Foundation, The William and Flora Hewlett Foundation, The Kresge Foundation, The Forrest and Frances Lattner Foundation, The Andrew W. Mellon Foundation, The Curtis and Edith Munson Foundation, The Overbrook Foundation, The David and Lucile Packard Foundation, The Summit Foundation, Trust for Architectural Easements, The Winslow Foundation, and other generous donors.

The opinions expressed in this book are those of the author(s) and do not necessarily reflect the views of our donors.

SAFE PASSAGES

Safe Passages

HIGHWAYS, WILDLIFE, AND HABITAT CONNECTIVITY

EDITED BY

Jon P. Beckmann
Anthony P. Clevenger
Marcel P. Huijser
Jodi A. Hilty

ISLANDPRESS

Washington | Covelo | London

Design and typesetting by Karen Wenk

Library of Congress Cataloging-in-Publication Data

Safe passages : highways, wildlife, and habitat connectivity / edited by
Jon P. Beckmann . . . [et al.].
p. cm.
Includes bibliographical references and index.
ISBN-13: 978-1-59726-653-6 (cloth: alk. paper)
ISBN-10: 1-59726-653-1 (cloth: alk. paper)
ISBN-13: 978-1-59726-654-3 (pbk.: alk. paper)
ISBN-10: 1-59726-654-X (pbk.: alk. paper)
1. Wildlife crossings. 2. Habitat conservation. 3. Roads—Environmental aspects.
4. Animals—Effect of roads on. I. Beckmann, Jon P.
SK356.W54S34 2010
639.9'6—dc22
2010001416

Printed on recycled, acid-free paper

Manufactured in the United States of America
10 9 8 7 6 5 4 3 2 1

CONTENTS

FOREWORD

The book in your hand is amazing, a revelation. In less than a decade, road ecology has coalesced and spread widely through North America and now reaches across much of the globe. Vehicles, roads, plants, animals, and water intensively interact across the land. Exploring this relationship between our road system and our natural environment, road ecology also catalyzes a rich array of solutions for diverse transportation-and-environment challenges.

The pages ahead bulge with useful case studies and solutions for society. A range of players—academics, citizens, nonprofit organizations, and transportation professionals—play key roles and share new successes.

Road networks are essentially permanent; few roads disappear. The Inca route along the Andes and arrow-straight Roman roads in hilly terrain are well used today. Road quality changes, becoming smoother or rougher and wider or narrower. But what will use this persistent road system generations ahead? The number of walkers, bikers, cars, buses, and trucks often varies widely, not only hour by hour but decade by decade. Both the mix of users and the speed of movement fluctuate. I just walked the 2-kilometer Promenade Plantee in Paris, which was transformed "overnight" from an elevated rail line to a glorious series of outdoor vegetated rooms and passages and surprises and views for appreciative walkers above the city hubbub. Society can quickly change usage and speed on our inertia-laden road system by altering road surface and verge, and indeed the moving objects themselves.

Today's transportation-and-environment issues are partly a legacy of a road system created in large part before the rise of modern ecology, so impact mitigation and compensation are core objectives. Newer issues—a billion vehicles, coalescing streams of tailpipe emissions, permeating traffic noise, wide tarmac ribbons, runoff pollution, and fragmenting habitats—also pose giant challenges. Yet pilot solutions have emerged for each problem. Multiplied for the scale of the road system, these should produce a mammoth benefit for both nature and us.

For example, cut traffic noise with quieter road and tire surfaces. Have discrete wildlife crossing zones (e.g., 100–1000 meters only), like those for school children, targeted with a concentrated dose of approaches so drivers care, and are hesitant to speed. Reduce driving with a tight array of known techniques and novel technologies. Allocate 10 percent of ongoing maintenance and improvement transportation funds to enhance water, wildlife, and walkers. Erect lightweight wildlife overpasses with light soils and attractive drought-resistant vegetation. The list goes on.

Wildlife, especially mammals—big/little, herbivore/predator, rare/abundant, nocturnalist—are the stars of this volume. The arrangement of land uses and distinctive habitats, plus the zone of habitat degradation near roads and traffic, strongly determines whether and where diverse wildlife live as well as move. Most movements are away from a busy road, few are parallel to and near it, and some are toward the road, "a bad place," which must be crossed to reach a good place on the other side. Safe passages are solutions for the bad places. Some wildlife readily learn, adjust behavior, and even adapt. But others, the most sensitive specialist species, are stuck with a bleak future.

Mobility and accessibility seem central to people and animals. They welcome more, never less. Increasing a highway system or network increases accessibility for drivers. But this includes to remote areas, where people degrade rapidly disappearing remote habitat, an impoverishment of our planet.

More roads also lessen connectivity, both for wildlife and for nature walkers across the land. A landscape fragmented by busy roads has small wildlife populations subject to local extinction, and may be miserable to walk across. Strategically closing a low-usage road, especially a spur road, in every town and county would have a huge cumulative benefit for wildlife. When a new highway, Carretera de los Tuneles, was proposed near Barcelona, the Ministry of Transport designed it, gave the plans (by law) to the Ministry of the Environment, received the latter's conclusion that erecting barriers that subdivide the land is inappropriate, and built vegetated overpasses for local residents and wildlife to cross. Safe passages across roads address the big picture, the road system plus its surrounding landscape.

Consider roadsides, the key linkage between road and surrounding land. Woody vegetation along most roadsides would provide an enormous amount of new habitat, doubtless increasing wildlife populations far more than any loss to road-kill. More importantly, such roadside vegetation significantly narrows "the bad place," thus increasing the probability of successful wildlife movement across roads and reducing the threat of local small-population extinctions. Equally interesting, safer roads is one of the several human benefits of woody roadsides if this does not bring animals onto roads. A growing literature emphasizes that a narrower perceived width of the road ahead reduces traffic speed, thus pointing to fewer vehicle crashes and human fatalities per kilometer of road. Moreover woody roadsides sequester carbon. Narrower roads in town are pedestrian friendly "for 7-year-olds and 70-year-olds"; across the land they reduce transportation's carbon footprint. In short, new thinking on roadside design is needed.

Still, big dark clouds threaten—and opportunities beckon. Greenhouse gases make daily headlines, while both clean freshwater and easily accessible oil become globally scarcer and expensive. Our extensive road network degrades water quantity and quality across the land, from stormwater pollutants to heated ditch water, altered groundwater, and straightened streams. Transportation is arguably the leading player in the oil and greenhouse gas crunches. Society's solutions to the dark clouds should be combined with serious enhancements for water and wildlife, rather than pondered *in vacuo*. Looming clouds are opportunities—at exactly the right few-decades time scale as solutions for wildlife and nature.

Start with the two major goals applying road ecology for society: (1) improve the natural environment alongside every road segment; and (2) integrate roads with a sustainable emerald network, and with near-natural water conditions, across the landscape. Novel, as well as normal, solutions follow. For instance, instead of locating wildlife underpasses/overpasses where animals now cross roads, or try to, or crash into vehicles, install the structures where wildlife corridors between large green areas will cross roads a century ahead. That is, place safe passages to sustain the land's future emerald network.

The editors and authors of *Safe Passages* have given us a goldmine of insights, case studies, and solutions that move the bar noticeably higher. With growing support from diverse interested parties, transportation projects and plans in every jurisdiction can increasingly provide big benefits to wildlife and water and walkers across the land.

No single solution or cookbook will solve the accumulation of transportation-and-environment issues. We can now outline the theatre, and even parts of the stage—but the players will create the evolving play ahead. Leaders with new ideas, alliances, and solutions will play lead roles.

Richard T. T. Forman
Harvard University
November 6, 2009

PREFACE

As human activities continue to spread across the globe, infrastructure such as roads facilitates this expanding human footprint. New roads also inevitably lead to higher rates of human access in areas that were previously relatively more remote. Roads, both a result of the expanding footprint and a driver in human expansion, are a leading cause of habitat fragmentation and the resulting loss of connectivity throughout the world and particularly in North America. The United States alone contains approximately 6.4 million kilometers (4 million miles) of public roads, and 4.2 million kilometers (2.6 million miles) of those are paved. These roads now cover over 1 percent of the total land area within the United States.

Road ecology, a relatively new subdiscipline of ecology, centers on understanding the interactions between road systems and the natural environment. The seminal book on road ecology for North America, *Road Ecology: Science and Solutions*, was published in 2003 (Forman et al. 2003). Since that time, the field has been rapidly advancing both in terms of new research and in offering new approaches to decrease the impact of surface transportation systems on biodiversity. Shaping transportation plans and projects originally fell under the purview of transportation professionals, but increasingly many other entities are engaging from conservation organizations to other agencies and to communities.

There are no professional societies or scientific journals dedicated solely to road ecology at this time. Therefore, scientific information is scattered across a variety of disciplines: civil engineering, conservation biology, landscape ecology, wildlife management, and many others. Currently there are only two regular national or international venues for gathering and sharing information in North America. On a biennial basis, the International Conference on Ecology and Transportation allows over 300 researchers, practitioners, and policy makers to meet and share their latest findings. On an annual basis, the Transportation Research Board of the National Academies of Science hosts its annual meeting in Washington DC with more than 10,000 attendees. Recognizing the importance of this growing field, the TRB has a standing committee dedicated to ecology and transportation. However, few land managers, wildlife biologists, and conservationists attend the TRB meeting.

An emerging facet in highway–wildlife mitigation is the rapidly increasing rate of participation by the conservation community in transportation planning and projects. As the science of road ecology has evolved, so has the sophistication of environmental nongovernmental organizations (NGOs) to seek outcomes that incorporate the needs of wildlife, including habitat connectivity. These efforts have resulted in the development of a broad range of highway mitigation projects across North America. The outcomes of the partnerships between public agencies and environmental stakeholders offer insights into highway project implementation and design that minimize the impacts to wildlife, particularly as related to their ability to move across landscapes.

These collective efforts span a range of scales from site-specific wildlife crossing structures to statewide planning for habitat connectivity to national legislation. The case studies include not only the roadway design, but also how transportation systems should be evaluated in the context of a landscape. This means that such conservation efforts may include a mix of public and private land conservation in concert with transportation mitigation. Despite the relative success of these efforts, there is a surprising paucity of venues to present these new approaches in road ecology. This information is essential to share with others so that practical conservation solutions can be replicated across the continent. This book describes the ingredients of working in successful partnerships to achieve common highway mitigation goals to enhance wildlife conservation even when agencies and conservationists may have differing missions, goals, and objectives.

There are other resources on road ecology available elsewhere, and this book does not attempt to duplicate those valuable efforts. Rather, the purpose of this book is to offer a practical handbook of tools and examples that may assist individuals and organizations thinking about or engaging in reducing road–wildlife impacts, with particular focus on providing insight into habitat connectivity across highways for wildlife, both terrestrial and aquatic. We focus on highways because they have the largest ecological footprint, receive the highest use by motorists, and are generally better funded than lower-volume roads to implement mitigation solutions. In this book, we define a highway as any road that is part of the interstate and transcontinental highway system, multilane limited access freeways and expressways, and other paved road corridors that serve local areas.

The first part of the book begins with an abbreviated overview (chapter 1) of the importance of connectivity as related to roads. The three following chapters (chapters 2 through 4) review and synthesize current methods of planning approaches and technologies for mitigating the impacts of highways on both terrestrial and aquatic species. We hope that this will provide transportation and resource management professionals and other stakeholders with a common "toolbox" of potential measures for mitigating the effects of highways on both terrestrial wildlife and aquatic species. In the second part of the book the contributing authors explore the different facets of public participation in highway–wildlife connectivity mitigation projects. Public involvement has increased across North America, national legislation has changed, and there are more opportunities for nongovernmental organizations, communities, and interested individuals to become involved. This second part starts with chapter 5 on progressive planning and what a leading state did to incorporate transportation with concerns for biodiversity and habitat connectivity. Chapter 6 explores public participation from both transportation agency and public interest perspectives. Our intent is to stimulate better understanding and communication between transportation agencies and public interest groups to arrive at mutually desired highway mitigation outcomes in future transportation plans and projects.

The third part of the book provides a series of case studies from a variety of partnerships throughout North America highlighting successful implementation of ecological and engineering solutions on the ground. The case studies in chapters 7 through 13 elucidate the cooperative efforts that are emerging as a result of transportation agencies, land and wildlife management agencies, and nongovernmental organizations finding common ground. These examples illustrate varied approaches to developing partnerships, the rationale, unique circumstances that hindered or assisted implementation, the outcomes, and lessons learned from each project. Additionally, each case study chapter discusses any new standards that were developed for the participating agencies as a result of the project.

The fourth and final part of the book comprises four chapters that describe some recent innovative highway–wildlife mitigation developments. Chapter 14 reviews a project that employed citizen-based science using cutting edge Web technology. Chapter 15 describes how a local tax initiative to support wildlife connectivity including road crossing structures came into being. Chapter 16, a review of available and emerging technologies, may assist those interested in pursuing road–wildlife mitigation projects in the future. Finally, in chapter 17, the editors synthesize emerging themes and lessons from the book. They discuss recommended information needs and future directions as well as improvements of public–private partnerships.

Many of these recommendations came from break-out sessions of a highways and wildlife workshop held March 29–30, 2007, at the Western Transportation Institute at Montana State University. This workshop was cosponsored by American Wildlands, the Wildlife Conservation Society, and the Yellowstone to Yukon Conservation Initiative. Many of the contributing authors to this book attended this workshop.

This book is intended to be a resource for local, state, and national entities addressing road–wildlife mitigation issues. We expect this book to be useful to a diverse audience, ranging from transportation professionals to conservation activists, land managers, fish and wildlife agency personnel, and policy makers—brought together by a common interest in the conservation of species and their habitats by reducing the impacts of roads. The content may appeal to a broad audience of academic and practicing scientists; agency biologists; county, state, and federal planners; private, state, and federal land managers; private sector transportation consultants; graduate and advanced undergraduate students interested in road ecology; conservation organizations participating in transportation planning and projects; local, regional, and federal policy makers, and local citizens engaging in road–wildlife mitigation projects. We hope this book can be a helpful resource as more individuals and organizations focus their attention on ameliorating the impacts of roads on wildlife mortalities, landscape connectivity, aquatic passageways, and other conservation priorities.

We owe thanks to the many people who contributed to this book. These include all the chapter authors as well as the many voices of individuals who attended and contributed at the 2007 highways and wildlife workshop held in Bozeman, Montana. In particular, we thank the organizers of the workshop, Rob Ament, Neil Darlow, Josh Burnim, and Jodi Hilty, and we especially thank Rob Ament for his early leadership in conceptualizing and promoting this book and his support through the writing of this book. Thanks also to BMW—North America, Coalition for Sonoran Desert Protection, Defenders of Wildlife, I-90 Wildlife Bridges Coalition, Henry P. Kendall Foundation, Southern Rockies Ecosystem Project, Turner Foundation, Western Transportation Institute, and the Wilburforce Foundation.

We much appreciate support from The Wildlife Conservation Society and the Western Transportation Institute for support in time, space, and equipment to allow the editing of this book. We thank Island Press and in particular Barbara Dean and Erin Johnson for their wonderful help in guiding us through the publication process. We thank Kathryn Socie, Jeff Burrell, Melissa Richey, David House, and Amy Beckmann for reviewing various chapters. Last and most important, we thank our families for supporting us given the long and often late hours it took to complete this book.

Jon P. Beckmann
Anthony (Tony) P. Clevenger
Marcel P. Huijser
Jodi A. Hilty

PART I

Current Practices

Traffic and roads are strongly implicated in many of the major environmental problems we face in North America. Transportation and resource management agencies must understand the negative impacts of roads on terrestrial and aquatic resources if they are to design effective measures to alleviate these impacts. Populations of terrestrial and aquatic organisms need to be connected; this is a central priority for their conservation. Transportation and land management practitioners must devise ways to maintain or restore connections for species and habitats under both existing conditions and forecasted scenarios of growing transportation systems and climate change. Sound project plans and wise decision making require an awareness of the most current research and management practices to best ameliorate the effects of roads.

This first part of the book begins by looking at roads well beyond their right-of-way, describing their ecological effects in a landscape context. We review and synthesize the most current management concepts, technologies, and practices used by agencies today to protect terrestrial and aquatic species from impacts of roads. This "toolbox" of management practices provides a solid foundation for planning and implementing measures aimed at reducing wildlife mortality and the fragmentation effects of roads

1

on fish and wildlife populations. The tools discussed in the first part of the book are later highlighted in a series of case studies from a variety of partnerships occurring across North America. The examples in the third part of the book demonstrate the successful implementation of the ecological and engineering solutions described in this first part.

Chapter 1

Connecting Wildlife Populations in Fractured Landscapes

JON P. BECKMANN AND JODI A. HILTY

I see . . . an America where a mighty network of highways spreads across
our country.

Dwight D. Eisenhower (www.fhwa.dot.gov)

President Eisenhower is credited with leading the creation of the national
system of interstate highways across the United States in the 1950s, the primary basis of today's commerce and travel. The system was originally conceived by Eisenhower as a necessary part of defending the country and was
called the national highway defense system. He simultaneously touted the
importance of the interstate system for communication and the transport of
goods. Since its inception, this highway system has continued to grow,
with states adding to the network as well as broadening already existing
roadways. As of 2008, there were more than 74,000 kilometers (46,000
miles) of roads in the interstate highway network, making it both the
largest highway system in the world and the largest public works project in
history (U.S. Department of Transportation 2008). During most of the
twentieth century, road construction focused on the ease of terrain, soil,
and other location considerations along with logistics and cost. It was not
until late in the twentieth century that projects started to consider the needs
of wildlife in relation to road construction and maintenance (e.g., Forman

et al. 2003). As a result, many roadways pass through what were once the best habitats for numerous species, such as in valley bottoms and along streams or rivers.

Today, there are more vehicles on the roads than in the 1950s, and both the technology of roads and automobile design enable vehicles to move faster than in the past. At the same time, many species of wildlife have rebounded in number, such as elk, deer, and other ungulates. Increasing cars and car speeds and increasing numbers of ungulates—the group of animals most commonly reported in accidents—have resulted in more animal–vehicle collisions. These wildlife–vehicle collisions are probably the main reason that society has been increasingly interested in the need to address wildlife–vehicle collisions and the adverse effects roads have on wildlife populations (Trombulak and Frissell 2000, Forman et al. 2003). In 2007, 1 to 2 million traffic accidents involving large mammals in the United States caused an estimated US$8.3 billion in vehicle damage (Huijser et al. 2007). In some U.S. states, six to eight cents of every insurance dollar goes toward paying for wildlife-related claims (Lowy 2001). Beyond vehicular damage, an astounding 26,000 people are injured or killed in the United States each year due to vehicle collisions with wildlife (Huijser et al. 2007). According to the Insurance Corporation of British Columbia, the cost to society of a human fatality is C$4.17 million, while an injury costs approximately US$97,000 (Sielecki 2000).

Europeans were first to recognize and start resolving the problem of wildlife–vehicle collisions by commissioning studies of the problem, building wildlife crossings, and reducing traffic speeds to reduce vehicle collisions. In the 1980s, North America saw an increase in interest in studies and construction of wildlife crossings such as those in Florida that are discussed in chapter 10 (Forman et al. 2003). In 2003 the seminal book *Road Ecology: Science and Solutions* (Forman et al. 2003) formalized this new subdiscipline of ecology. Today, increasing research, general media, and visible projects continue to move the nascent field of road ecology forward. This means that now is the time to bring emerging science, policy, and innovation into standard transportation planning, design, and construction.

Creating projects to minimize or reverse the negative impacts of roads on wildlife, however, is challenging. Given funding constraints and design limitations, it is unlikely that a project will benefit all possible species; therefore, species must be prioritized. Where new roads are being constructed, there are increasing opportunities to consider wildlife early and throughout the decision-making process. Unfortunately, given that many of the mitigation projects for wildlife are for existing roads, most efforts require the daunting task of redesigning roadways to restore connectivity.

This chapter offers a brief review of why roads can be a challenge for wildlife in general; how roads impair connectivity for wildlife populations, including the broader toll of roads on wildlife; and the importance of connectivity. For those interested, many of these concepts are covered in depth elsewhere (e.g., Forman et al. 2003, Hilty et al. 2006), but an overview of these challenges is important in framing the case studies and opportunities presented throughout this book.

The Problem of Roads for Wildlife

Human activities impact much of the world and continue to expand (Sanderson et al. 2002). Rapid population growth, an increase in extractive industries, uncontrolled and unplanned development, and new transportation infrastructure are threatening many of our natural resources and the persistence of wildlife populations (e.g., Western Governor's Association 2008). Many human-made linear infrastructures such as railroads, power lines, and petroleum pipelines intensify habitat degradation (Primack 2006), but roads have the most widespread and lasting impacts (Spellerberg 1998, Davenport and Davenport 2006). Roads serve as the arteries of this ever-expanding human footprint. Networks of roads also inevitably lead to increased human access to areas that were once more remote and undisturbed. Increased access and corresponding impacts can have potentially negative consequences for wildlife. Human activities ranging from logging and petroleum drilling to hiking, camping, and illegal poaching can negatively impact wildlife, and as road networks increase these activities also increase across a much larger expanse of the globe (e.g., Dyer et al. 2002, Roever et al. 2008). This is particularly true when recreational vehicles, such as ATVs and snowmobiles, even further expand human activities into previously remote areas. Roads are also a leading cause of habitat fragmentation and the resulting loss of connectivity for wildlife populations throughout the world, including North America.

The United States alone contains approximately 6.4 million kilometers (4 million miles) of roads, with 4.2 million kilometers (2.6 million miles) of those being paved (Forman et al. 2003, U.S. Federal Highway Administration 2008). Roads now cover more than 1 percent of the total land area within the United States, and roads influence the ecology of at least one-fifth of the land area of the entire country (Forman 1999, Cerulean 2002). In the United States alone, 4.8 million hectares (11.9 million acres) of land have been directly lost due to road construction (Trombulak and Frissell 2000). This means that roads alter ecosystem processes and species distributions. They can also serve as a vector to introduce new, invasive species into adjacent habitats.

The amount of use that roads receive generally corresponds to the level of effects on wildlife. In 1999, approximately 90 percent of all trips taken by Americans were made in vehicles (Turrentine et al. 2001), which equates to more than 200 million vehicles driving 8 trillion kilometers (5 trillion miles) in the United States (Ritters and Wickham 2003). Of all the miles of roads, highways such as the interstate systems have the largest ecological footprint and receive the most use by motorists. Such high-use roads often create the most significant barriers to connectivity between wildlife populations (Transportation Research Board 2002a). Further, traffic speed, as discussed in later chapters, impacts wildlife in that wildlife–vehicle collisions are more likely to occur where automobiles are moving faster, such as on highways.

While the field of road ecology has expanded in recent years to document the consequences of roads on wildlife, relatively little information is available about how wildlife species navigate lands bisected by roadways and how they cross roads (Transportation Research Board 2002, Forman et al. 2003). Such gaps in knowledge impede prudent management and conservation. With increasing awareness and knowledge of transportation's impacts on wildlife, wildlife and land management agencies have the opportunity to make more informed decisions about where and how roads are designed and retrofitted to better accommodate wildlife needs.

Toll of Roads on Wildlife

The development, maintenance, and ongoing use of roads have profound impacts on the world's biodiversity from amphibians and ungulates to birds and even vegetation (e.g., Forman and Alexander 1998). Roads and associated traffic can impact wildlife populations in four ways: (1) decrease habitat amount and quality; (2) increase mortality from collisions with vehicles; (3) limit access to resources; and (4) fragment wildlife populations into smaller and more vulnerable subpopulations (Jaeger et al. 2005).

Decreased Wildlife Habitat Amount and Quality

The creation and expansion of roads decreases existing natural habitat and can lower the quality of remaining habitat adjacent to roads. The area impacted includes the lanes of road and also the area of vegetation that is maintained alongside the road, which can extend anywhere from a meter to 10 or more meters (32.8 feet) away from the edge of the road. The road and associated vegetation management often create an abrupt edge in once continuous habitat, such as creating a clearing in once continuous forest habitat. Physical and biological effects occur at such edge habitats. The edge climate may be warmer and drier, for example, and this can lead to changes in species composition. Depending on the ecosystem type and species, these so-called edge effects can permeate hundreds of meters into adjacent habitat (Reijnen et al. 1995, 1996). Species that are sensitive to edge habitat, especially forest interior species, decrease in density and/or may be less likely to survive due to competition with exotic species, edge predators, and overall poor habitat quality (Laurance 2004, Laurance et al. 2004, Bollinger and Gavin 2004).

In addition to direct habitat loss, road construction can also contribute to indirect habitat loss. Indirect habitat loss occurs where species no longer occupy otherwise sufficient habitat that is adjacent to roads because of behavioral responses to roads (e.g., Berger 2007). For example, researchers have documented that grizzly bears (*Ursus arctos horribilis*) are less likely to occupy regions with higher road densities (Mattson 1992)

Roads can also be a source of pollution, degrading adjacent core habitat areas. In California, species dependent on nitrogen-poor serpentine soil are negatively affected by car pollution that deposits nitrogen. The introduced nitrogen enables generalist species to out-compete serpentine soil specialists (Weiss 1999). Light and noise are other sources of pollution spilling into nearby habitats. The combination of lighting, noise, runoff of pollutants, and high human activity inhibits the occurrence of big and small species in adjacent habitats (Jaeger et al. 2005). Grizzly bears in the Canadian Rocky Mountains, for example, tend to avoid areas with high vehicle use, traffic noise, and human disturbance (Gibeau et al. 2002, Chruszcz et al. 2003).

Impacts of Vehicle Collisions on Wildlife

Mortality from collisions with vehicles does not appear to pose a significant threat to robust populations, but road mortality can be devastating to small or dwindling populations (Bennett 1991). For example, vehicle–bird collisions are a major source of mortality for endangered Florida scrub jays (*Aphelocoma coerulescens*) such that road-related mortality is a significant threat to their population (Mumme et al. 2000). In addition road mortality has affected populations of a number of iconic species in North America, including Florida panther (*Puma concolor coryi*, Maehr et al. 1991), cougars (Beier 1995), pronghorn (*Antilocapra americana*, Berger et al. 2007), grizzly bears (Gunther et al. 2004), black bears (*Ursus americanus*, Brody and Pelton 1989, Beckmann and Berger 2003, Hebblewhite et al. 2003, Beckmann and Lackey 2004), and a variety of more common species such as white-tailed deer (*Odocoileus virginianus*, Widenmaier and Fahrig 2006), raccoons (*Procyon lotor*, Clark et al. 1989), and red fox (*Vulpes vulpes*, Hardy et al. 2006). With an estimated 1 million vertebrates killed every day on roads in the United States, death tolls on wildlife are astounding (Lowy 2001). A five-state study concluded that 15,000 reptiles and amphibians, 48,000 mammals, and 77,000 birds die each month due to collisions with vehicles (Havlick 2004).

The presence of roads can be advantageous for some species. For example, some moose (*Alces alces*) in Grand Teton National Park and surrounding areas appear to benefit from roads. Female moose have begun to realize that humans can be a shield of sorts against one of their major predators, the grizzly bear. Some female moose give birth closer to roads usually avoided by grizzly bears, a behavior that is thought to be a strategy to increase calf survival (Berger 2007). Other species that benefit from roads are scavengers. Increased levels of carrion due to collisions with vehicles have not only benefited ravens (*Corvus corax*), but other scavengers such as crows (*Corvus brachyrhynchos*), coyotes (*Canis latrans*), and turkey vultures (*Cathartes aura*), to name a few (e.g., Knight et al. 1995). In rare cases, vegetation associated with road right-of-ways may be important to conservation of wildlife, especially in landscapes in which road-associated habitats may be the only natural or seminatural habitat in an otherwise intensely altered landscape (e.g., Huijser and Clevenger 2006). For most species of wildlife, roads are an additional hazard that they must negotiate in an increasingly humanized world.

Roads Limit Access to Resources

Many species require access to different habitats throughout their life cycle or even throughout the year. Species such as salamanders may require ponds or rivers for one part of their life cycle and live another part of their life in terrestrial habitat. Roads that run adjacent to rivers or that circle ponds and lakes either limit movement altogether or can be a major source of mortality for such populations. Other species such as ungulates move from summer to winter ranges. In particularly harsh winters, they may need to move even further to find scarce resources. Roads can impede these migrations. For example, Interstate 80 in southern Wyoming can be a complete barrier to pronghorn migration when animals are trying to move further south to avoid deep snow during severe winters (Berger et al. 2007). In fact, several hundred pronghorn died as a result of their inability to move south of the Interstate 80 barrier during harsh winters in the 1970s and 1980s (Johnson 1988).

Fragmenting Wildlife Populations into Smaller and More Vulnerable Subpopulations

Road creation can be a direct source of habitat fragmentation and loss, bisecting a continuous population and creating two or more less-connected subpopulations (Spellerberg 1998, Epps et al. 2005, McRae and Beier 2007). Many organisms have evolved in naturally discontinuous habitats. However, roads, like many other human-altered environments, have changed the parameters of habitat configuration on the landscape by massively expanding levels of habitat fragmentation. For example, roads potentially contribute more to fragmentation of forest habitats than certain arrays of clearcuts (Reed et al. 1996).

For many populations of rare, low-density, and/or wide-ranging species of wildlife the single greatest threat to long-term persistence is the continued fragmentation of their habitat that isolates small populations (Hilty et al. 2006). This is particularly true for those species that require (1) large expanses of land for their daily, seasonal, or annual ecological needs (i.e., large home range); (2) migratory movements between seasonal habitats for either food or mates; and/or (3) dispersal to connect isolated (naturally or otherwise) subpopulations both genetically and demographically. One group of species most vulnerable to habitat fragmentation includes large carnivores that require a large amount of area to maintain a viable population. The fragmentation of these naturally low density populations into smaller subpopulations can lead to the smaller, more isolated subpopulations going extinct (Weaver et al. 1996, Wydeven et al. 2001, McRae et al. 2005).

For wide-ranging carnivores such as wolverines (*Gulo gulo*), grizzly bears, and black bears that roam the Greater Yellowstone Ecosystem, continued expansion of human activities—primarily road construction and associated urban and rural sprawl—may compromise these species' long-term ability to survive in the region. This fact is exemplified by data collected by the Wildlife Conservation Society. In an ongoing study in the Greater Yellowstone Ecosystem, researchers found several wolverines killed by cars; a death toll that is significant since the population of wolverines in the Greater Yellowstone Ecosystem is probably fewer than 100 individuals (R. Inman, pers. comm.). This wolverine study also documented a male wolverine dispersing over 500 miles from Grand Teton National Park in northwest Wyoming south into northern Colorado during April and May 2009 (fig. 1.1). Such observations demonstrate the vulnerability of such species to roads and that managing for connectivity must be done at a continent-wide scale for such carnivores and likely for other taxa as well.

The tolls of roads on wildlife are further exemplified in the struggle to maintain a viable population of grizzly bears in the U.S. northern Rocky Mountains. While an occasional road-kill may remind drivers of wildlife

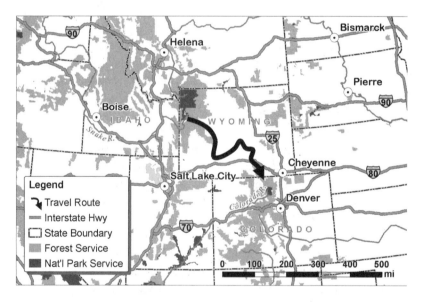

FIGURE 1.1. This map depicts the travel route of a dispersing male wolverine tracked by Wildlife Conservation Society biologists. M56 was captured near Grand Teton National Park and traveled approximately 500 miles during April and May 2009 into northern Colorado. Because wolverines exist at naturally low densities, maintaining viable wolverine populations in the U.S. contiguous states requires ensuring that the landscape remains permeable to enable such movements. (Permission to use by Robert Inman)

that roam the region, it is nearly impossible for the casual driver to see the road from the perspective of wildlife. Every day about 3,000 vehicles zoom across western Montana on Interstate 90, and that number reaches approximately 7,000 vehicles per day between Missoula, Montana, and the Idaho border (Montana Department of Transportation 2008). Grizzly bear movement studies indicate that Interstate 90 may largely be a barrier for grizzlies (Chris Servheen, United States Fish and Wildlife Service, pers. comm.). Grizzly bears were listed as a federally endangered species in the United States under the Endangered Species Act in the 1970s. Their long-term well-being hinges on a couple of key factors, including maintaining good core habitat areas and ensuring that these core habitat areas stay connected in a manner that keeps bears and humans largely separated. For these bears to have the best opportunity of surviving where they currently occur, many scientists suggest that subpopulations in the northern U.S. Rockies should be connected and not be completely isolated from one another. If so, connecting grizzly bear populations across Interstate 90 will be necessary.

A further example of the negative impacts of roads on wildlife populations is that of black bears in the Lake Tahoe Basin in the northern Sierra Nevada in western Nevada. In this region, the single largest source of mortality for both adult black bears and cubs is collisions with vehicles. The number of bears killed by vehicles annually has increased more than 20-fold over a decade and a half (Beckmann and Berger 2003, Beckmann and Lackey 2008). During the late 1980s, on average less than 0.5 bear/year was killed by vehicles (Goodrich 1990, 1993). In contrast, during the twelve years from 1997 to 2009, 129 bears (an average of 10.75 bears/year) were killed by motor vehicles. This is a striking number given that Nevada's bear population is estimated at less than 300 bears in the entire state. The high levels of bear mortality, particularly from roads in the northern Sierra Nevada, are currently exceeding birth rates of bears in these urban regions (Beckmann and Lackey 2008). As a result, roads are creating sink areas (i.e., regions where mortality rates exceed recruitment rates in a population) for bears in these urban centers. If connectivity to adjacent habitat in California is severed by roads or other unnatural barriers, Nevada could lose its bear population (see Beckmann and Lackey 2008). There exists opportunity to inform where and how development, including the road network, expands, and some designs could significantly benefit such species through maintaining connectivity and decreasing potential human–wildlife conflict. Examples of such successes are provided in the following chapters of this book.

Why Connectivity Is Important

Roads are one of the major causes of habitat fragmentation that are disconnecting once continuous habitat. We know connectivity is important, but what is it exactly? Connectivity is arguably one of the very core tenets of conservation. We offer a few definitions to help clarify the expanse of terms often used interchangeably in the context of connectivity. First, for the purposes of this book, we define connectivity as a measure of the extent to which plants and animals can move between habitat patches. Further, connectivity includes highway crossing structures for wildlife, corridors, greenbelts, linkages, ecological networks, and other elements that are the potential means for achieving connectivity (Hilty et al. 2006). One of the challenges of maintaining and restoring connectivity is that it can occur at many scales. Some organisms, such as the wolverine (see earlier example), need connectivity at continental scales, while others, such as elk (*Cervus elaphus*) and bighorn sheep (*Ovis canadensis*) may require site-specific corridors, such as to cross a particular road safely. The reality is that regardless of the scale one chooses, much of the practice of maintaining and restoring connectivity occurs at smaller, localized scales such as sections of road where animal movement occurs on a regular basis.

The concept of connecting fragmented populations has largely stemmed from two bodies of ecological theory that have contributed much to landscape ecology and conservation biology. Island biogeography and metapopulation theory both emphasize the importance of connectivity. The theory of island biogeography originated to describe species richness and composition on islands of varying sizes and varying distances from the mainland (MacArthur and Wilson 1967). In recent years, island biogeography has been applied to terrestrial landscapes where suitable habitat patches represent islands and poor habitat is equated with the sea or an uninhabitable void. Strongly isolated and small habitat patches, according to island biogeographic theory, are likely to contain fewer species than patches that are larger and have higher connectivity (Dunning et al. 1995). Similarly, habitat fragmentation will lead to a loss of species in smaller, more isolated patches that cannot maintain as many species (MacArthur and Wilson 1967). Based on this theory, many studies have linked fragmentation to a number of deleterious environmental consequences (e.g., Tewksbury et al. 2002a, Damschen et al. 2006). In fact, a number of different field studies have demonstrated that we are losing species from regions such as national parks and protected areas, and these studies correlate species loss to variables such as the isolation of parks from other parks and increasing levels of human development in surrounding areas (e.g., Newmark 1995, Parks and Harcourt 2002). The biological effects of fragmentation can vary from a decline in species requiring large amounts of connected habitat to virtual replacement of native biota by exotic species (Crooks and Soulé 1999).

The application of island biogeography theory to terrestrial systems has been largely subsumed by metapopulation theory where the mainland (source) and island (recipient) situation constitute one scenario (McCullough 1996). Metapopulation theory seeks to explain how different levels of connectivity in an array of local populations (demes) affect the long-term persistence of the entire population of a species. The metapopulation is the sum of its constituent spatially separated demes. Increased connectivity should increase persistence among the demes and thus the metapopulation such as by providing access to necessary resources, recolonization of patches where populations have gone extinct, and maintenance of genetic diversity.

Connectivity, based in these above described theories, is a key component of conserving biodiversity for several reasons. Connected populations generally have a higher likelihood of surviving. Connectivity between populations also provides greater flexibility for a species to respond to changing environmental conditions, such as climate change, compared to populations that exist in isolated patches. Field research and modeling exercises have shown that well-connected populations are more resilient to environmental changes and natural disturbances, such as drought or fire (Tewksbury et al. 2002b, Damschen et al. 2006).

Connectivity can help maintain genetic diversity, another variable that helps population persistence. Low levels of genetic connectivity and interchange among populations can in some instances lead to inbreeding within the reduced and isolated populations. Inbreeding can cause the entire population to become more susceptible to disease or other environmental stressors, thus having demographic effects by lowering reproductive rates and increasing mortality rates. These demographic factors can potentially cause populations to vanish or become locally extinct. If small subpopulations within a larger population are able to maintain connectivity through dispersal and other processes, these harmful inbreeding effects can be reduced or even eliminated (Pannell and Charlesworth 2000, Hanski and Gaggiotti 2004, Hilty et al. 2006).

Subpopulations with some level of connectivity to other subpopulations also mean a larger total population size and overall range of a given species, and their connectivity enables breeding between subpopulations. Such connectivity provides opportunity for recolonization of ranges when subpopulations go extinct (Hanski 1999). Examples of species that exist naturally as smaller subpopulations that are often linked by dispersal are mountain goats (*Oreamnos americanus*), bighorn sheep, marmots (*Marmota* spp.), and pika (*Ochotona princeps*). These species are typically found in alpine habitat that is patchily distributed, an archipelago of high elevation islands creating natural subpopulations that are somewhat connected, or a metapopulation scenario. Many of these species make seasonal altitudinal movements that may be affected by unnatural barriers such as roads, which can influence gene flow, connectivity among populations, and ultimately survival of the metapopulation (Epps et al. 2005, Rice 2008).

Maintaining continuous habitats can also buffer communities from changes. Smaller populations and habitat patches can be more susceptible to the introduction of new diseases that affect wildlife and exotic plant species. These exotic plants can alter resource availability for wildlife through competition with native plants and/or through altered system processes such as the frequency of fire (Tyser and Worley 1992, Lonsdale and Lane 1994, Hansen and Clevenger 2005). Roads bisecting once continuous habitat are known vectors of non-native, invasive animal species and disease. Roads can also alter the spatial proportion of a given patch of habitat such that the edge-to-area ratio changes (Reed et al. 1996). The results of these changes can be dramatic, particularly for species that are more vulnerable to predation at the edge of habitats, such as is the case with many ground nesting birds that are preyed on by predators (Delgado et al. 2001, Niehaus et al. 2003). Similarly, an increasing edge-to-area ratio in habitats along roads can result in some passerine birds being more susceptible to nest parasitizers such as brown-headed cowbirds (*Molothrus ater*; Niemuth and Boyce 1997, Ortega and Capen 1999).

The need for connectivity is intensified as we recognize and document not only the effects of human-induced habitat fragmentation but that of climate change. We know that historical climate change altered species occurrence and distribution across the globe. This means that climate change today, which is occurring more rapidly than at any time in documented history, is likely to lead to shifts in where species occur. In fact, such shifts are already being documented around the world (Parmesan and Yohe 2003, IPCC 2007). Some of the most commonly recommended tools to help species adapt to changing climates are to enhance and maintain connectivity, and maintain and expand protected areas of continuous intact habitat (Thomas et al. 2004, Heller and Zavaleta 2009). Populations of species will likely need the flexibility to move between various locales in order to find food, shelter, mates, and other resources during this time of climate change (e.g., Koteen 2002). Increasing resiliency through connectivity is therefore even more important today because of the changing climate.

Maintaining or Restoring Connectivity: Factors to Consider

Any project that is looking to maintain or restore connectivity including transportation-related projects should consider a number of variables. First, an appropriate planning team should be assembled and clear goals including the identification of focal species should be made explicit. It is unlikely that a designed crossing system will serve all species in the region, so those focal species that most need restoration of connectivity should be identified. The decision of focal species is ideally made by using existing science or science expertise in the region. Understanding which species are most likely to benefit from connectivity and the type of mitigation options available will be necessary. Identifying the needs of specific focal species drives the type of mitigation or planning needed to maintain or restore connectivity. For a hypothetical example, grizzly bears may utilize overpasses whereas cougars may be more likely to use underpasses. Also, clarity on what core habitats will be connected through the project is important. Identifying pathways between core habitats that are secure from human development and activities that may alter their connectivity value is important. Second, appropriate data sets should be assembled and evaluated because in most cases additional site specific data collection is necessary to inform the best project design. This step should also explicitly identify what areas are presently barriers to focal species, such as developed areas or human disturbance, and also consider how the species' movements may be affected by future scenarios of changing land use and human activity. Understanding current and future obstacles to wildlife movement ensures that efforts do not inadvertently maintain or create dead-end corridors. For example, if an urban subdivision is created some years later next to a wildlife crossing structure, the conservation efforts and transportation investments will be nullified. Well designed corridors should keep human-wildlife conflicts to a minimum by directing wildlife away from areas of human activities. Once these types of information have been collected and options for protecting connectivity are highlighted, cost-benefit analyses should be conducted. The concept behind a cost-benefit analysis is to aim to achieve the highest level of ecological connectivity while minimizing the economic cost and maximizing human and wildlife safety. Also any project should engage local communities from the beginning, and social acceptance from surrounding communities must be weighed throughout the decision-making process. Local support for a project, as several examples in this book point out, often determines a project's success (Hilty et al. 2006).

Conclusions

Today, conservationists, scientists, communities, and departments of transportation are increasingly aware that highways have negative impacts on wildlife through both direct mortality and the disruption of population connectivity. Connectivity is an important tool for maintaining wildlife populations in many situations. This means that planning for connectivity of focal wildlife populations both in retrofitting and in designing new roads should be strongly considered. Research and conservation efforts are demonstrating mechanisms to decrease road-kill and increase connectivity for different species and in different ecosystems. To address the issue of connectivity as related to existing or proposed roads requires looking systematically at the problem, defining a goal, and establishing a team of interested parties to design a project that will address the issue. As the science of road ecology has evolved, so has the sophistication of various stakeholders to seek outcomes that incorporate the needs of wildlife. These efforts have resulted in the development of a broad range of highway mitigation projects across North America. The wildlife conservation issues coupled with the substantial human and economic costs resulting from wildlife–vehicle collisions have caused scientists, engineers, and transportation practitioners to consider a number of mitigation tools for reducing the conflict between roads and wildlife and restoring or maintaining connectivity of wildlife populations. The following chapters will discuss some of these tools and partnerships.

Chapter 2

Wildlife Crossing Structures, Fencing, and Other Highway Design Considerations

ANTHONY P. CLEVENGER AND ADAM T. FORD

The linear nature of surface transportation systems creates a suite of concerns for transportation professionals as they seek to ameliorate the impacts of their projects on environmental resources. Roadways and railroads slice through a series of habitats and hydrological features, impacting wildlife and motorist safety, aquatic resources, habitat connectivity, and many other environmental values (Forman and Alexander 1998).

The effects of roads on wildlife populations have been the focus of many studies in recent years. Roads affect populations in numerous ways, from habitat loss and fragmentation to constituting physical barriers to animal movement and traffic-related mortality. Calls for new solutions to these problems are increasingly heard from the public, environmental scientists, the transportation community, and decision makers (Transportation Research Board 2002a).

Wildlife crossing structures are being designed and incorporated into road construction and expansion projects to help restore or maintain animal movements across roads (Spellerberg 2002, Forman et al. 2003). Engineered wildlife crossings are designed both to allow animals to cross roads and to reduce hazards to motorists and wildlife. Crossing structures are typically combined with high fencing, and together these measures have proved to reduce road-related mortality of wildlife and connect populations

(Foster and Humphrey 1995, Clevenger et al. 2001a, Huijser et al. 2007). An increasing number of crossings have been built in North America and worldwide in the last decade (McGuire and Morrall 2000, Davenport and Davenport 2006, Bissonette 2007). Anticipated population growth and ongoing transportation infrastructure investments in most regions, coupled with the resounding concern for maintaining large-scale landscape connectivity, are generating interest in wildlife crossings as conservation tools. The role of wildlife crossing structures and landscape corridors allowing wildlife populations to adapt to changing climate scenarios is gaining increasing attention (Heller and Zavaleta 2009).

As road networks continue to grow and expand throughout North America, transportation agencies, land managers, and local decision makers need to know the most effective approaches to designing safe roadways for motorists and wildlife. In this chapter we review and synthesize current knowledge of North American wildlife crossing systems as related to design, monitoring, and performance criteria. The chapter provides information on how to increase the effectiveness of established designs and recommends ways to design for particular species and species groups in different landscapes. These guidelines can be used for wildlife crossings on new or existing roads, highway expansions (e.g., two-lane to four-lane) and bridge reconstruction projects. The review is not meant be exhaustive but captures the most current literature, knowledge, and data with regard to the current practices in wildlife crossing mitigation.

Planning

The following section provides a general explanation of the planning process and how wildlife habitat connectivity needs can be incorporated into transportation projects.

Avoidance and Minimization of Impacts

Before initiating project planning for wildlife habitat connectivity, the first step is to determine whether impacts can be avoided or minimized. Avoidance eliminates impacts to wildlife habitat, whereas minimization reduces the impact on wildlife to levels that can often avoid or reduce the need for mitigation. For example, can alternative routes for highway expansion be found in areas that do not contain endangered species habitat? Development of a transportation system without adequate considerations for avoidance and minimization can result in long, drawn-out negotiations with regulatory agencies. Transportation agencies feel that the demonstration of avoidance and minimization in transportation planning is consistent with the mandated streamlining provision of the most recent U.S. transportation bill, a Safe, Accountable, Flexible, Efficient Transportation Equity Act: A Legacy for Users (SAFETEA-LU).

The U.S. Federal Highway Administration and state transportation agencies have recognized the need to consider environmental issues early in the planning process (Brown 2006). The success of streamlining the environmental process during project development and environmental studies phase has relied on getting environmental considerations into the process at the systems planning stage so that coordination leads to resolution prior to the project development and environment stage. System planning is the time when connectivity for wildlife should be evaluated. At the systems level, the cumulative effect of roads on the landscape can be evaluated and measures can be taken to demonstrate avoidance and minimization of environmental impacts to important ecological areas. These considerations then can be applied to the entire system rather than individual projects. Critical habitat linkages can be identified and prioritized at this stage of planning and then coordinated with the planning of future projects.

Scaled Habitat Connectivity Planning

Systems-level and project-level approaches are two different scales of habitat connectivity planning and means of incorporating measures to reduce the effects of roads on wildlife populations. Project-based approaches are most common with transportation agencies, although systems-level approaches encompassing entire states and provinces have become more common in the last few years (Smith 1999, Singleton et al. 2002, Kintsch 2006).

Project Level

Mitigating roads for wildlife conservation is most economical during road expansion or upgrade projects. Thus funding for road mitigation measures such as wildlife crossing structures is most likely to originate from specific transportation projects that address multiple transportation management concerns, one of which may reduce vehicle collisions with wildlife and provide safe passage across busy roadways. This project-based approach is concerned with proximate objectives (i.e., those within the transportation corridor and occasionally lands adjacent to it). A project-level focus may not necessarily consider how the wildlife crossing structure(s) fit into the larger landscape and regional wildlife corridor network. Wildlife crossings should not lead to ecological "dead-ends" or "cul-de-sacs," where wildlife have nowhere to go. Crossings must link to a larger functional landscape and habitat complex that allows wildlife to disperse, move freely, and meet their daily and life requisites. This not only requires large spatial-scale considerations but should also incorporate future (or projected) land-use change into the planning process.

Systems Level or Landscape Level

Wildlife crossings may also emerge from a systems-level analysis of transportation management concerns and priorities over a much larger area than transportation corridor projects. Rather than seeking to place a specific crossing structure (± 1 mile), the systems perspective identifies which stretches of highway should require mitigation (± 10–100 miles) and how intensive the mitigation should be. Key wildlife crossing areas may also be identified from a regional landscape assessment of wildlife connectivity needs around a state-/provincial-wide road system or regional transportation corridor. This landscape-focused approach can be viewed as the inverse of the project-level, or corridor-focused approach. With the right information it is possible to identify key habitat linkages or zones of important connectivity for wildlife that are bisected by transportation corridors. Linkages and potential wildlife crossing locations can be prioritized based on future transportation investments, scheduling, and ecological criteria. This helps to strategically plan mitigation schemes at a regional or ecosystem level.

This landscape-level approach, which is institutionalized in most of Europe, is gaining appeal with North American transportation agencies. In the United States, the overlay of two state agency maps—Statewide Transportation Improvement Program plans with comprehensive wildlife conservation plans from natural resources agencies (Smith 1999, White 2007)—facilitate the integration and coordination of spatially explicit transportation and wildlife habitat conservation plans at the state level. A recent policy by the Western Governors' Association to "protect wildlife migration corridors and crucial wildlife habitat in the West" sets a management directive to coordinate habitat protection and land use management for wildlife across jurisdictional boundaries (Western Governors' Association 2008). Of particular note was the section of the report produced by the Transportation Infrastructure Working Group, which makes detailed recommendations on ways to integrate future transportation planning with wildlife habitat conservation at the systems level.

There are substantial benefits from the systems-level analysis. By establishing a formal, broad-scaled planning process, it is possible to readily address stakeholder concerns, prioritize agency objectives, and incorporate landscape patterns and processes into the planning and construction process. It also helps ensure that well-meaning project-level efforts take into consideration the larger ecological network in the surrounding region. This results in more streamlined projects that save transportation agencies money over the long term.

Planning Resources

Identifying where to locate wildlife crossing structures requires adequate tools and resources to identify the most suitable sites at the project and systems levels. The following sections describe many of the resources that can help define the important wildlife habitat linkages across roads and important areas for road mitigation.

Maps and Data

Many resources are available today that facilitate the identification of wildlife habitat linkages and movement corridors (Chetkiewicz et al. 2006). Most electronic resources are geographic information system (GIS) based, are readily available from government or nongovernmental agencies, and can be downloaded from their respective Internet sites, for example, state/provincial or national geospatial data clearinghouses (Transportation Research Board 2004). Some basic map and data resources for planning wildlife connectivity and crossing mitigation include the following:

- Aerial photos
- Land cover vegetation maps
- Topographic maps
- Landownership maps
- Wildlife habitat maps
- Wildlife movement model data

- Field research data
- Road-kill data
- Road network data

Table 2.1 describes each resource and how it can be used for project-level and systems-level planning of wildlife habitat connectivity and highway mitigation.

Use of these resources in combination with road network and traffic data is an ideal place to start identifying the intersections of high probability habitat linkages and roads. Combining multiple resources will provide greater accuracy in identifying habitat linkages and finalizing site selection for wildlife crossing structures. Most of the resources listed in table 2.1 work best at the more localized, project level; however, some can be used or adapted for larger, systems-level assessments.

Geographic Information System Layers

GIS analysis is widely used in transportation and natural resources management today (Transportation Research Board 2004). Analyses can be done in multiple spatial scales ranging from the project scale to larger landscapes and regions (Beier et al. 2008). Many of the map and data resources listed earlier are available in digital format and can be overlaid and analyzed in ArcView/ GIS or ArcMap. Basic GIS layers useful for identifying habitat linkages and sighting wildlife crossings at the corridor level include the following:

- Digital elevation model (DEM; characterizes topography, preferably ≤ 30 m resolution)
- Water or hydrology (includes all lakes, ponds, rivers, streams)
- Vegetation or ecological land classification system (general habitat types)
- Wildlife habitat suitability (species-specific habitat map)
- Built areas (areas of human development and activity)
- Roads (network of all paved and unpaved roads)

How to Site Wildlife Crossings

Generally systems-level habitat linkage assessments are not suitable for identifying specific locations for wildlife crossings due to differences in

TABLE 2.1. Data layers and maps for planning wildlife connectivity and crossing mitigation

Map/Data Type	Project Level	Landscape Level
Aerial photos	Photos can be used to help identify vegetation types and human developments. Photos come in many scales and image formats (ortho-photos, color infrared, black and white). Some images are high resolution (5 mb). Readily available from local and state/provincial government agencies.	Typically not practical to use for large landscape-scale assessments of linkage zones. Landsat TM satellite imagery and other remotely sensed imagery are good substitutes for working at a state/provincial scale. Satellite imagery should be available at most local and state/provincial government agencies.
Land cover vegetation maps	These maps help identify general vegetation types such as deciduous versus coniferous forests, shrublands, grassland/marshes, rock, and ice. Land cover maps are more general and include physical (built areas) and biological information. Readily available from local and state/provincial government agencies and their Web sites.	Maps are available for large-scale habitat and corridor network planning. The scale is much larger and resolution lower, nonetheless important resource to use in large-scale planning endeavors. Readily available from local and state/provincial government agencies and their Web sites.
Topographic map	Information on slopes, ridgelines, valley bottoms, drainages, and other main topographic features are valuable for identifying wildlife habitat corridors. Roads, powerlines and other human developments are usually found on these maps. Readily available from local and state/provincial government agencies.	Like land cover vegetation maps above, topo maps are available for state-/provincial-wide mapping exercises; however, resolution is lower quality. Readily available from local and state/provincial government agencies.

TABLE 2.1. Continued

Map/Data Type	Project Level	Landscape Level
Landownership map	Coordinating management of lands adjacent to roads is key to successful mitigation. Maps that identify adjacent land use management such as public/crown lands, designated reserves, municipal and private lands are needed for planning corridors and crossings. Readily available from local and state/provincial government agencies.	Also available for large-scale planning endeavors. Generalized vegetation and land-use types are provided. Readily available from local and state/provincial government agencies.
Wildlife habitat map	Generally developed from combination of biophysical maps (vegetation maps being one) and models of habitat suitability for certain wildlife species or groups. They identify key habitat types for the species they are prepared for. Some are very accurate and derived from site-specific studies, while others are less accurate relying on extrapolated information. Readily available from local and state/provincial government agencies and nongovernmental organizations.	Some states have prepared (or are preparing) statewide habitat connectivity maps (e.g., FL, WA, CA). U.S. state agencies have prepared "comprehensive wildlife conservation plans" that identify statewide, key habitats for wildlife conservation. These should be readily available from most, if not all, state natural resource agencies today.
Wildlife movement model	Similar to wildlife habitat maps but more specific to where wildlife are most likely to move through the landscape. These are based on either expert opinion or empirical studies that integrate species ecology and landscape suitability. Generally available from wildlife agencies conducting the modeling research.	Generally not available for large-scale exercises unless designed specifically for that purpose. Least-cost path and circuit theory modeling may be promising methods at this scale.

TABLE 2.1. Continued

Map/Data Type	Project Level	Landscape Level
Field research data	Supplemental data in the form of telemetry points or population surveys can help guide the location selection for connectivity and crossing structures. Generally available from wildlife agencies conducting research in the project area.	Not generally available for state-/provincial-wide work; however, local data can be extrapolated to larger landscapes to aid in habitat corridor planning.
Road-kill data	Many state/provincial transportation agencies collect location-specific data on wildlife species killed on their roads, either through carcass collections or collision reports. These data are primarily collected for large mammals and rarely for small and medium-sized fauna. These data can be used to identify road-kill hotspots, but do not provide information on where wildlife are successfully crossing the roadway.	Data are readily available from state/provincial transportation agencies, usually collected by districts and then stored in a state-/provincial-wide database. These data can be used to identify most critical sections of state/provincial highway for accidents with large mammals (primarily elk and deer).
Road network	Municipal and state/provincial governments have digital information on all road types in their jurisdiction.	Road data from state/provincial to national scale can be obtained from the U.S. Census Bureau geospatial database or GeoConnections in Canada.

design considerations (e.g., broad-scale movement patterns of large carnivores versus local topographic and engineering concerns). However, a linkage assessment can help prioritize and identify where wildlife–road conflict areas occur over a large area (Beier et al. 2008). Once identified, this is a good starting point for initiating discussions with transportation and regulatory agencies about mitigation plans in the short and long term.

Specific placement of wildlife crossings is generally determined at the project level or after a thorough field survey as part of a larger systems-level assessment. Regardless of the method, considerations of wildlife crossing placement begin by determining the wildlife species or groups of concern. Once the focal species or group is identified, many of the resources listed earlier can be used to identify the best locations for wildlife crossing mitigation. Methods to identify those locations are briefly described following here.

Field Data

Road-kills. Intuitively road-kill data would be best suited for determining where wildlife crossings should be placed. However, research suggests that the locations where wildlife are struck by vehicles may have little in common with where they safely cross roads (Clevenger et al. 2002a). Many factors associated with roads and adjacent habitats are the causes of wildlife–vehicle collisions (Malo et al. 2004, Ramp et al. 2005), and these factors may not influence where wildlife safely cross roads. Use of road-kill data alone provides a very limited scope of wildlife movement areas and should be combined with habitat linkage mapping or movement models (see later discussion). If reducing road-kill and increasing habitat connectivity are project objectives, then identifying the location of safe wildlife crossings will be an important consideration in planning crossing structures.

Radio and satellite telemetry. Telemetry has been commonly used to describe successful road crossing locations, usually through intensive monitoring of wildlife movements. More accurate crossing data are now being obtained using global positioning system monitoring devices and satellite-based telemetry (van Manen et al. 2001, Waller and Servheen 2005). Compared to radio-based methods, satellite methods allow for more frequent and more accurate relocation data while the animal is collared.

Capture-mark-recapture. By live-trapping and marking individuals and monitoring their movements via translocation or natural movements across roads, the distribution and population density of wildlife can be identified. This approach is most common among smaller fauna but is becoming less popular as more noninvasive survey methods are being developed (MacKenzie 2005, Schwartz et al. 2006, Long et al. 2008).

Snow tracking. In areas that receive regular snowfall, animal crossing locations can be identified via transects adjacent and parallel to the road or road surveys carried out by driving slowly along the road edge (Clevenger et al. 2002a, Barnum 2003).

Track beds. Beds of sand or other tracking media laid out parallel to sections of roadway to "record" animal movements across roads have been used to estimate the number of animal crossings before road expansion and the construction of wildlife crossings (Hardy et al. 2007). These data can be used to determine the duration of monitoring to detect a proportional change in crossing rates after construction.

Remote cameras. Camera systems along roads have their own inherent problems of operation and have not proven reliable for obtaining information on where animals actually cross roads (Clevenger et al. 2008). These problems are due to cameras' limited range of detection. However, camera data can be used to provide information on wildlife distribution and relative abundance by using camera "traps" (Long et al. 2008). Camera sampling stations can be placed within the study area (road corridor) using a grid or stratified sampling approach that will provide the best results per unit effort. Animal distributions can be modeled using presence-only data from cameras (Guisan and Zimmermann 2000, Pearce and Boyce 2006). Determining relative abundance is more problematic, as it is difficult to identify individual animals detected by cameras (but see Rowcliffe et al. 2008).

Genetic sampling. Similar to camera traps, noninvasive genetic sampling of hair for DNA may be practical if used in a high-density grid pattern or when focusing efforts at a smaller scale of resolution (e.g., medium-sized mammals). Genetic sampling may only be able to provide information on the general location of potential wildlife crossing structures. Unlike data from camera systems genetic sampling and DNA analysis can provide minimum estimates of local population size and can identify individuals, their sex, and their genetic relatedness (Manel et al. 2003, Schwartz et al. 2006, Kendall et al. 2008).

Geographic Information System Models

GIS-based movement model. Landscape-scale GIS-based models have been used to identify key habitat linkages, evaluate habitat fragmentation resulting from human activities, and discover areas of where highways are permeable to wildlife movement (Crooks and Sanjayan 2006). Models that simulate movements of wildlife tend to use "resource selection functions" that map habitat quality. The models have rules for simulated movements based on habitat quality and how animals are able to travel through the landscape. The data used to generate a GIS-generated "habitat surface" for these models are based on animal distribution data, usually obtained by radiotelemetry locations, but can also be derived from other methods to survey animal populations, such as DNA sampling, sooted track plates, acoustic surveys, or scat-detection dogs (Pearce and Boyce 2006). Regardless of how the simulated movement or habitat linkage models are developed, the accuracy of the models' ability to predict crossing locations needs to be tested with empirical field data (e.g., road-kill locations, telemetry location data, field observations, survey data, etc.). We describe many of the popular GIS-based techniques for identifying habitat linkage modeling methods in box 2.1.

No Data

Often transportation and natural resource agencies lack easily accessible empirical data for planning the location of wildlife crossing structures. Usually decisions regarding design and location need to be made in a few months, leaving no time for preconstruction studies. When this is the case, there are several options to consider.

Expert-based habitat model. Expert information can be used to develop simple, predictive, habitat linkage models in a relatively short period of time (Store and Kangas 2001, Yamada et al. 2003). Expert information may consist of models based on the opinion of experts or qualitative models based on the best available information obtained from the literature. Several methods have been used to quantitatively analyze expert opinion data, but the analytical hierarchy process (Saaty 1990) is popular among environmental biologists (Dodson Coulter et al. 2006). Expert opinion has been successfully used to identify key habitat linkages across roads and to site wildlife crossings (Clevenger et al. 2002b). The advantages are as follows: (1) it is quick and easy to carry out, (2) legitimacy can be quite high if a consensus model is employed by participants, (3) the method can be statistically sound and biologically robust for identifying and prioritizing critical habitat linkages, and (4) GIS software to assist in linkage identification is readily accessible. Software for the analytical hierarchy process is freely available on the Internet, and was designed by analytical hierarchy process authority Thomas L. Saaty. Major limitations of expert-based modeling are that it works best in the context of a narrow taxonomic focus, and like all

BOX 2.1. GEOGRAPHIC INFORMATION SYSTEM–BASED MODELING TECHNIQUES
USED TO IDENTIFY HABITAT LINKAGES

Least-cost path (Epps et al. 2007): Can be quite arbitrary as people tend to select a protected or wilderness area as a core and then ask how an animal leaving
the core gets to the destination. Unrealistic because for animals to follow a
least cost corridor, they require global knowledge of landscape structure and
must be totally rational in their decision making. See Majka, D., J. Jenness, and
P. Beier. 2007. CorridorDesigner: ArcGIS tools for designing and evaluating corridors. Available at http://corridordesign.org.

Resource-selection functions (Manly et al. 2002): models the probability of
an animal using an area on the landscape as a function of environmental variables. Useful for predicting areas of high-quality habitat adjacent to roads. Can
be used at both fine and broad scales to assess habitat suitability.

Circuit theory (McRae et al. 2008): Uses a binary node (habitat) and linkage
in a network perspective. Similar to graph theory in its networked approach,
and includes directionality and degree of connectivity between nodes. Can be
used to quickly analyze large landscapes and datasets.

Graph theory (Bunn et al. 2000; Minor and Urban 2008): Similar to circuit
theory, graph theory uses a network (nodes and edges) approach to assessing connectivity. Scenario building can be used to show the most important
linkages needed to maintain regional habitat connectivity. See the FUNCONN
GIS tool box. Available at http://www.nrel.colostate.edu/projects/starmap/
funconn_index.htm.

Percolation theory (O'Neill et al. 1992): Grid-cell-based model of landscapes, often used with a simplistic binary view of habitat suitability. More
realistic and explicit modeling techniques are available to provide more detailed
and powerful assessments of connectivity.

Brownian bridges (Horne et al. 2007): Used to infer the most likely distribution of animals between known locations (e.g., movement patterns between
two sequential telemetry locations). Has been used to identify local-scale road
crossing areas.

Customized approaches: Includes spatially explicit, individual-based (Clevenger and Wierzchowski 2006; Graf et al. 2007), statistical models from genetic
data (Cushman et al. 2006) and complex simulations requiring supercomputers
(Hargrove et al. 2005). Very powerful but requires high levels of expertise in
statistical techniques and software programming.

models, results are most reliable when validated with field data. When using expert-based modeling it is important to consider "who" is invited to participate as an expert and how transparent the process is when it comes to finding broader support for the findings of the model.

Rapid assessment. A rapid assessment process has been used that involves gathering experts from the area of concern (Ruediger and Lloyd 2003). This process differs from the foregoing in that there is no quantitative analysis of expert opinion or modeling. Through consensus, participants delineate where key corridors are located on a given section of highway. The advantages are similar to the expert-based model; however, rapid assessments can have a broad taxonomic focus. There are two main shortcomings: (1) criteria are rarely used for the selection of potential linkage areas, and (2) a lack of decision rules or weighting makes it difficult to identify and prioritize the most critical linkages in a statistically and biologically meaningful way. As such, large sections of highway may be deemed critical when actually a smaller subset and most ecologically important linkages are not readily identified. For rapid assessment results to be applied by agencies they should be validated with rigorously collected field data.

Local knowledge. Historically, local knowledge has been important for wildlife biologists conducting research or managing habitats for wildlife. Long-term residents can provide valuable information about where and how wildlife moves across the land. In highly fragmented landscapes where crossing locations are limited, local knowledge can help guide the planning of wildlife crossings. Local participation in project planning is not only good public relations but, more importantly, provides stakeholders with input and participation in the project. Recently, local knowledge and public participation have been formalized through citizen-scientist programs that encourage active participation by the local community in wildlife movement and road mortality data collection (see chapter 14).

Compatibility of Adjacent Land Use

Wildlife crossings are only as effective as the management strategies developed around them that incorporate all the key landscape elements (humans, terrain, natural resources, and transportation). Wildlife crossings are in essence small and narrow, site-specific habitat linkages or corridors. For these measures to fulfill their function as habitat connectors, mitigation strategies must be contemplated at two scales. Site-level or local-scale impacts from development, or human disturbance adjacent to crossing structures can impede wildlife use (Clevenger and Waltho 2000, Dodd et al. 2007a). Similarly, alteration of landscape elements at a broader regional scale could impede or obstruct movements toward the structures, preventing animals from using them entirely, thus rendering them ineffective. This is a second scale of concern that must be recognized if local-scale measures are to be effective. This will require coordination between land management and transportation agencies, and to some degree municipal planning organizations. Put one way, if the transportation agency provides for functional wildlife crossings, but the land management agency fails to properly manage the land away from the transportation corridor, transportation agency funds will be wasted and mitigation will be ineffective. Likewise, if the land is being managed to ensure regional scale connectivity across a highway and the transportation agency does not provide a functional system of wildlife crossings to link the bisected habitats in the area of concern, then efforts of the land management agency will be worthless.

In developing recommendations for wildlife crossings, temporal and spatial contexts of ecosystems should be considered. Mitigating highways for wildlife is a long-term process that will last for many decades and affect individuals and populations alike (Opdam 1997). Thus highway mitigation strategies developed around land-use planning should not terminate with the construction process, but need to be proactive at both local and regional scales to ensure that crossing structures remain functional over time.

Mitigation is for the long term, the lifespan of the crossing structures being put in place, which is seventy-five to eighty years on average. In developing mitigation plans an effort needs to be made to visualize and comprehend the growing infrastructure and how the physical and biological elements of the natural landscape will adapt, endure, and remain viable. This is a considerable challenge for all that are involved in the planning of transportation projects with wildlife conservation concerns.

Design

Just as important as finding the correct location to build wildlife crossings is to have them properly designed to meet their performance objectives. Questions arise as to the size of the crossing and how species-specific behaviors should be incorporated into the crossing structure design. These concerns are offset by the logistics of the project, which include costs of the structure, available material and expertise, and physical limitations of the site (e.g., soil, terrain, hydrology). The various stakeholders involved in the crossing structure design process can then find themselves searching through published and gray literature on the design, performance, and cost of the project. As project managers attempt to incorporate the designs and lessons from other jurisdictions, several specific questions arise:

- What do wildlife crossings look like?
- Where were they built?
- For what species were they designed?
- What types of roads and highways were they installed on?
- What environmental settings were they built in (national park/forest, wildland–urban interface, urban, rural agricultural, etc.)?
- Were they successful?
- What documentation is available regarding specific design and construction cost?
- What are the practicalities of each design? Were they overdesigned? (i.e., were successful but could have been built more cheaply). Were they underdesigned? (i.e., they performed poorly and wildlife used them less than expected).

Here we provide examples of what tools and practical applications are available today for designing wildlife crossings in transportation projects. Rather than presenting a complete list of technical designs or methods used, we describe the most common crossing structure types currently in use.

Function of Wildlife Crossings and Associated Measures

Wildlife crossing mitigation has two main objectives: (1) to connect habitats and wildlife populations and (2) to increase motorist safety and reduce mortality of wildlife on roads (fig. 2.1).

Objective 1: Facilitate Connections between Habitats and Wildlife Populations

To achieve this goal crossings are designed to allow movement of wildlife above or below the road, either exclusively for wildlife use, mixed wildlife–human use, or as part of other infrastructure (e.g., creeks, canals). Crossing structures come in a variety of shapes and sizes, depending on their specific objective; they can be divided into eleven different design types (fig. 2.1).

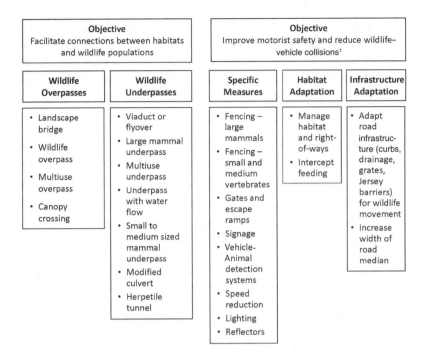

Objective
Facilitate connections between habitats and wildlife populations

Wildlife Overpasses	Wildlife Underpasses
• Landscape bridge • Wildlife overpass • Multiuse overpass • Canopy crossing	• Viaduct or flyover • Large mammal underpass • Multiuse underpass • Underpass with water flow • Small to medium sized mammal underpass • Modified culvert • Herpetile tunnel

Objective
Improve motorist safety and reduce wildlife–vehicle collisions[1]

Specific Measures	Habitat Adaptation	Infrastructure Adaptation
• Fencing – large mammals • Fencing – small and medium vertebrates • Gates and escape ramps • Signage • Vehicle-Animal detection systems • Speed reduction • Lighting • Reflectors	• Manage habitat and right-of-ways • Intercept feeding	• Adapt road infrastructure (curbs, drainage, grates, Jersey barriers) for wildlife movement • Increase width of road median

FIGURE 2.1. Types of measures used to reduce the impacts of roads on wildlife. (Adapted from Iuell et al. 2005)
[1]See Huijser et al. (2007) for more information.

- Four wildlife crossings are above-grade (over-the-road); seven are designed for below-grade (under-the-road) wildlife movement.
- Two of the eleven crossings are designed for both wildlife and human use (multiuse); nine are exclusively for wildlife use.
- Unique wildlife crossings include canopy crossings for arboreal wildlife, underpasses that accommodate movement of water and wildlife, adapted walkways at canal bridges, and below-grade tunnels designed for movement of amphibians and reptiles.

Objective 2: Improve Motorist Safety and Reduce Wildlife–Vehicle Collisions

Traffic-related mortality of wildlife can significantly impact some wildlife populations; particularly those that are found in low densities, are slow reproducing, and travel over large areas. Common and abundant species like deer, elk, and moose, however, can present serious problems for motorist safety. Many mitigation measures have been designed over the years to reduce collisions with wildlife, but few actually perform well or have been rigorously tested (Romin and Bissonette 1996, Huijser et al. 2007). Mitigation measures can be generally categorized as three types:

1. Specific mitigation measures designed to improve motorist safety and reduce collisions with wildlife
2. Mitigation measures that require habitat alterations in or near roads
3. Mitigation measures that require modifications to the road infrastructure

Objectives 1 and 2 should work together and can be integrated to provide for safe movements of wildlife across road corridors by reducing motor vehicle accidents with wildlife. Wildlife crossings typically require one or more types of specific measures designed to improve motorist safety and reduce wildlife–vehicle collisions (e.g., fencing, escape gates, and ramps). Other techniques used to increase motorist safety and reduce collisions with wildlife, such as specific measures (signage and animal-detection systems) and the adaptation of habitats and road infrastructure, are not within the scope of this chapter.

Spacing of Wildlife Crossings

Landscape connectivity is the degree to which the landscape facilitates wildlife movement and other ecological flows (Taylor et al. 1993). No two landscapes function the same for wildlife movement. Terrain, habitat type, levels of human activity, and climate are some factors that influence wildlife movements and ecological flows. Therefore the spacing of wildlife crossings on a given section of roadway will depend largely on the variability of landscape elements, terrain, and the juxtaposition of critical wildlife habitat that intersects the roadway. In landscapes that are highly fragmented with little natural habitat bisected by roadways, fewer wildlife crossings will be required compared to relatively intact, less-fragmented landscapes.

Wildlife crossings are permanent structures embedded within a dynamic landscape. With the lifespan of wildlife crossing structures around seventy to eighty years, the location and design of the crossings need to accommodate the changing dynamics of habitat conditions and their wildlife populations over time. How can we reconcile the dynamic environmental processes of nature with static physical structures on roadways? Environmental change is inevitable and will occur during the lifespan of the crossing structures. There are some basic principles for management to consider:

- *Topographic features*: Wildlife crossings should be placed where movement corridors for the focal species are associated with dominant topographic features (riparian areas, ridgelines, etc.). Sections of roadway can be ignored where terrain (steep slopes) and land cover (built areas) are unsuitable for wildlife and their movements.
- *Multiple species*: Crossings should be designed and managed to accommodate multiple species and variable home range sizes. A range of wildlife crossing types and sizes should be provided at frequent intervals, along with necessary microhabitat elements that enhance movement of smaller fauna, (e.g., root wads for small mammal cover). Unlike the physical structure of wildlife crossings, microhabitat elements are movable and can be changed over time as conditions and species distributions change.
- *Adjacent land management*: How well a wildlife crossing structure performs is entirely dependent upon the land management that surrounds it. Transportation and land management agencies need to coordinate in the short and long term to ensure that tracts of suitable habitat adjacent to the crossings facilitate movement to designated wildlife crossings.
- *Larger corridor network*: Wildlife crossings must connect to and be an integral part of a larger regional corridor network. They should not lead to "ecological dead-ends." The integrity and persistence of the larger corridor network are not the responsibility of the transportation agency, but of neighboring land management agencies and municipalities.

These basic principles will help guide determination of how many wildlife crossings may be necessary and how to locate them in order to obtain the greatest long-term conservation value. There is no simple formula for determining the recommended distance between wildlife crossings; as mentioned earlier each site is different. Planning will largely be landscape and species specific. Bissonette and Adair (2008) used an allometric approach to formulate the spacing of wildlife crossing structures based on the reported home range of each species. While this approach shows promise in developing broad principles for the frequency of crossing structure placement, we think that engineering, funding, topographic, and other local environmental factors generally outweigh information derived through allometry.

TABLE 2.2. Spacing interval (per kilometer) of wildlife crossings for large mammals for existing and future projects

Number of Crossings	Road Length Km (mi)	Average Spacing Km (mi)	Location (Reference)
17	27 (17)	1 / 1.6 (1/1.0)	SR 260, Arizona (Dodd et al. 2007a)
24	45 (27)	1 / 1.9 (1/1.2)	Trans-Canada Highway,[a] Banff, Alberta (Clevenger et al. 2002a)
8	12 (7.5)	1 / 1.5 (1 / 0.9)	Trans-Canada Highway,[b] Banff, Alberta (Parks Canada, unpubl. data)
32	51 (32)	1 /1.6 (1 / 1.0)	Interstate 75, Florida (Foster and Humphrey 1995)
42	90 (56)	1 /2.14 (1/1.3)[c]	US 93, Montana (Marshik et al. 2001)
16	24 (15)	1 / 1.5 (1 / 0.9)	Interstate 90, Washington (Wagner 2006)
4	24 (15)	1 / 6.0 (1 / 3.8)	US 93 Arizona (McKinney and Smith 2006)
82	72 (45)	1 / 0.9[c] (1 / 0.5)	A-52, Zamora, Spain (Mata et al. 2005)

[a]Phase 1, 2, and 3A reconstruction.
[b]Phase 3B reconstruction.
[c]Includes crossings for small and large mammals.

The spacing interval of some wildlife crossing projects designed for large mammals are found in table 2.2. Listed are several large-scale mitigation projects in North America (existing and planned). The spacing interval varies from one wildlife crossing per 1.5 kilometers (0.9 mile) to one crossing per 6.0 kilometers (3.8 miles). The projects listed indicate that wildlife crossings are variably spaced but average about 1.9 kilometers (1.2 miles) apart.

Species Groups

Planning and designing wildlife crossings will often be focused on a certain species of conservation interest (e.g., threatened or endangered species), a specific species group (e.g., amphibians), or abundant species that pose a threat to motorist safety (e.g., deer, elk, moose). In this section we refer to

North American wildlife and species groups when discussing the appropriate wildlife crossing designs. The eight groups mentioned in the following list are general in composition. Their ecological requirements and how roads affect them are described along with some sample wildlife species for each group.

1. *Large mammals* (ungulates [deer, elk, moose, pronghorn], carnivores [bears, wolves, cougars]): Species with large area requirements and potential migratory behavior; large enough to be a motorist safety concern; traffic-related mortality may cause substantial impacts to local populations; susceptible to habitat fragmentation by roads.
2. *High mobility medium-sized mammals* (bobcat, fisher, coyote, fox): Species that range widely; fragmentation effects of roads may impact local populations.
3. *Low mobility medium-sized mammals* (raccoon, skunk, hare, groundhog): Species with smaller area requirements; common road-related mortality; relatively abundant populations.
4. *Semiarboreal mammals* (marten, red squirrel, flying squirrel): Species that are dependent on forested habitats for movement and meeting life requisites; common road-related mortality.
5. *Semiaquatic mammals* (beaver, river otter, mink, muskrat): Species that are associated with riparian habitats for movement and life requisites; common road-related mortality.
6. *Small mammals* (ground squirrels, voles, mice): Species that are common road-related mortality; relatively abundant populations.
7. *Amphibians* (frogs, toads, salamanders): Species with special habitat requirements that may include habitat complementation (e.g., upland forests and lowland wetlands); relatively abundant populations at the local scale; populations are highly susceptible to road mortality.
8. *Reptiles* (snakes, lizards, turtles): Species with special habitat requirement; road environment tends to attract individuals; relatively abundant populations.

Wildlife Crossing Design Types and Objectives

Earlier the eleven different wildlife crossing design types were introduced (see fig. 2.1). The intended use and function of each design are described in the following sections.

Landscape Bridge

Landscape bridges are the largest wildlife crossing structures that span highways. Landscape bridges are generally more than 100 m (330 feet) wide. Their large size enables the restoration of habitats, particularly if designed and integrated so there is habitat continuity from one side to the other. These are primarily intended to accommodate the movement of a broad spectrum of wildlife from large mammals to reptiles, and even many invertebrates (Forman et al. 2003). High- and low-mobility medium-sized mammals, small mammals, and reptiles will utilize structures, particularly if habitat elements are provided on overpasses. Semiarboreal, semiaquatic, and amphibious wildlife may use structures that are adapted for their needs. Types of vegetation and placement can be designed to enhance crossings by bats and birds.

Wildlife Overpass

Next to landscape bridges, wildlife overpasses are the largest crossing structures to span roadways. They are primarily intended to move large mammals; however, small and medium-sized fauna will use wildlife overpasses if the right habitat elements are provided. Wildlife overpasses are generally 50 to 70 m (165–230 feet) wide, while some may be as narrow as 40 to 50 m (130–165 feet). The guidelines for landscaping of wildlife overpasses are identical to those for landscape bridges already described, but for a smaller structure. Wildlife overpasses should be closed to public and any other human use or activities. Roads should not be on or near wildlife overpasses because they will hinder wildlife use of the structures.

Multiuse Overpass

Although multiuse overpasses have a design similar to that of a wildlife overpass, the management objective of these structures is to allow use by both wildlife and humans. These overpasses are smaller than wildlife overpasses because of their mixed use. They may be adequate for movement of some large mammals, but not all species will regularly use structures with frequent human use and activity. Small- and medium-sized mammals tend to utilize these structures, particularly generalist species common in human-dominated environments. Structures may be adapted for semiarboreal species. Semiaquatic and amphibious species may use these if the structures are located within these animals' habitats. Multiuse overpasses are generally 15 to 25 m (50–82 feet) wide, while some may be as narrow as 10 m (32 feet).

Canopy Crossing

Canopy crossings are unique above-grade crossing structures designed to link forested habitats separated by roads. They are designed for semiarboreal and arboreal species whose movements are strongly impacted by roads, limiting movements and potentially fragmenting habitat. Canopy crossings allow for movements between forests over many road types and widths. Structures can be designed to meet the needs of particular focal species. Relatively few canopy crossings have been constructed to date, mostly in Europe and Australia (Goosem 2004).

The design and materials used will depend on the site and species for which they are intended. Structures consist of anchoring thick ropes or cables to trees or permanent fixtures (signage beams, light posts, etc.), allowing animals to move between tree canopies situated on opposite sides of the road. Over small roads (or railways) ropes or cables can be installed between trees. For multilane highways and roads with wide clearance (i.e., large distance between trees) more permanent and stable fixtures are required.

Viaduct or Flyover

A viaduct is the largest of the underpass structures wildlife use; however, they are usually not built specifically for wildlife movement. The large span and clearance of most viaducts allow for use by a wide range of wildlife. Structures can be adapted for amphibians and semiaquatic and semiarboreal species. Viaducts are an alternative to constructing underpasses on cut-and-fill slopes, which tend to have a greater barrier effect to wildlife and less habitat connectivity compared to viaducts. Viaducts with support pillars help keep habitats intact and nearly undisturbed. Viaducts also help restore or maintain hydrological flows and the biological diversity associated with riparian habitats. They are commonly used structures for crossing wetland habitats. A range of dimensions exist from long structures with low vertical clearance for wetlands to short structures with high clearance spanning deep canyons.

Large Mammal Underpass

Not as large as most viaducts, these are the largest and most common underpass structures designed specifically for wildlife movement. They are primarily designed for large mammals but their use by some large mammals will depend largely on how the structure may be adapted for the animals' specific crossing requirements. Small- and medium-sized mammals (including carnivores) generally utilize these structures, particularly if cover is provided along walls of the underpass by the use of brush or stump walls. These underpass structures can be readily adapted for amphibians and semiaquatic and semiarboreal species. Large mammal underpasses are generally at least 10 m (32 feet) wide and 4 m (13 feet) high, while some smaller structures may be 7 m (23 feet) wide and 4 m (13 feet) high.

Multiuse Underpass

With a design similar to that of a large mammal underpass, a multiuse underpass allows use by both wildlife and humans. These structures are generally smaller than large mammal underpasses because of their mixed use, although they may be adequate for movement of some large mammals. Small- and medium-sized mammals will utilize these structures, particularly generalist species common in human-dominated environments (e.g., urban or periurban habitats). The structures may be adapted for semiarboreal species. Semiaquatic and amphibious species may use them if located within their habitats. Multiuse underpasses are generally at least 7 m (23 feet) wide and 3.5 m (11.5 feet) high, while some smaller structures may be 5 m (16.5 feet) wide and 2.5 m (8.2 feet) high.

Underpass with Water Flow

These underpass structures are designed to accommodate dual needs of moving water and wildlife. They are generally located in wildlife movement corridors given their association with riparian habitats; however, some may be only marginally important. Structures aimed at restoring proper function and connection of aquatic and terrestrial habitats should be situated in areas with high landscape permeability, known wildlife travel corridors, and minimal human disturbance. These underpass structures are frequently used by several large mammal species, yet usage will depend largely on how a structure may be adapted for their specific crossing requirements. Small- and medium-sized mammals (including carnivores) generally utilize these structures, particularly if riparian habitat is retained or cover is provided along walls of the underpass by the use of logs, brush, or root wads. These underpass structures can be readily adapted for amphibians and semi-aquatic and semiarboreal species. Travel paths are generally at least 3 m (10 feet) wide, and vertical clearance is 4 m (13 feet) high, while some smaller structures may have travel paths at least 2 m (6.5 feet) wide and 3 m (10 feet) vertical clearance.

Small- to Medium-Sized Mammal Underpass

Like herpetile tunnels (see later section), these are the smallest wildlife crossing structures. They are primarily designed for small- and medium-sized mammals, but wildlife use will depend largely on how a particular structure may be adapted for specific crossing requirements and cover needs. Small- and medium-sized mammals (including carnivores) generally utilize these structures, particularly if they provide sufficient cover and protection and are properly spatially aligned within their habitats. Similarly, they can be of value to semiaquatic mammals and amphibians if they are located in or near the habitats of these species.

Modified Culvert

This wildlife crossing is adaptively designed for movement of small and medium-sized wildlife associated with riparian habitats or irrigation canals. Designs to adapt canal bridges for wildlife crossing can take many forms. Dry platforms or walkways are typically constructed on the lateral interior walls of the bridge and above the high-water mark. Ramps from adjacent habitat and dry ground lead to the dry walkways inside the drainage structure.

Adapting drainages and canals for wildlife use is an easy and cost-effective means to provide wildlife passage associated with wetlands and other habitats that are inundated year-round or seasonally. Few modifications are needed to adapt canal bridges for wildlife passage. Any work to adapt a bridge for wildlife passage should not impede or reduce the bridge's hydrologic capacity or function. Wildlife walkways should run along both sides of the canal bridge. Walkways can be placed on one side of the bridge interior in situations where wildlife habitat is primarily on one side of the bridge. There is generally little human activity in these areas, but to ensure performance and function a modified culvert should have minimal human disturbance.

Amphibian and Reptile Tunnels

The main conflicts with amphibians and reptiles are where roads intercept periodic migration routes to breeding areas or areas where young are produced (ponds, lakes, streams, or other aquatic habitats). For some species the migration to these critical areas, including the dispersal of juveniles to upland habitats, is synchronized each year. This large movement event results in a massive migration of individuals in a specific direction during a short period of time. Amphibian and reptile tunnels should be located in these key sections of road that intercept their movements year after year.

Although they are designed specifically for passage by amphibians and reptiles, other small- and medium-sized vertebrates may use these tunnels as well. There are many different tunnel designs to meet the specific requirements of each species or taxonomic group. Amphibians and reptiles tend to have special requirements for wildlife crossing design since they are unable to orient their movements to locate tunnel entrances. Walls or fences play a critical function in intercepting movements and directing animals to the crossing structure (Langton 1989).

Large tunnels provide greater airflow and natural light conditions; however, smaller tunnels with grated slots for ambient light and moisture can be effective. Grated tunnels are placed flush with the road surface. Requirements for tunnel design and microhabitat differ among amphibian and reptile taxa (see Lesbarrères et al. 2003). Hesitancy and repeated unsuccessful entry attempts at tunnels are believed to be due to changes in microclimatic conditions, particularly temperature, light, and humidity, that animals perceive as localized climate degradation. Larger tunnels (ca. 0.9 meters [3 feet] diameter) permit greater airflow and increased natural light at tunnel exits. Smaller tunnels can be effective if they are grated on top, increasing natural light and moisture. Sandy soil (sandy loam) should be used to cover the bottom of the tunnel to provide a more natural substrate for travel.

Amphibians have been documented using tunnels that range in length from 6.7 m (22 feet) (spotted salamanders *Ambystoma maculatum*, Massachusetts) to 40 m (125 feet) (Lausanne, Switzerland). The effectiveness of long tunnels spanning four-lane highways has not been tested. Shorter tunnels are better for amphibian movement.

Drift fencing. Because amphibian and reptiles generally do not avoid roads and have biased, directed movements while migrating, guiding walls or drift fences should be installed to direct movement toward the tunnel (Woltz et al. 2008). These walls should angle out from each end of the tunnel at approximately 45 degrees, at 1.25 feet high and be made of concrete, treated wood, or other opaque material. Guiding walls/fences made of translucent material or wire mesh are not recommended because some amphibians and reptiles try to climb over them instead of moving toward the tunnel.

The bottom section of a guiding wall or fence should be secured to the ground, not leaving any gaps. The guiding wall/fence should be tied into the tunnel entrance, avoiding any surface irregularities that might impede or distract movement toward the tunnel entrance.

Large Fauna Fence Design and Substitutes

Fencing is a key part of a mitigation plan involving wildlife crossings. Fences and wildlife crossings have been around for many years; however, relatively little is known about effective fence designs and other innovative solutions to keep wildlife away from roads. Wildlife can become trapped inside the fenced area; thus measures need to be in place to allow animals to safely exit the highway area. Steel swing gates, hinged metal doors, or earthen ramps or jump-outs are some commonly used methods. Small and medium-sized mammals can get through most fencing for large mammals. Different fencing types and designs are needed to keep these animals from reaching the roadway.

Typical fencing for large mammals consists of 2.4-meter (7.9 feet)-high page wire fencing material with wooden or steel posts spaced at 5-meter intervals. To keep digging animals from going under the fence and reaching the roadway, an apron of fencing material is spliced to the fence. The apron is approximately 1 meter long and buried at a 45-degree angle away from the fence and roadway.

Performance of Wildlife Crossings

Basic guidelines have been developed to monitor the function of wildlife crossings and assess their conservation value (Forman et al. 2003). The criteria used to measure their function or conservation value, however, will depend on the intended purpose of the wildlife crossing(s), the taxon of interest, and the biological level of organization most relevant to answering monitoring and research goals (Clevenger 2005).

Monitoring needs to be an integral part of a highway mitigation project even long after the measures have been in place. Mitigation is costly, requiring an important investment of public funds. Post-construction evaluations are a judicious use of public infrastructure funds and can help agencies save money in future projects (see Adaptive Management section). Monitoring and research can range from a simple, single-species population with the highway corridor to more complex ecological processes and functions within regional landscapes. Like wildlife corridors, crossing structures should allow for the following five ecological functions:

1. Reduced mortality and increased movement (genetic interchange) within populations
2. Meeting biological requirements such as finding food, cover, and mates
3. Dispersal from maternal or natal ranges and recolonization after long absences
4. Redistribution of populations in response to environmental changes and natural disturbances (e.g., fire, drought); movement or migration during stressful years of low reproduction or survival
5. Long-term maintenance of metapopulations, community stability, and ecosystem processes

From these five functions it is possible to set performance objectives, determine the best methods to monitor, develop study designs, and address the management questions associated with the project objectives (table 2.3; Clevenger et al. 2008).

Determining measures of mitigation success or performance is a complex task. It needs to be scientifically defensible and done a priori (Roedenbeck et al. 2007). Short-term and long-term monitoring and research are needed to measure the performance criteria over time. Note that these functions increase in complexity and the costs and time required to properly monitor whether they are being facilitated (see table 2.3). Not all ecological functions may be of management concern for transportation and land management agencies, particularly those at the more complex end of the scale. A noninvasive genetic sampling method was used to assess population-level benefits (level 2, table 2.3) of twenty wildlife crossings on the Trans-Canada Highway in Banff National Park, Alberta (fig. 2.2; Clevenger and Sawaya, 2010).

TABLE 2.3. Levels of conservation value for wildlife crossing systems as measured by ecosystem function achieved, level of biological organization targeted, type of connectivity potential, and cost and duration of research required to evaluate status

Level	Ecosystem Function (simple to complex)	Level of Biological Organization	Level of Connectivity[a]	Cost and Duration – Research[b]
1a	Movement within populations and genetic interchange	Genetic	Genetic	Low cost—short term
1b	Reduced mortality due to roads	Genetic and species/population	Genetic and species/population	Low cost—short term
2	Biological requirements of finding food, cover, and mates	Species/population	Demographic	Moderate-to-high cost—long term
3	Dispersal from maternal ranges and recolonization after long absences	Species/population	Functional	Moderate-to-high cost—long term
4	Populations to move in response to environmental changes and natural disasters	Ecosystem/community	Functional	High cost—long term
5	Long-term maintenance of meta-populations, community stability, and ecosystem processes	Ecosystem/community	Functional	High cost—long term

[a]Genetic: predominantly adult male movement across road barriers; Demographic: genetic connectivity with confirmed adult female movement across road barriers; Functional: genetic and demographic connectivity with confirmed dispersal of young females that survive and reproduce.
[b]Based on studies of large mammals. Cost and duration will largely be dependent upon area requirements, population densities, and demographics.

FIGURE 2.2. Wolf passing through barbed wire, hair-sampling system on one of the wildlife overpasses in Banff National Park, Alberta. (Photo: Banff Wildlife Crossings Project/WTI-MSU)

Simpler techniques using remote cameras can be used to detect animals using wildlife crossing structures; however, information about number of individuals, gender, and genetic relationships cannot be obtained (fig. 2.3; Ford et al. 2009).

Adaptive Management

Adaptive management consists of deriving benefits from measured observations from monitoring to inform decision making with regard to planning and design of subsequent phases of a project (Holling 1978, Walters 1986). An example of adaptive management would be changing the design of wildlife crossing structures on subsequent phases of highway reconstruction after obtaining empirical data from the use of structures from earlier phases. Microhabitat elements within wildlife crossings may require changes if monitoring shows they do not facilitate movement of smaller

FIGURE 2.3. Remote camera and track pads used in monitoring wildlife use at a crossing structure. (Photo: Robert Long)

wildlife. Monitoring of fencing may identify deficiencies that lead to revised design or materials used for construction in future phases. Preconstruction data on local species occurrence and wildlife movements may lead to slight changes in the locations and types of wildlife crossing structures (e.g., small to medium-sized culverts) should monitoring reveal previously undocumented unique populations or important habitat linkages. Whatever the case may be, monitoring ultimately provides management with sound data for mitigation planning, helps to streamline project planning, and saves on project costs.

Adaptive management of the project design from monitoring results requires regular communication between the wildlife monitoring project coordinator and the sponsoring transportation agency. Close coordination between research and management will allow for timely changes to project design plans that reflect the most current results from monitoring activities.

Conclusions

There is undoubtedly an increasing trend in acceptance and implementation of mitigation measures by transportation agencies in North America. A qualitative evaluation of state and provincial transportation agencies will find that roughly one-third of agencies in the United States have either institutionalized mitigation practices or are moving in that direction. Only a decade ago little interest was shown by transportation agencies, academia, and even some land management agencies in adopting measures to reduce the impacts of roads on wildlife populations. Fifteen years ago, it was unlikely that transportation engineers and planners would be in the same room with environmental managers and biologists to discuss ways of meeting each other's regulatory needs and agency objectives. Transportation and ecology were completely isolated disciplines and ran on separate tracks.

Today we find there are numerous tools available and policies in place that promote highway planning that ensures habitat connectivity for wildlife and safe passage. Here in North America, federal, state, and provincial transportation agencies have recognized that early stakeholder involvement and identification of issues and areas of concern are essential if their projects are to be streamlined, cost-effective, and supported by local communities. Recent developments in state- and province-wide GIS-based information for transportation planning and mapping priority habitat conservation needs now provide an unprecedented opportunity to coordinate ecological and transportation networks at biologically significant scales (Brown 2006). Integrating these plans helps ensure that habitat conservation and connectivity concerns appear at the beginning of the planning process and guide transportation and land management actions. Mapping ecological and transportation corridors will help clarify stakeholder concerns, prioritize agency objectives, and incorporate landscape patterns and processes in the planning and construction process (Forman 1987).

A growing body of knowledge has developed in the last decade regarding wildlife response to crossing structure design and placement (Forman et al. 2003, Bissonette 2007). This information has been obtained through systematic monitoring and research of wildlife crossing use by a range of wildlife. Transportation and resource management agencies are starting to include pre- and post-mitigation research as part of mitigation schemes (Dodd et al. 2007a, Clevenger et al. 2008). This allows agencies to better evaluate the performance of their mitigation investments and inform decision making with regard to planning and design of mitigation on future projects.

Chapter 3

Reducing Wildlife–Vehicle Collisions

Marcel P. Huijser and Pat T. McGowen

Wildlife–vehicle collisions affect human safety, property, and wildlife. The most visible effects to drivers are dead animals on or alongside the road (fig. 3.1). While many vehicles can continue to be driven after a collision, some have to be left along the roadside until a tow truck can transport the damaged vehicle to a repair shop (fig. 3.2). Both damaged vehicles and animal carcasses, at least the ones that are easily visible to drivers, are typically removed from the road or roadside within a few days after a collision has occurred. Therefore drivers, and the public in general, tend to underestimate the number of collisions with large mammals.

In the United States the total number of large mammal–vehicle collisions is estimated at 1 to 2 million and at 45,000 in Canada annually (Conover et al. 1995, Tardif & Associates Inc. 2003, Huijser et al. 2007). In the 1990s in the United States, these collisions were estimated to cause 211 human fatalities, 29,000 human injuries, and over $1 billion in property damage annually (Conover et al. 1995). These numbers increased even further in the last decade in both the United States and Canada (Tardif & Associates Inc. 2003, Huijser et al. 2007; Huijser et al. 2009, fig. 3.3). Most of the reported wildlife–vehicle collisions relate to white-tailed deer (*Odocoileus virginianus*) and mule deer (*O. hemionus*). A major auto insurance company in the United States that covered 17.5 percent of the market

FIGURE 3.1. An elk killed as a result of a collision with a vehicle. (© Marcel Huijser)

FIGURE 3.2. Vehicle damage as a result of a collision with a white-tailed deer. (© Marcel Huijser)

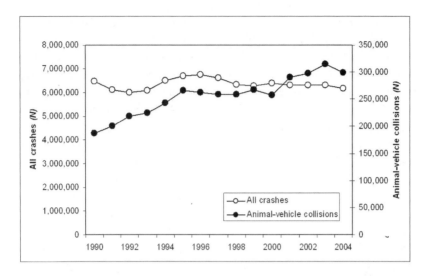

FIGURE 3.3. The number of reported crashes (all types) per year remained relatively constant in the United States between 1990 and 2004 while the number of reported animal–vehicle collisions per year increased by about 50 percent. Note that the number of reported collisions is only a fraction of the actual number of collisions. The actual number of collisions with large mammals in the United States is estimated at 1 to 2 million per year (see text).

share in 2006–2007 reported that of all claims related to wildlife–vehicle collisions 99.2 percent related to deer, 0.5 percent to elk (*Cervus elaphus*), and 0.3 percent to moose (*Alces alces*) (total sample size of the claims is approximately 180,000; Dick Luedke, State Farm Insurance, pers. comm.).

The substantial increase in wildlife–vehicle collisions between 1990 and 2004 was related to deer population size as well as the number of vehicles and the location and time these vehicles were used. Deer population sizes, especially those of white-tailed deer, grew substantially over the last century in the United States (Côté et al. 2004). This population growth is especially apparent since the 1960s (Porter and Underwood 1999, Côté et al. 2004). This increase in population size was triggered by better protection, a matrix of habitat providing cover (forests) and food (agriculture, silviculture), the loss or decline of their natural predators, and more recently, reduced hunting pressure by humans through an increase in refugia (private land and urban areas), and a decrease in hunters (Porter and Underwood 1999, Brown et al. 2000a, Enck et al. 2000, Rooney 2001, Côté et al. 2004, Ditchkoff et al. 2006). Currently, white-tailed deer numbers are believed to be higher than they have ever been in the past several hundred years, causing or contributing to a series of problems, including wildlife–vehicle collisions (Waller and Alverson 1997, Porter and Underwood 1999, Rooney 2001, Côté et al. 2004). At the same time the total number of kilometers driven in the United States increased 38 percent from 3,450,278 million kilometers (2,144,362 million miles) in 1990 to 4,770,344 million kilometers (2,964,788 million miles) in 2004 (FHWA 2008). During the same time period, the lane kilometers in the United States increased only 3.6 percent from 12,954,189 kilometers (8,051,081 miles) in 1990 to 13,417,163 kilometers (8,338,821 miles) in 2004 (FHWA 2008). In addition, the time of day people travel, partially because of a twenty-four-hour economy, and partially to avoid peak hours, may expose them to a greater risk of wildlife–vehicle collisions (peak between 5:00 and 7:00 a.m. and 6:00 and 10:00 p.m., Huijser et al. 2007). Furthermore, travel in rural areas and in the urban–rural interface also increases the exposure of drivers to wildlife–vehicle collisions (Huijser et al. 2007). Finally, while the reconstruction of rural two-lane roads typically leads to safer roads through wider lanes, shoulders, and clear zones, wildlife–vehicle collisions appear to increase rather than decrease after road reconstruction, probably as a result of an increase in design speed (Vokurka and Young 2008).

In most cases the animals die immediately or shortly after the collision (Allen and McCullough 1976). In some cases it is not just the individual animals that suffer. Road mortality may also affect some species at the population level (e.g., Zee et al. 1992, Huijser and Bergers 2000, Fahrig and Rytwinski 2009), and some species may even face a serious reduction in population survival probability as a result of road mortality, habitat fragmentation, and other negative effects associated with roads and traffic (Proctor et al. 2005, Klar et al. 2006, Grift et al. 2008, see also chapter 1). In addition, some species also represent a monetary value that is lost once an individual animal dies (Romin and Bissonette 1996, Conover 1997).

Over forty types of mitigation measures aimed at reducing animal–vehicle collisions have been implemented or described (see reviews in Hedlund et al. 2004, Knapp et al. 2004, Huijser et al. 2007). Examples include warning signs that alert drivers of potential animal crossings, wildlife warning reflectors or mirrors, wildlife fences, and animal detection systems. However, the effectiveness of these mitigation measures and the quantity and quality of the data on the effectiveness of these mitigation measures vary greatly.

This chapter describes 41 different types of mitigation measures or dif-

TABLE 3.1. Estimated effectiveness of individual mitigation measures aimed at reducing collisions with large, wild ungulates through influencing driver behavior, and whether the measures result in an absolute barrier for large, wild ungulates

Mitigation Measure	Estimated Effectiveness (%)	Absolute Barrier?
Public information and education	?	No
Standard warning signs	0	No
Enhanced wildlife warning signs	?	No
Seasonal wildlife warning signs	26	No
Animal detection systems (ADS)	87	No
ADS linked to on-board computer	87	No
On-board animal detectors	?	No
Increase visibility (roadway lighting)	?	No*
Increase visibility (vegetation removal)	38	No*
Increase visibility (wider striping)	?	No
Increase visibility (reflective animal collars)	?	No
Increase visibility (reduce snowbank height)	?	No
Reduce traffic volume	?	No
Temporary road closure	?	No
Reduce vehicle speed	?	No
Wildlife crossing assistants	?	No

*Measure may increase barrier effect of road but does not present an absolute physical barrier.

ferent combinations of mitigation measures aimed at reducing collisions with large, wild ungulates in the United States and Canada, most notably deer (*Odocoileus* spp.). In addition, we discuss their estimated effectiveness and whether they present an absolute barrier to large, wild ungulates. We distinguish between measures aimed at influencing driver behavior (*n* = 16) and measures aimed at influencing animal behavior (*n* = 25) (tables 3.1 and 3.2). If more than one study reported on the effectiveness of a mitigation measure, we calculated the average percent reduction in ungulate–vehicle collisions (tables 3.1 and 3.2).

Mitigation Measures Aimed at Influencing Driver Behavior

Mitigation measures aimed at influencing driver behavior include public information and education, various types of permanent warning signs,

TABLE 3.2. Estimated effectiveness of individual mitigation measures aimed at reducing collisions with large, wild ungulates through influencing animal behavior, and whether the measures result in an absolute barrier for large, wild ungulates

Mitigation Measure	Estimated Effectiveness (%)	Absolute Barrier?
Deer reflectors and mirrors	0	No*
Deer whistles	0	No*
Olfactory repellents	?	No*
Deer flagging models	?	No*
Hazing	?	No*
Escape paths from road corridor	?	No
Deicing alternatives	?	No
Intercept feeding	?	No
Minimum nutritional value right-of-way vegetation	?	No
Carcass removal	0	No
Increase median width	?	No
Population culling	50	No
Relocation	50	No
Antifertility treatment	50	No
Habitat alteration away from road	?	No*
Fence (including dig barrier)	87	Yes
Boulders in right-of-way	?	Yes
Fence with gap and warning signs	0	No*
Fence with gap and crosswalk	40	No*
Fence with gap and animal detection system	82	No*
Fence with underpasses	87	No*
Fence with overpasses	87	No*
Fence with under- and overpasses	87	No*
Long bridges	100	No
Long tunnels	100	No

*Measure may increase barrier effect of road but does not present an absolute physical barrier.

seasonal warning signs, animal detection systems, measures that increase the visibility for drivers, measures that reduce traffic volume, temporary road closures, reduced vehicle speed, and wildlife crossing assistants (i.e., humans that stop traffic so that animals can cross the road safely; see table 3.1). The effectiveness of these mitigation measures is reviewed in the following sections.

Public Information and Outreach

Many transportation agencies and insurance companies provide information to the public on the number of animal–vehicle collisions, particularly with deer, the road sections that have the most collisions, the time of day, and how to respond when large ungulates are on or close to the road in order to avoid a collision or limit the severity of a potential collision (Kansas DOT 2004, Wildlife Collision Prevention Program 2008, Maine DOT 2009). However, this information requires a constant state of alertness and awareness of drivers over often considerable distances, and no studies were identified which showed that public information and outreach have resulted in fewer deer–vehicle collisions. Our preliminary conclusion is that, although public information and outreach may be valuable for other reasons (e.g., awareness of the problem, generating public support for mitigation measures), public information and outreach are unlikely to result in a substantial reduction in deer–vehicle collisions.

Standard and Enhanced Warning Signs

Standard black on yellow deer warning signs are probably the most widespread road side mitigation measure aimed at reducing wildlife–vehicle collisions. Over the last decades, the dominant practice for transportation agencies was to install wildlife warning signs when (variable) thresholds were reached for ungulate–vehicle collisions (Knapp and Witte 2006). However, such permanently visible signs do not appear to be effective in reducing wildlife–vehicle collisions, even when accounting for possible differences in road, traffic, and landscape parameters (Rogers 2004, Meyer 2006). Enhanced warning signs are also permanently visible to drivers, but compared to the standard deer warning signs these signs are more noticeable, for example, through the addition of bright orange flags, a permanently flashing amber warning light, a sign that is lighted in its entirety, or displays on dynamic or variable message signs (DMS/VMS). The effectiveness of such enhanced warning signs is not fully clear. In a field study, DMS displays that warned for wildlife–vehicle collisions resulted in significantly lower vehicle speeds compared to DMS that were turned off, with the greatest speed reductions occurring during the dark hours and during the weekend (Hardy et al. 2006). Similarly, lighted animated deer crossing signs reduced vehicle speed by 4.8 kilometers per hour (3.0 miles per hour) compared to the same signs when they were turned off (Pojar et al. 1975).

Interestingly, the presence of deer carcasses in addition to the signs resulted in a much greater reduction in vehicle speed: 12.6 kilometers per hour (7.9 miles per hour) (lights turned off) and 10.0 kilometers per hour (6.2 miles per hour) (lights turned on). In a driving simulator study, the average vehicle speed with a standard deer warning sign was 99.6 kilometers per hour (61.9 miles per hour) (Hammond and Wade 2004). An enhanced sign that had a flashing light on top of a standard deer warning sign, but with the light turned off, resulted in similar speeds of 99.5 kilometers per hour (61.8 miles per hour). When the flashing light was switched on however, the enhanced sign resulted in significantly lower vehicle speed of 95.9 kilometers per hour (59.6 miles per hour), a reduction of 3.7 kilometers per hour (2.3 miles per hour). In another study, three different types of enhanced warning signs all resulted in significantly lower vehicle speed (range 115.9–120.0 kilometers per hour [72.0–74.6 miles per hour]) compared to a standard deer warning sign (123.3 kilometers per hour [76.6 miles per hour]) (Stanley et al. 2006). In addition, participants in this driving simulator study were exposed to a group of deer crossing the road. The participants started braking earlier with the enhanced warning signs compared to the standard deer warning sign, but the difference was significant for one and not all three types of enhanced warning signs (Stanley et al. 2006). These results all suggest that enhanced warning signs can lead to lower vehicle speed, greater speed reduction compared to standard deer warning signs, and decreased driver reaction time. However, despite these encouraging data, enhanced warning signs have not been shown to significantly reduce the number of deer–vehicle collisions (Pojar et al. 1975, Stanley et al. 2006).

Seasonal Warning Signs

Seasonal wildlife warning signs are only present at certain times of the year, such as during a seasonal migration. As such they present a more time-specific warning signal to drivers than permanent warning signs. Temporary warning signs with reflective flags and permanently flashing amber lights at locations used by mule deer during their seasonal migration reduced collisions by 51 percent (range 41.5–58.6 percent for individual test areas) compared to control areas (Sullivan et al. 2004). The signs reduced the percentage of speeders from 19 percent to 8 percent during their first season of operation, but the effect was less pronounced during the second season, perhaps due to driver habituation. In another study, enhanced deer warning signs (black on yellow sign showing a deer and a car symbol, combined with a black on orange sign stating HIGH CRASH AREA) deployed between October and January during the peak of collisions did not reduce deer–vehicle collisions (Rogers 2004).

Animal Detection Systems

Road-based animal detection systems use sensors to detect large animals that approach the road. Once a large animal is detected, warning signals are activated that urge drivers to slow down and be more alert. Compared to permanent and seasonal warning signs, animal detection systems provide an extremely time-specific warning signal to drivers. While permanent warning signs need to be installed at the right location, and seasonal warning signs need to be installed at the right location and at the right time, animal detection systems can, in principle, be installed anywhere as long as the warning signs are only visible when an animal has been detected. However, it is extremely important for an animal detection system to be reliable; it must detect all or nearly all large animals that approach the road, and it may not produce too many false warning signals (Huijser et al. 2009a). If an animal detection system is too unreliable, it can erode driver confidence in the system, and consequently, result in an ineffective system. Therefore reliability norms have been suggested (Huijser at al. 2009a). These suggested norms are based on interviews with three stakeholder groups: employees of transportation agencies, employees of natural resource management agencies, and drivers. Five out of nine systems tested for their reliability in a test-bed proved to meet the suggested norms for the reliability of animal detection systems (Huijser et al. 2009a). This suggests that if a reliable animal detection system is installed, its effectiveness in reducing collisions with large mammals is mostly dependent on driver response. Similar to other types of warning signs drivers may respond to the warning signals of animal detection systems by lowering the speed of their vehicle, by becoming more alert, or a combination of the two. Animal detection systems have shown variable results with regard to speed reduction: substantial decreases in vehicle speed of greater than or equal to 5 kilometers per hour (3.1 miles per hour), minor decreases in vehicle speed of less than 5 kilometers per hour (3.1 miles per hour), and no decrease or even an increase in vehicle speed (review in Huijser et al. 2009b). This variability of the results appears to be related to the type of warning signs, whether the warning signs are accompanied with advisory or mandatory speed limit reductions, road and weather conditions, whether the driver is a local resident, and perhaps also cultural differences in driver response in different regions (review in

Huijser et al. 2009b). The effectiveness of animal detection systems in reducing collisions with large ungulates has been estimated at 82 percent (Mosler-Berger and Romer 2003) and 91 percent (Dodd and Gagnon 2008). Road-based animal detection systems may also communicate the warning to on-board computers in vehicles that are approaching the area in which a large mammal has been detected. Such an application can be expected to be at least as effective, as long as roadside warning signals are still present to alert drivers that may not have such communication devices in their vehicle. Vehicle-based animal detection systems may be developed based on pedestrian detection systems (e.g., Omar and Zhou 2007, Bauer et al. 2008). However, detecting large animals while driving at high speeds may be a challenge to this type of detection technology.

PROS FOR ANIMAL DETECTION SYSTEMS COMPARED TO WILDLIFE CROSSING STRUCTURES

Animal detection systems have the potential to provide wildlife with safe crossing opportunities anywhere along the mitigated roadway, but wildlife crossing structures (see chapter 2) are usually limited in number, and they are rarely wider than about 50 meters (164 feet). Animal detection systems are less restrictive to wildlife movement than fencing or crossing structures; they allow animals to continue to use existing paths to the road or to change them over time. In addition, animal detection systems can be installed without major road construction or traffic control for long periods. Finally, once they are mass produced, animal detection systems are likely to be less expensive than wildlife crossing structures.

CONS FOR ANIMAL DETECTION SYSTEMS COMPARED TO WILDLIFE CROSSING STRUCTURES

Although the available data on the effectiveness of animal detection systems with regard to collision reduction are encouraging, animal detection systems currently are not as "tried and proven" as wildlife crossing structures (chapter 2). Currently, animal detection systems only detect large animals (e.g., deer, elk, or moose). Relatively small animals are not detected, and drivers are not warned about their presence on or near the road. In addition, wildlife crossing structures can provide cover (e.g., vegetation, living trees, tree stumps) and natural substrate (e.g., sand, water) allowing better continuity of habitat. Furthermore, some types of animal detection systems are only active in the dark, and animals that cross during the daylight may not be protected. Finally, animal detection systems usually require the presence of poles and equipment in the right of way, sometimes even in the clear zone, presenting a safety hazard of their own.

Increase Visibility: Roadway Lighting

Increased visibility for drivers through roadway lighting may allow drivers to better see wildlife on or near the road and react accordingly. A study in Colorado that involved alternately turning roadside lights on and off for one-week periods did not result in a change in deer–vehicle collisions, but drivers did significantly reduce speeds when a deer decoy was present (Reed and Woodard 1981). A study of the installation of roadway lighting, without fencing, along a section of the Glenn Highway near Anchorage, Alaska, showed a 65 percent reduction in moose–vehicle collisions (Mc-Donald 1991 cited in Biota Research and Consulting Inc. 2003). However, it was not clear if the reduction was the result of animals avoiding the lighted area or the result of increased visibility to drivers.

Increase Visibility: Vegetation Clearance

Increased visibility for drivers through the mowing or cutting of vegetation may allow drivers to better see wildlife on or near the road and react accordingly. A study of the effect of vegetation removal on moose–train collisions in Norway found that clearing vegetation across a 20 to 30 meter (70–100 feet) swath on each side of the railway reduced moose–train collisions by 56 percent (Jaren et al. 1991). Clearing of vegetation along roadsides in Sweden resulted in a 20 percent reduction in moose–vehicle collisions (Lavsund and Sandegren 1991). However, such reductions may not only be the result of increased visibility to drivers, they may also have caused animals to spend less time near the transportation corridor. On the other hand, regrowth of shrubs and trees may attract browsers such as moose, perhaps increasing animal presence and the probability of collisions.

Increase Visibility: Wider Striping, Reflective Animal Collars, Reduce Height of Snow Banks

Some mitigation measures are aimed at increasing the visibility of animals to drivers. For example, drivers may see a break in the pattern of painted highway striping (particularly if it is wide) when an animal crosses it, and thus be warned of its presence on the highway. In addition, wide striping may make the driver perceive a narrower roadway, potentially resulting in lower vehicle speed. However, the effectiveness of this measure is not yet known (Maine DOT et al. 2004). Similarly, collars with reflective tape on some of the free ranging wood bison (*Bos bison*) in northeast British Columbia, Canada, have been considered to increase their visibility to drivers, but this measure has not been implemented so far (Conrad Thiessen, Ministry of Environment, British Columbia, pers. comm.). Finally, it has been suggested to reduce the height of snow berms to increase visibility of moose to drivers, but again effectiveness data are lacking.

Reduce Traffic Volume

Modeling efforts showed a relationship between traffic volume and the number of wildlife–vehicle collisions (Jaarsma and Willems 2002, Langevelde and Jaarsma 2004, Meyer and Ahmed 2004). Traffic calming on more minor rural roads and shifting traffic onto a few major highways have been suggested (Jaarsma and Willems 2002). However, higher traffic volume is not necessarily associated with a reduction in wildlife–vehicle collisions because low-volume roads may present less of a barrier to animals than high-volume roads, and the quality of the habitat adjacent to the road may not be affected as much (Huijser et al. 2000, Jaarsma and Willems 2002, Seiler 2003). Better insight in the effect of traffic volume on wildlife–vehicle collisions and the barrier effect is needed.

Temporary Road Closure

During periods of high animal movement, roads are sometimes closed temporarily. However, temporary road closures have been implemented primarily on low-volume roads during the night, and temporary road closures are typically for amphibians or reptiles that may display mass seasonal migration rather than for large wild ungulates (Delaware Water Gap National Recreation Area 2003, National Park Service 2006, Shawnee National Forest 2006, Todaro 2006). Since most deer–vehicle collisions occur in the dark or in twilight conditions, and since most deer–vehicle collisions occur at a predictable time (October–November) (Huijser et al. 2007), it is likely that seasonal or night road closures for large ungulates would substantially reduce the number of collisions. However, the traffic volume on roads with substantial numbers of collisions with large ungulates may be too high to consider seasonal or night road closure.

Reduced Vehicle Speed

Lower vehicle speed allows drivers and wildlife more time to respond to each other's presence so that a collision can be avoided. For example, in Yellowstone National Park, roads with lower design speeds and lower posted speed limits (≤45 miles per hour [≤72 kilometers per hour]) had fewer wildlife–vehicle collisions than roads with higher design speeds and higher posted speed limits (55 miles per hour [88 kilometers per hour]) (Gunther et al. 1998). However, simply reducing the posted speed limit is dangerous as a substantial difference between the design speed of a road and the posted speed limit can lead to great differences in speed. Some drivers will observe the new and lower posted speed limit while others will drive the speed they perceive as safe given the design of the road. This phenomenon is referred to as speed dispersion and is associated with an increase in overall crashes. Therefore it is considered good practice to not simply lower the posted speed limit, but to have the posted speed limit match the design speed of a road and verify this by measuring the actual vehicle speed (i.e., operating speed). A few examples of measures aimed at reducing vehicle speed are presented later in the chapter. In Jasper National Park in Alberta, Canada, on the Yellowhead Highway the speed limit for the road was 90 kilometers per hour (56 miles per hour). In 1991, the speed limit was reduced from 90 kilometers per hour (56 miles per hour) to 70 kilometers per hour (43 miles per hour) on three road sections, and the speed limit was enforced (Bertwistle 1999). A speed study at two of the road sections showed that less than 20 percent of the vehicles actually obeyed the lower speed limit, and collisions with large ungulates appeared to have increased compared to those in control road sections. Changes in the basic road design (e.g., lane and shoulder width, sight distances) and other traffic calming methods (e.g., speed bumps, traffic circles, curb extensions, sidewalk extensions, raised medians, and rumble strips) are likely to be more effective. Currently such measures are typically used in residential neighborhoods or on a highway approaching a town, and rarely on major highways in rural areas. However, in Tasmania, Australia, concrete barriers were installed with a "give way" sign that constricted traffic to a single lane in the center of the road close to road sections that had a concentration of collisions with eastern quolls (*Dasyurus viverrinus*) and Tasmanian devils

(*Sarcophilus laniarius*) (Jones 2000). After installation, median vehicle speed in the center of the road section dropped by about 20 kilometers per hour (12.3 miles per hour), and wildlife road mortality was reduced.

Wildlife Crossing Assistants

In areas such as national parks traffic jams occur as drivers stop their vehicles on the road to view wildlife. Agency personnel will direct traffic for safety reasons and to keep the road clear for emergencies. However, agency personnel or volunteers will also keep people at a certain minimum distance from the animals allowing them to move and cross the road (Cochran 2006). However, the effectiveness of this strategy in reducing wildlife–vehicle collisions is not known.

Mitigation Measures Aimed at Influencing Animal Behavior

Mitigation measures aimed at influencing animal behavior include measures directed at scaring ungulates away from the road and road corridor, alerting them to approaching traffic, reducing the attractiveness of the road or road sides, increasing the attractiveness of areas away from the road, providing them with a resting area when crossing multiple lanes of traffic, pathways that allow animals to escape from the road corridor (e.g., with high snow accumulation), population size reduction efforts, physical barriers that keep animals off the road (with or without safe road crossing opportunities), and elevating or tunneling roads (see table 3.2).

Deer Reflectors and Mirrors

Deer mirrors and reflectors are designed to reflect vehicle headlights off the roadway and into the surrounding right of way and keep deer and other large ungulates from the road. Most studies testing the effectiveness of mirrors and/or reflectors on reducing collisions with large ungulates found that they had no effect, mixed results, or inconclusive results (Waring et al. 1991, Ford and Villa 1993, Reeve and Anderson 1993, Pafko and Kovach 1996, Barlow 1997, Gulen et al. 2000, Cottrell 2003, Rogers 2004). However, some studies did find a significant reduction in deer–vehicle collisions with reflectors (Schafer and Penland 1985: 88 percent; Pafko and Kovach 1996: 50–97 percent in rural areas). However, Pafko and Kovach (1996) also found an increase in collisions in metropolitan areas. Others reported on problems with low light reflection intensities (Sivic and Sielecki 2001) and maintenance (replacing missing reflectors, keeping reflectors clean) (Sielecki 2004). Studies testing the influence of reflectors on animal behavior found little or no evidence of inducing avoidance behavior (Zacks 1986, Waring et al. 1991, D'Angelo et al. 2006). A study of wildlife warning reflectors in four colors (red, white, blue-green, and amber) found them to be ineffective at altering white-tailed deer behavior so that deer–vehicle collisions might be prevented (D'Angelo et al. 2006). However, reflectors produced a weak fleeing response in kangaroos (Ramp and Croft 2002), and one study found that deer initially responded to reflectors with alarm and flight but then became habituated to the light reflection (Ujvari et al. 1998).

Deer Whistles

Deer whistles and other audio animal warning devices are designed to alert wildlife of oncoming traffic and keep deer and other large ungulates from the road. Deer can hear ultrasonic frequencies up to at least 30 kHz (D'Angelo et al. 2007). However, most studies report no behavioral changes in animal behavior or the number of collisions (Romin and Dalton 1992, Bender 2001, Ujvari et al. 2004).

Olfactory Repellents

Olfactory repellents are designed to keep deer and other large ungulates from the road. Experimental scent marking with scent of ungulate predators, and other components, showed an 85 percent reduction in moose–train collisions in Norway. These results may be questionable given the short treatment distances of 500 meters (1,640 feet), which yielded small and variable sample sizes for the number of collisions (Andreassen et al. 2005). Captive trials of a synthetic scent repellent showed no repellency of caribou or black-tailed deer (Brown et al. 2000b, Shipley 2001). Another synthetic canine predator odor had aversive effects on one species of marsupial but attracted another (Ramp and Croft 2002). Captive trials of a putrescent whole egg repellent initially altered caribou feeding behavior, but feeding times and amount eaten eventually returned to pretreatment levels (Brown et al. 2000b). Habituation of the animals and the effort associated with continuous replacement of the scent sources may make this mitigation measure ineffective and impractical.

Deer Flagging Models

White-tailed deer raise their tails to expose the white underside of their tail when fleeing (Caro et al. 1995). Painted wooden silhouette models of deer with painted or actual deer tails (deer flagging models) were designed to invoke a fleeing response in real deer (Graves and Bellis 1978). However, the researchers found the models to be ineffective in deterring deer from the roadway (Graves and Bellis 1978). As with many other devices designed to scare ungulates away from the road, habituation is likely.

Hazing

Hazing or aversive conditioning of large ungulates involves scaring wildlife away from the road by frightening them, for example, by using human presence and movement, lights, water sprays, pyrotechnics, cannons, guns, and helicopters (DeNicola et al. 2000, Peterson 2003, Kloppers et al. 2005, VerCauteren et al. 2006). In a survey, three of forty-three natural resource agencies reported the use of hazing as a method to reduce deer road mortality (Romin and Bissonette 1996). One of the three agencies reported success using this method. Hazing was not considered effective in ungulate–vehicle collision mitigation efforts in Alaska (Thomas 1995) and British Columbia (Sielecki 2004). Similarly, green and blue lasers were found to be ineffective as frightening devices to disperse deer at night as the deer saw and followed the laser light and appeared to be more curious than frightened. Similarly lights and water sprays have only limited effectiveness in scaring deer (DeNicola et al. 2000). Hazing with sounds (e.g., pyrotechnics, cannons, guns, and helicopters) may offer a temporary solution for dispersing animals, but noise can be a problem in areas with human presence. Again, habituation appears a problem when trying to scare ungulates away from roads.

Escape Paths from Road Corridor

In Alaska, the Alaska Moose Federation creates escape paths from the road corridor through clearing and compacting snow (Peninsula Clarion 2008). Moose that would otherwise not be able to leave the road corridor because of high snow berms can now escape from the road corridor by using these paths. The escape paths are combined with supplementary feeding away from the road corridor. It is unclear whether escape paths also guide moose toward the road corridor and what the effectiveness of the escape paths is in reducing moose–vehicle collisions.

Deicing Alternatives

The principal deicers used by transportation agencies are chloride-based salts such as sodium chloride (NaCl), calcium chloride ($CaCl_2$), and magnesium chloride ($MgCl_2$), and acetate-based deicers such as potassium acetate, sodium acetate, and calcium magnesium acetate (Xianming Shi and Laura Fay, Western Transportation Institute, Montana State University, pers. comm.). The use of chloride salts in winter maintenance can attract wildlife to the road and the right of way and may increase collisions with large ungulates, especially in areas without natural salt licks (Fraser and Thomas 1982, Miller and Litvaitis 1992, Brownlee et al. 2000). A moose study in New Hampshire found that all of the animals' home ranges converged in an area with salt (NaCl) deposits formed by runoff from the road (Miller and Litvaitis 1992). Reducing the amount of salt or using alternative deicers (without salts), in combination with providing alternative salt sources away from the road, may reduce the attractiveness of the road and the right of way and, as a consequence, reduce the number of wildlife–vehicle collisions (Grosman et al. 2009). Lithium chloride, a gastrointestinal toxicant, was found to effectively discourage captive caribou from eating treated food and may prove useful as an alternative to NaCl (Brown et al. 2000b). CaMg-acetate has also been recommended as a deicing alternative (Groot Bruinderink and Hazebroek 1996). On the other hand, trials with calcium chloride were considered unsuccessful in Jasper National Park, Canada (Bertwistle 1997). While deicing alternatives are needed for a variety of reasons, including water quality concerns, there are concerns with the use of some alternatives as well.

Intercept Feeding

Intercept feeding involves strategically placing supplemental food sources in order to lure animals away from a road. A study that tested the effectiveness of intercept feeding in reducing mule deer–vehicle collisions in Utah found that intercept feeding may have reduced collisions by as much as 50 percent (Wood and Wolfe 1988). Similarly, in Norway a combination of scent marking, forest clearing, and supplemental feeding, may have reduced moose–train collisions (Andreassen et al. 2005). Providing salt sources away from the road reduced moose–vehicle collisions in Quebec, Canada (Grosman et al. 2009), but, attempts at discouraging animals from road salt using intercept mineral baiting were unsuccessful in Jasper National Park, Canada (Bertwistle 1997). While intercept feeding may be effective under certain conditions on the short term, intercept feeding is typically labor intensive and may cause animals to depend on supplemental food, and it may increase their population size (Wood and Wolfe 1988).

Minimize Nutritional Value of Right-of-Way Vegetation

Roadside vegetation can attract wildlife to roads and may increase wildlife–vehicle collisions (Putman 1997). Minimizing the nutritional quality of the right-of-way vegetation may reduce ungulate presence and, as a consequence, reduce ungulate–vehicle collisions. The nutritional quality of the right-of-way vegetation may be minimized by planting unpalatable species, reducing forage quality, mowing or cutting, or applying noxious chemicals (Groot Bruinderink and Hazebroek 1996, Putman 1997, Forman and Alexander 1998, Wells et al. 1999, Evink 2002, Rea 2003, Riley and Sudharsan 2006). In Norway, vegetation removal along a railway line (20–30 meters [66–98 feet] on each side) resulted in a 56 percent reduction in moose–train collisions (Jaren et al. 1991). However, the timing and frequency of mowing and cutting must be carefully evaluated as re-growth may be very attractive to ungulates (Rea and Gillingham 2001, Rea 2003). It is difficult to predict the effects of different types of vegetation management. In addition, the effect on native species that may now mainly occur in right-of-ways has to be considered.

Carcass Removal

Carcasses of animals killed by traffic can attract carnivores and scavengers. Although large wild ungulates are typically not attracted to or deterred by carcasses; the timely removal of road-killed animals is considered good practice and is not only safer to humans but also keeps birds and mammalian carnivores and scavengers from spending substantial time on or close to the road and reduces their exposure to potential collisions.

Increase Median Width

A wide median may allow animals to cross only one direction of traffic at a time. However, it is unclear whether a wide vegetated median reduces wildlife–vehicle collisions or whether the habitat in the median actually attracts animals to a dangerous environment (Clevenger and Kociolek 2006).

Population Size Reduction; Culling

Reviews suggest that a reduction of deer population size across a relatively wide area can result in fewer deer–vehicle collisions (Waring et al. 1991, Craven et al. 2000, Iowa Department of Natural Resources 2005). However, actual data on the effectiveness of population reduction programs on wildlife–vehicle collisions are scarce. One of the few examples is a field test in Minnesota which showed that a deer population reduction program reduced winter deer densities by 46 percent and deer–vehicle collisions by 30 percent (Doerr et al. 2001). Deer population size can be reduced through culling, relocation, antifertility treatment, or habitat alteration away from the road. Culling needs to be targeted at young animals and does rather than bucks, and combined with the declining number of hunters and potentially dangerous situations in urban settings, a shift from recreational hunting to shooting by professionals may be expected (Enck et al. 2000, Riley et al. 2003, Porter et al. 2004). Culling efforts are more likely to result in a substantial reduction in deer population size if the herd size is relatively small to begin with and if it is a closed population that does not allow influx of animals from nearby places. The effort has to be repeated periodically as the deer population will grow back to the same levels if the habitat conditions remain similar (i.e., it is not a one-time-only measure). In addition, the fertility of white-tailed deer is density dependent (Swihart et al. 1998) suggesting that as population density is reduced, increased effort is needed to keep the deer density at the lower level. Furthermore, wildlife culling may be met with strong public opposition, and a public relations campaign should be considered along with a proposed culling effort. The legal and ecological feasibility of baiting should be carefully evaluated (Brown and Cooper 2006, Van Deelen et al. 2006). Culling may not be possible, desirable, or effective on private lands, in remote areas or areas with relatively little human disturbance, or in certain urban and suburban settings (Brown et al. 2000a, Côté et al. 2004).

Population Size Reduction; Relocation

Deer relocation involves the capture, transport, and release of deer (i.e., moving deer to another location). It is typically considered if population reduction is required but culling is not an option. The effectiveness of relocation efforts can be seriously diminished if it is an open population that allows the individuals from neighboring populations to fill the gaps or that allows the relocated individuals to return (Cromwell et al. 1999). The effort will have to be repeated periodically as the deer population will grow back to the same levels (growth, immigration, potentially including individuals that were relocated) if the habitat conditions remain similar; it is not a one-time-only measure. In addition, relocated deer can experience relatively high mortality from capture-related causes, and in one study half of the relocated deer dispersed from their release site (Cromwell et al. 1999, Beringer et al. 2002), and relocation can result in the spread of infectious diseases (Miller and Kaneene 2006). Furthermore, relocated individuals may compete with individuals that are already present at the release site, or they may contribute to the growth and overpopulation at the release site and the negative effects associated with overpopulation (Côté et al. 2004). In summary, wildlife relocation is generally not recommended (Craven et al. 1998).

Population Size Reduction: Antifertility Treatment

Antifertility treatment can limit the size of a deer population (Turner et al. 1992, Turner et al. 1996, McShea et al. 1997, Waddell et al. 2001, Naz et al. 2005). This measure is typically considered or applied where killing (through hunting) is illegal (private lands, legislation) or impractical (suburban areas, public pressure), and where a relatively small and closed deer population is concerned (e.g., 200 or fewer breeding females) (Boone and Wiegert 1994, Seagle and Close 1996, Turner et al. 1996, Rudolph et al. 2000, Kirkpatrick and Rutberg 2001, Merrill et al. 2003, Porter et al. 2004). Some drugs have proven to be effective for up to one or two years, but repeated application is typically needed. The reversibility of antifertility treatment can be considered an advantage (if reproduction is necessary later) as well as a disadvantage (continuing treatment required) (Turner et al. 1996, Kirkpatrick et al. 1997). Depending on the drug, fertility control and immunocontraceptives can disrupt normal reproductive behavior and can cause physical problems with the reproductive system, abscesses and inflammations, weight gain, changes in general behavior, and changes in the sex ratio in the herd (Nettles 1997, McShea et al. 1997). Antifertility treatment requires considerable effort and cost (see review in Huijser et al. 2009b) and is generally considered only in very specific circumstances.

Population Size Reduction: Habitat Alteration Away from the Road

Deer population size depends on the quantity and quality of their habitat. An abundance of food and cover, in combination with an absence of predators and hunting, allows for relatively high population densities and potentially also a relatively high number of deer–vehicle collisions (Porter and Underwood 1999, Brown et al. 2000a, Enck et al. 2000, Rooney 2001, Côté et al. 2004, Ditchkoff et al. 2006). In general, good feeding habitat for deer includes young forests (e.g., in harvested areas that have been replanted or that have naturally regenerated), agricultural lands (grasslands or alfalfa fields, especially if they are fertilized and irrigated, and croplands), lawns and gardens (including golf courses), and riparian habitat. Good cover is provided by forests or shrubland. When there is a matrix of good cover with good feeding habitat, deer population densities are typically relatively high. Habitat alteration may include reducing edge habitat by having larger patches of cover and feeding habitat or reducing the quality and quantity of the available food (Pettorelli et al. 2005). Reducing the quality of the available food may be achieved by certain mowing or cutting practices, allowing for natural succession to more mature forests (where applicable) with different grass–herb and shrub vegetation on the forest floor, and reducing or stopping irrigation and the use of fertilizers (Gill et al. 1996). Reducing the quantity of the available food can be achieved by allowing the natural succession to more mature forests (where applicable) with less grass–herb and shrub vegetation on the forest floor, or making prime feeding habitat unavailable to the deer (e.g., through the use of wildlife fencing) (Gill et al. 1996, Darimont et al. 2005). Some of the measures discussed earlier may take a long time to take effect, while other measures may require a change in land-use practices. If the habitat is negatively affected on a large scale within a short time period, population control may be required to avoid potential starvation or dispersal in response to the reduction in habitat quality and availability.

Wildlife Fencing and Crossing Opportunities for Wildlife

Fencing is one of the most commonly applied measures to keep wildlife from entering the road corridor and is one of the primary components of highway mitigation for wildlife (chapter 2). Wildlife fences in North America typically consist of 2.0- to 2.4-meter (6.5–8 feet) high woven wire mesh or cyclone fence material. In some cases electric fencing is used, which may not need to be as high (e.g., 1.3 meters [4.3 feet] for deer) (Seamans and VerCauteren 2006). Wildlife fencing has reduced ungulate–vehicle collisions by 78.5 to 99 percent (Reed et al. 1982, Lavsund and Sandegren 1991, Clevenger et al. 2001a, Dodd et al. 2007a, Ward 1982, Woods 1990, Sielecki 1999). If the purpose of a fence is not just to keep large ungulates from entering the fenced road corridor, but also to be a barrier to medium-sized mammal species or even amphibians, smaller-sized mesh fencing or screens may be partially dug into the ground (Feldhamer et al. 1986, Clevenger et al. 2001a, Huijser et al. 2008). In addition, the effectiveness of wildlife fencing depends on fence maintenance and repair of holes (Feldhamer et al. 1986). Fencing is recommended when traffic volume is so high that animals almost never succeed in crossing the road and high traffic mortality is known to be a major threat to the population viability of the species concerned (Jaeger and Fahrig 2004). Fencing is discouraged when the population is stable or increasing or if animals need to access resources on both sides of the road, unless the fencing is combined with safe crossing opportunities for wildlife (Jaeger and Fahrig 2004). Good practice always combines absolute barriers, such as wildlife fencing, with safe crossing opportunities for wildlife. Wildlife crossing opportunities that have been implemented in combination with wildlife fencing include gaps in fence combined with warning signs, gaps in fence combined with crosswalks and warning signs (Lehnert and Bissonette 1997), gaps in fence combined with animal detection systems, and wildlife under- and overpasses (Clevenger et al. 2002a, chapter 2). These different types of safe crossing opportunities vary in their effectiveness in reducing wildlife–vehicle collisions (see table 3.2). In addition to providing safe crossing opportunities for wildlife, good practice also combines absolute barriers with opportunities that allow animals to escape from the road corridor (wildlife jump-outs or escape ramps) should they happen to end up between the fences (Bissonette and Hammer 2000, Huijser et al. 2008). Other design and implementation considerations relate to fence ends, access, road landscape aesthetics, and the like (Huijser et al. 2008).

Boulders

Large boulders have been placed in the right of way, outside of the clear zone, as an alternative to wildlife fencing. Large boulders are thought to make it hard for animals, especially ungulates, to walk across an area and approach the road. Boulders have been used for this purpose along State Route 260 in Arizona and are thought to be effective (Terry Brennan, U.S. Forest Service, pers. comm.; Norris Dodd, Arizona Game and Fish Department, pers. comm.).

Long Bridges and Tunnels

Long bridges that elevate roadways and road tunnels under the landscape are defined as structures that are at least several hundreds of meters long, sometimes many kilometers. Long bridges leave the landscape and ecological processes under the roadway intact, whereas long tunnels leave the landscape and ecological processes above the road intact. Tunnels may include "cut and cover" strategies, where a "roof" is constructed over the road that may be revegetated. Drilled tunnels do not disturb the landscape above the road during construction. Long tunnels and bridges are primarily constructed because of the nature of the terrain (e.g., through a mountain, across a floodplain), but in some cases they are constructed to avoid areas that are ecologically very sensitive and where no alternatives are available. If the nature of the terrain permits, animals can move freely over long tunnels or under long bridges, and because the animals are physically separated from traffic, ungulate–vehicle collisions are eliminated. However, because of their relatively high costs (Huijser et al. 2009b) these structures are rarely specifically implemented to reduce wildlife–vehicle collisions.

Conclusions

Long bridges and tunnels, wildlife fencing in combination with underpasses and overpasses, and animal detection systems, with or without wildlife fencing, are among the most effective measures to reduce collisions with large ungulates (see tables 3.1 and 3.2). Long bridges and tunnels are likely to eliminate ungulate–vehicle collisions for the road sections concerned, but the reduction of wildlife–vehicle collisions is rarely the primary reason for their construction. On the other hand, wildlife fencing in combination with underpasses and overpasses are often implemented to reduce wildlife–vehicle collisions. While this measure does not eliminate these types of collisions, it reduces them substantially (greater than 80 percent). The recommended spacing, type, and dimensions of wildlife under- and overpasses are discussed in chapter 2. The effectiveness of animal detection systems may be similar to wildlife fencing in combination with wildlife under- and overpasses, but the estimate of the effectiveness of animal detection systems is not nearly as robust. Therefore, animal detection systems should still be considered experimental, and the estimates on their reliability and effectiveness may change substantially as more data become available.

Animal detection systems may substantially reduce the number of ungulate–vehicle collisions, but since they allow large animals to cross the road at grade, they will never completely eliminate ungulate–vehicle collisions. Animal detection systems can be aesthetically displeasing, especially large components like solar panels (if they are used as a power source). Wildlife crossing structures have greater longevity and lower maintenance and monitoring costs (Huijser et al. 2009b). Additional considerations are discussed earlier in this chapter.

Perhaps the most important difference between implementing either wildlife fencing in combination with wildlife under- and overpasses or animal detection systems relates to the goals of the project. If the success of a project largely depends on reducing the number of ungulate–vehicle collisions substantially, the use of wildlife fencing in combination with under- and overpasses is probably advisable. On the other hand, if a project aims to contribute to the body of knowledge on the reliability and effectiveness of animal detection systems, regardless of how much ungulate–vehicle collisions may be reduced, at least in the short term, one may consider the implementation and monitoring of an animal detection system.

Acknowledgments

We would like to thank Rob Ament , Jon Beckmann, Tony Clevenger, Julie Fuller, Amanda Hardy, Jodi Hilty, Angela Kociolek, Dick Luedke, and Dan Smith for their help with preparing various parts of this manuscript.

Chapter 4

Safe Passages for Fish and Other Aquatic Species

MATTHEW D. BLANK

Fish and other aquatic species require access to a variety of habitats for maintenance of their life histories (Rieman and McIntyre 1993, Kahler and Quinn 1998). Barriers to their movement can spatially isolate populations, increase genetic isolation, and decrease the long-term viability of species (Winston et al. 1991, Morita and Yamamoto 2002, Wofford et al. 2005). This means that fish and many other aquatic species must move, in some cases very long distances, to access habitats and maintain viable populations.

Waterways intersect frequently and a variety of passage structures may enable aquatic continuity under roads. These structures are often grouped as culverts or bridges. Culverts come in a variety of shapes, materials, and sizes and were created to provide drainage by allowing water to pass under a road. A common type of culvert is a round, corrugated metal pipe. Other examples of culverts include concrete box culverts, pipe arches, and open-bottom arches that can span an entire stream channel. Bridges are structures that fully span a waterbody, sometimes requiring piers to support the span. Culverts will typically have a layer of soil or aggregate between the culvert and the road surface. Most bridges do not have any aggregate between the support structure and the road surface.

Improperly designed, constructed or maintained culverts can create barriers to the movement of fish and other aquatic species (Votapka 1991,

Warren and Pardew 1998, Sagar et al. 2003, Gibson et al. 2005). Data presented at the National Fish Passage Summit held on February 15 and 16, 2006, in Denver, Colorado, indicated that there are over 2.5 million barriers to fish from culverts, dams, and canals on streams and rivers across the United States (USFWS 2003).

In some areas of North America considerable strides have been made to quantify road–stream crossings and the number of potential barriers to fish movement (Taylor 2001, Tchir et al. 2004, Gibson et al. 2005, Whatcom County Public Works 2006, USDA–Northern Region 2008, and others). Figure 4.1 shows the percentage of culverts that are thought to be barriers to fish relative to the total number assessed in various regions of North America. The data show the number of culverts that were believed to be barriers ranged from 50 percent of the total inventoried to as high as 82 percent, with an average of 66 percent. These percentages group total, partial, and temporal barriers together.

This chapter describes how roads affect aquatic connectivity with a focus on fish, although aquatic connectivity is clearly important for all aquatic organisms and the cycling of energy and nutrients through an ecosystem (Coe 2004, Williams 2006). Fish are perhaps the most visible and important aquatic species economically, and road mitigation research and design have centered on them. Discussion of aquatic connectivity for other aquatic species is also dispersed throughout the chapter.

Passage structures create potential barriers to fish movement. First this

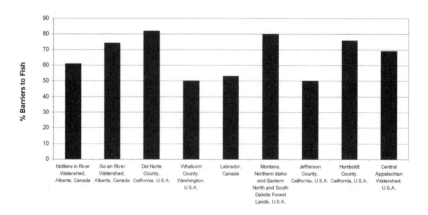

FIGURE 4.1. Summary of results from a sampling of culvert barrier inventories for various parts of North America.

chapter discusses common approaches used to assess whether a passage structure is a barrier to movement and the degree that it limits movement. Current designs that effectively provide fish and other aquatic species passage are also reviewed. Approaches to retrofitting existing crossings are described next. In the last section, the broader context of how roads fragment both surface and subsurface hydrologic flows is considered.

Although the focus is on road–stream crossings, also called passage structures in this chapter, the information applies to many physical structures that are placed in a stream or river system. Irrigation diversion structures are an example of a structure that is ubiquitous in many parts of North America (especially the arid and semiarid regions west of the Mississippi River) and the world. Like improperly designed, constructed, or maintained road–stream crossings, these structures can also be barriers to aquatic movement. The same holds true for dams.

Assessment of Passage Structures

This section addresses two key questions and related issues: Why and how do road crossings create barriers to fish and aquatic organism movement? How does one determine whether a structure is a barrier and the extent or severity of the barrier? Figure 4.2 shows a range of road crossings, including structures that are barriers to passage and others that are passable and more ecologically sensitive.

It is the static nature of the physical structures, such as culverts, placed in a dynamic stream environment that is the main reason why barriers develop over time at road–stream crossings. Rivers and streams are dynamic systems as they continually adjust their shape, both plan and section, to balance the erosive power of moving water with the capacity of the stream to transport sediment and woody debris. A rigid, immobile structure restricts the natural flow, variability, and processes of a stream system.

For example, a culvert may be sized to pass the bankfull flow of a stream channel by designing the culvert width to equal or exceed the stream channel bankfull width and by constructing a natural stream bed through the culvert. Bankfull flow is commonly defined as the flow level that begins to overtop the stream channel banks and access the floodplain (Rosgen 1996). When flow rates are contained within the bankfull channel, the culvert will maintain flow depths, velocities, and sediment transport similar to the natural channel. However, when higher flow events occur, the natural channel will allow the water to flow over the stream's banks and onto the

FIGURE 4.2. Photos show a range of passage structures. Upper left photo is a concrete box culvert with a large outlet drop that presents a barrier to upstream fish movement. Upper right photo is a squash corrugated metal pipe with an outlet drop that blocks juvenile trout, yet allows some adult trout to access upstream habitats. The middle photo shows a passage structure that provides passage for fish and function for stream processes. The lower left photo shows a large culvert battery on a large river system. The lower right photo shows four small corrugated metal pipes that restrict stream continuity and create barriers to upstream fish passage under some flow conditions.

floodplain, thus dissipating its energy. No floodplain access exists in the culvert example, so the water flow is restricted. The restricted flow causes increased water velocity in the culvert. The increased velocity in turn increases the erosive power of the water, which can create scouring of stream substrate downstream of the culvert barrel — a condition called the shotgun effect. The scour at the outlet creates a plunge pool and, over time, lowers the stream bed profile immediately downstream of the crossing. Eventually, the outlet drop below the culvert will grow large enough to be a leap barrier to upstream migrating fish and other aquatic species.

Types of Physical Barriers

Improperly designed, constructed, and maintained road–stream crossings can create the following types of physical barriers:

- Leap barrier
- Velocity barrier
- Depth barrier
- Debris accumulation and/or dam barrier
- Others

Figure 4.3 presents examples of common physical barriers created by improperly designed, constructed or maintained passage structures.

Factors That Create Barriers

The physical factors influencing fish passage through a culvert are discussed as they would be encountered by a fish ascending a culvert. The first potential obstacle to passage is presented on the approach to a culvert. Obstacles there may include a jump that must be overcome to reach the culvert. Jumping abilities vary by species and by size (Hoar and Randall 1978). Recent studies involving brook trout (Salvelinus fontinalis) showed jumping performance to be strongly affected by jump height, water depth, and fish size (Brandt et al. 2005, Kondratieff and Myrick 2006). This research showed that, although larger fish could leap greater heights, smaller fish were capable of leaping a greater number of body lengths than larger fish, and the probability of successfully leaping over an obstacle was inversely proportional to jump or leap height.

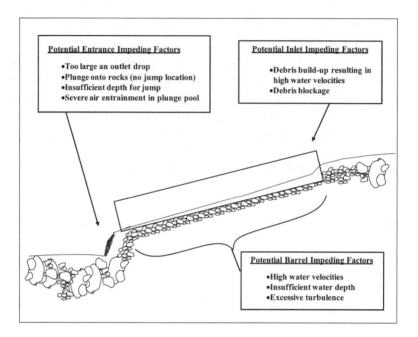

Potential Entrance Impeding Factors

•Too large an outlet drop
•Plunge onto rocks (no jump location)
•Insufficient depth for jump
•Severe air entrainment in plunge pool

Potential Inlet Impeding Factors

•Debris build-up resulting in
 high water velocities
•Debris blockage

Potential Barrel Impeding Factors

•High water velocities
•Insufficient water depth
•Excessive turbulence

FIGURE 4.3. Side view of a hypothetical culvert and adjacent stream channel with potential physical barriers to upstream passage by fish and other aquatic species. Water flow is right to left in the diagram.

The hydraulics and physical characteristics of the jump location appear to play a role in the ability to successfully leap an obstacle, and in the leaping behavior of some salmonids. Salmon and sea trout were observed using the vertical currents that formed on the downstream side of a plunging outfall in a natural falls (Stuart 1962). The hydraulics in this area also appeared to stimulate leaping behavior (Stuart 1962). Pacific salmon accelerated from the bottom of pools below natural falls by using burst swimming to the surface rather than leaping from near the surface itself (Lauritzen 2002). The ratio of leap height (or height of fall) to pool water depth also plays a role in successful leaps over natural falls and into manmade structures like culverts (Stuart 1962, Lauritzen 2002, Brandt et al. 2005).

Once inside a culvert, water depth becomes a factor in passage. If there is insufficient depth in the culvert, the fish cannot swim upstream even if other conditions are favorable to passage. Research from the Clearwater River drainage in Montana showed small brook trout and westslope cutthroat trout (*Oncorhynchus clarki lewisi*) negotiated water depths as low as 31 millimeters (1.2 inches) (average length of adults = 97 millimeters [3.8

inches]) (Burford et al. 2009). Researchers have suggested minimum water depths for passage of adult trout should be 80 millimeters (3.2 inches; (Saltzman and Koski 1971), 120 millimeters (4.7 inches; Lauman 1976) and 150 millimeters (5.9 inches; Baker and Votapka 1990).

Water velocity can be a critical factor for preventing upstream passage. High water velocities force fish to expend more energy swimming against the flow. In some cases the velocity of the water is greater than the swimming ability of the fish. Fish become exhausted as they attempt to pass through the swift water and are flushed back downstream. Velocity thresholds that define acceptable conditions for passage depend on the species and size of the fish. For example, several studies of trout have placed an upper threshold for water velocity at 1.2 meters/second (4 feet/second; Saltzman and Koski 1971, Lauman 1976, Belford and Gould 1989). Water velocities exceeding this threshold are likely to prevent or impede passage. Data collected by Warren and Pardew (1998) suggest that increased water velocity restricts fish passage for warm water fish species.

The direction of flow and its magnitude affect how certain fish species behave. For example, in a series of flume experiments designed to investigate swimming behavior of shovelnose sturgeon (*Scaphirhynchus platorynchus*), fish were observed to have strong flow orientation. At low velocities, the fish swam upstream and downstream through the flume using a mixture of head first and tail first alignments; however, as velocity was increased, fish eventually aligned themselves parallel to the main flow direction and maintained a tail first orientation in both upstream and downstream movements (White and Mefford 2002).

Adult Arctic grayling (*Thymallus arcticus*), juvenile coho salmon (*O. kisutch*) and juvenile chinook salmon (*O. tschawytscha*) have been observed utilizing the lower velocity regions that develop along the sides and bed of culverts (Travis and Tilsworth 1986, Kane et al. 2000, Pearson et al. 2005). Researchers speculated that fish were seeking to minimize their energy expenditure by traveling against the lower velocities. In some cases, fish may not have been able to pass without seeking and using the lower velocity areas.

Passage success can be affected by the length of a culvert and the availability of resting areas in the culvert. For example, trout passed more easily through a culvert when rocks were placed within it as compared to a culvert barrel free of rocks (Belford and Gould 1989). These rocks were thought to provide resting areas for the fish during passage upstream.

The inlet of a culvert can also present challenges to fish. Excessive sediment can deposit on the upstream end of a culvert (Kane and Wellen 1985).

This deposition may result when culverts constrict flow by having widths that are narrower than the upstream active or bankfull channel width, or when the culvert slope is less than the upstream channel slope (Kane and Wellen 1985). Sediment buildup at the culvert inlet can also steepen inlet slopes and increase water velocities (Behlke et al. 1991).

The inlet area presents another potential challenge the fish must overcome for successful passage through a culvert. When fish are swimming through high velocity areas, which can form at constricted culvert inlets with sediment buildup, they often use their burst speed (Behlke et al. 1991). Fish have a time to exhaustion related to swimming mode, and for burst swimming the time to exhaustion ranges up to 15 seconds (Hoar and Randall 1978). By the time they reach the inlet area the fish may already have used up much of their available energy and are nearing the exhaustion point.

There are other factors, such as air entrainment and turbulence, that affect swimming abilities and the potential for a successful passage (Stuart 1962, Powers 1997, Lauritzen 2002, Enders et al. 2003). Air entrainment is the mixing of air with water (as occurs where water plunges into a pool) resulting in a fluid mixture that is less dense than water alone. The reduced density of the mixture affects a fish's swimming performance and ability to overcome passage obstacles. Turbulence is the erratic fluctuations and motion in water. These factors are complicated, and their role in passage is not well known. In addition, these factors are easily confounded with velocity.

Fish Locomotion

Fish locomotion is an important biological factor that interacts with the physical environment and plays a key role in fish passage. Temperature, dissolved oxygen, motivation to swim upstream, gender, physical condition, disease, and sexual maturity all affect fish locomotion (MacPhee and Watts 1976, Wardle 1980, Baker and Votapka 1990, Bell 1991, Kynard and O'Leary 1993). Fish swimming is often discussed as three regimes: sustained swimming, prolonged swimming, and burst swimming (Hoar and Randall 1978).

- Sustained swimming is a spectrum of swimming activities and speeds that can be maintained for an indefinite period.
- Burst swimming is rapid movements of short duration and high speed, maintained for less than 15 seconds.

- Prolonged swimming covers a spectrum of speeds between burst and sustained and is often categorized by steady swimming with more vigorous efforts periodically. The swimming period lasts between 15 seconds and 200 minutes and if maintained will end in fatigue.
- Fatigue or exhaustion occurs when a fish collapses and can no longer maintain a given swimming speed.

One might make an analogy to human movement as follows: a human walks at a sustained speed, jogs at a prolonged speed, and sprints at a burst speed. Swimming ability is related to the species of fish and the size of the individuals within a given species (Hoar and Randall 1978).

Two modes of swimming ability, prolonged and burst, are unsustainable, and the mechanisms for switching from one to another are not very well understood (Castro-Santos 2005). The fact that two unsustainable modes occur means that there is not a single swim speed–fatigue time relationship. Maximum distance achieved depends on which mode the fish utilizes and the behavior and strategy used to shift from one mode to the next (Castro-Santos 2005). For example, American shad (*Alosa sapidissima*), alewife (*A. pseudoharengus*), and blueback herring (*A. aestivalis*) used a distance-maximizing strategy to pass through a flume with high velocities, whereas striped bass (*Morone saxtalis*), walleye (*Stizostedion vitrium*), and white sucker (*Catostomus commersonii*) did not (Castro-Santos 2005).

Behavioral Barriers

Another category of barrier is behavioral barriers (Clay 1995). Behavioral barriers are conditions, such as unnatural lighting or excessive darkness created by very long, small-diameter pipes that trigger a negative reaction in fish. These conditions may prevent fish from attempting passage even if there is not a physical impediment to their passage. Sound pressures created by impact or vibratory pile driving techniques that are often used to embed steel piles for bridge support may negatively affect fish and other aquatic species (Hawkins 2005).

Assessment Techniques

Interest in aquatic connectivity has grown in the last few decades, as has interest in techniques for evaluating structures as barriers to movement and

Direct Approach: Field experiments that measure fish movement directly and compare movement to flow conditions in a structure.

Indirect Approach: Approximate movement potential using thresholds, modeling, or comparisons between population characteristics measured upstream and downstream of a crossing.

•Tagging studies: mark-recapture, PIT tagging, or others (e.g., radio telemetry)

•Visual observations

•Video camera

•Regional screens based upon field and laboratory experiments that established thresholds such as maximum outlet drop height, maximum culvert slope, or slope × length criteria (surrogate for velocity and exhaustion)

•Hydraulic modeling

•Comparisons between upstream and downstream fish population characteristics

•Genetic differences

FIGURE 4.4. Assessment techniques for investigating fish and other aquatic species passage through structures.

the severity or degree to which a crossing acts as a barrier to aquatic species. Figure 4.4 depicts most of the available techniques for assessing road–stream crossings as physical barriers to fish and other aquatic species movement.

Road–stream crossings are typically categorized as one of four types: (1) total barrier (no passage regardless of species and flow), (2) temporal barrier (structure that is passable only at certain flow rates), (3) partial barrier (structure that is a barrier to a certain species, size-class, or life stage only), or (4) passable (Flosi et al. 2002).

Developing "flow" or "passage windows" is a method that refines the status of a potential barrier and defines thresholds for passage relative to flow rates in the stream and crossing structure (Reiser et al. 2006). Figure 4.5 depicts a passage window for a culvert in a tributary stream to the Yellowstone River. Flow rates below the lower threshold present a low-flow barrier, and flows above the upper threshold create a velocity barrier. Passage windows are defined for a certain size and species of fish. The example shown was developed for adult Yellowstone cutthroat trout and is for illustration of this concept only (Blank 2008).

An approach to performing quality assessments of a large number of passage structures is to first develop a regional screening tool based upon a subset of the total number of structures (USDA 2005). One way to accomplish this is to group all structures and locations by their physical, geomor-

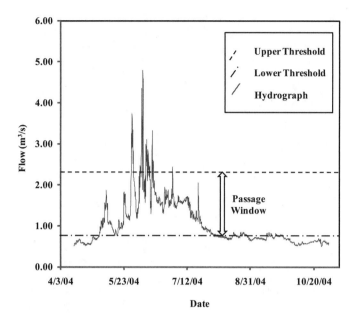

FIGURE 4.5. A measured hydrograph collected over the summer of 2004 in Mulherin Creek in Montana. A flow or passage window for adult Yellowstone cutthroat trout, estimated using a combination of hydraulic modeling and fish locomotion, is shown relative to the hydrograph.

phic, hydrologic, and biological attributes. A representative sample (subset) of each group should then be thoroughly evaluated using a direct measure of passage. An effective technique that works even through high-water periods uses passive integrated transponder (PIT) tags placed inside representative individuals of the aquatic species of interest, with detection antennae installed on the upstream and downstream ends of a structure. Although this method can be labor and equipment intensive, it provides very accurate information about the timing of natural movement relative to flow conditions (Solcz 2007). Results from the direct assessments can be used to determine or refine regional thresholds for passage structures within each group for various species and life stages. Passage can be inferred for other structures that were not directly assessed.

Placing individual road–stream crossings in the watershed context is a key part of the framework for addressing aquatic connectivity issues and prioritizing replacements. Understanding and addressing individual road–stream interfaces is essential; however, without expanding that understanding and placing it within the big-picture context, efforts to improve

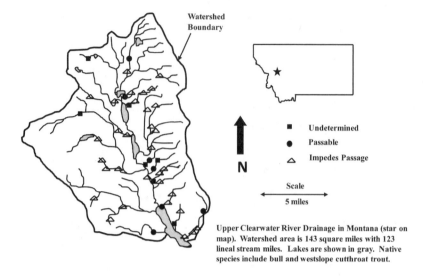

Upper Clearwater River Drainage in Montana (star on
map). Watershed area is 143 square miles with 123
lineal stream miles. Lakes are shown in gray. Native
species include bull and westslope cutthroat trout.

FIGURE 4.6. An example of a watershed-scale assessment of road–stream crossings
and their barrier status. The figure was modified for instructive purposes from the
original published in Burford et al. (2009).

aquatic connectivity will be piecemeal and less effective than a large-scale
watershed approach (see fig. 4.6, Burford et al. 2009). By looking at all
crossings collectively in combination with habitat quality, species distribution, population characteristics, and other site-specific considerations, decisions can be made regarding which replacements will have the most benefit
to the entire ecosystem and/or populations of key species targeted for restoration. Typically, improving crossing structures on the lower reaches of a
watershed first will provide a greater amount of aquatic connectivity than
replacing a barrier in a small tributary stream that has other barriers to passage downstream.

The watershed context can also help guide restoration and protection
of native species. For example, some conservation strategies will purposely
isolate native fish populations upstream of barriers to protect them from
harmful interactions with non-native species. In order to do this, knowledge regarding the size and quality of habitat in relation to barriers is critical to ensure there is sufficient habitat to maintain the long-term viability of
an isolated native population (Peterson et al. 2008).

Prioritization schemes and tools have been developed to guide removal
and replacement of culvert barriers (Washington Department of Fish and
Wildlife 2000, O'Hanley and Tomberlin 2005). One example prioritizes

crossings by calculating a priority index that incorporates the following criteria (Washington Department of Fish and Wildlife 2000):

- Proportion of fish run expected to gain access
- Species-specific production potential of gained habitat
- Amount of habitat gained
- Mobility needs
- Stock status of fish
- Cost of project

Prioritization schemes are necessary because there are many problem culverts and limited resources available for replacing them. Focusing restoration efforts on the biggest problems first is a prudent and effective approach.

Care must be taken when assessing culvert barriers because research has shown that different assessment methods can result in different results as to their barrier classification (Blank 2008, Burford et al. 2009). A basinwide study in Montana assessed trout passage through forty-five culverts during summer low flow using three different approaches: upstream and downstream fish population characteristics, hydraulic modeling using FishXing (Furniss et al. 2008), and direct passage experiments with marked trout. Agreement between FishXing results and direct passage results was low (17 percent) for a subset (n = 12) of culverts studied (Burford et al. 2009). Fish passage studies by Coffman (2005) and Rajput (2003) also show incongruence between different methods used to assess barriers to fish passage. Coffman (2005) assessed the accuracy of three predictive models by comparing their estimates of passability to passage assessments using a mark–recapture technique. The models failed to accurately predict passage approximately 50 percent of the time. The prediction success was improved after revising the predictive model parameters based on the data from the mark–recapture experiments. Finally, assessment approaches should strive to be as accurate as possible yet err on the conservative side. If a structure is limiting passage for the species of concern, it should be identified as a barrier.

Design of New Passage Structures

Considerable effort has been spent to identify and develop efficient and effective design methods and construction techniques that result in ecologically sensitive, robust road–stream crossings (USDA 2008). A range of possible design approaches includes techniques typically grouped as follows: hydraulic design, hydraulic simulation, and geomorphic design (Hotchkiss and Frei 2007). Historically, road–stream crossings were designed following a hydrologic approach, which sizes the passage structure to convey a design flood flow without consideration of the needs of aquatic species and stream continuity.

Hydraulic Design

Hydraulic design procedures combine the known swimming capabilities for a design fish with estimates of the hydraulic environment, including water depth and average velocities, within and near the road–stream crossing for a range of flows (Hotchkiss and Frei 2007, Behlke et al. 1991). Fish passage design flows, including a low- and high-passage flow, are estimated using flood frequency analyses or other hydrologic methods. Fish movement timing relative to typical annual hydrographs is estimated. A hydraulic model of the proposed road–stream crossing is developed, and the model is used to estimate velocity and depths within the structure for the identified range of hydrologic flows. A design fish, including species and size, is selected. Swimming abilities for the design fish are used to estimate passage through the modeled hydraulic environment. If the initial crossing configuration (shape, length, width, etc.) creates a hydraulic environment that exceeds the design fish's swimming capacity, the road–stream crossing configuration is modified, the model is rerun, and the comparison between the estimated hydraulic environment and fish locomotion is performed again. This iterative approach is repeated until the design constraints dictated by the design fish are met. Of course, the design must also factor in other engineering considerations such as road alignments, surface loads, soil and bedrock conditions, groundwater, and others.

Hydraulic Simulation

Hydraulic simulation is a technique that matches the hydraulic environment within the road–stream crossing to the hydraulic environment within the natural stream channel (Hotchkiss and Frei 2007). The approach creates a diversity of water depths and velocities within the crossing structure by embedding the crossing in substrate or adding other natural roughness elements.

Geomorphic Design

Geomorphic design approaches mimic natural stream channel characteristics through the road–stream crossing structure (Hotchkiss and Frei 2007, USDA 2008). Stream simulation is the name of the geomorphic design method developed by the U.S. Forest Service. Stream simulation designs create crossings that are wider than the stream channel and at grade with the streambed. They maintain a continuous substrate, with characteristics similar to those of the natural stream channel. This approach is the most conservative and relies upon the assumption that if the road–stream crossing maintains the same features of the natural stream channel, then passage for all aquatic species will be provided. Research on the effect of road culverts on salamander movement shows that culverts installed following stream simulation techniques will provide the maximum benefit to salamanders, which are considered to have limited mobility compared to fish (Sagar et al. 2003, Ward and Anderson 2008).

Cost Implications of Passage Structures

One important aspect of crossing design is the cost of the structure. To incorporate more ecological function, road–stream crossings will be sized larger. The larger the crossing structure, the more expensive it is. Therefore, the type of structure used must incorporate a cost–benefit analysis to determine what is feasible at a given road–stream crossing site. Each design approach should incorporate ecological function and need, and the relative cost should be compared for each approach (fig. 4.7).

Commonly Used Passage Structures

There are hundreds of different types and configurations of passage structures that can be used at road–stream crossing intersections (Hotchkiss and Frei 2007). Some of the more common culvert crossings are shown in figure 4.8. The figure shows common single-barrel shapes and two common multiple-barrel configurations. Depending on the size of the stream, the expected flows and the goals of the project, any of these configurations could be used for any of the design approaches.

As with culverts, there is also a wide range of bridge types. Because bridges typically span water bodies (with the use of support piers in some

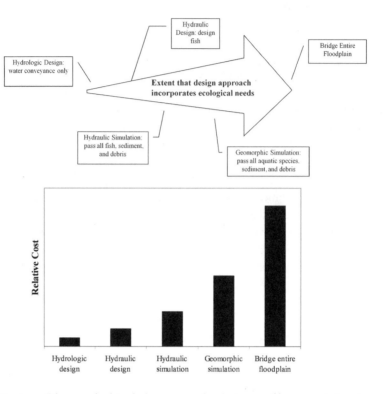

FIGURE 4.7. Diagram depicts design approaches in terms of how much they incorporate ecological needs (after Bob Gubernik). The relative cost for various approaches has been generalized for this chapter. There are situations where a hydrologic design approach may provide adequate ecological connectivity. For example, the hydrology in some regions may require sizing structures to be very large, which incorporates more ecological function.

cases) and allow proper stream functions, they are generally preferred over culverts from an ecological perspective.

Retrofitting and Other Mitigation Measures

In some situations, retrofitting a road–stream crossing is the only feasible approach to improving passage. Replacing a problem structure is almost always the preferred approach when compared to retrofitting; however, replacement may be cost prohibitive making retrofitting the only means of improving an inadequate structure. Another reason why retrofits may be

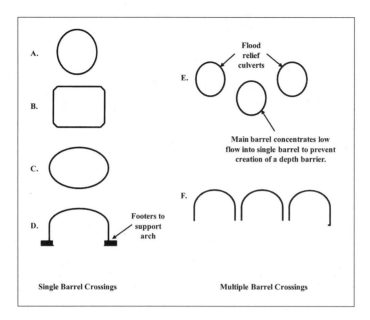

FIGURE 4.8. Illustration of typical shapes of culvert passage structures used for road–stream crossings: A, round corrugated metal pipe (CMP) or other material; B, square box culvert with chamfered corners; C, elliptical CMP; D, open-bottom arch; E, multiple round CMPs; F, multiple arches.

needed is that some road–stream crossing structures have a design life of 75 or more years. In these cases, the long-term viability of fish or other aquatic species populations that use the stream may be jeopardized if passage is not improved before replacement of the structure is required.

The general idea behind retrofitting structures is to slow the speed of the water, provide resting areas, and increase water depths. Retrofit approaches fall into two general categories: approaches that modify the interior of a crossing structure and those that modify the stream channel downstream of a crossing.

Many types of baffle systems have been developed over the past century to improve fish passage (Ead et al. 2002, fig. 4.9). Baffling systems can be added to new installations during design and construction or to an existing crossing structure. At times crossings must be placed on steep gradients (greater than 4 to 6 percent slopes). In these situations, baffles can be added to the structure to slow down the water and increase its depth or to hold and/or collect substrate in the structure.

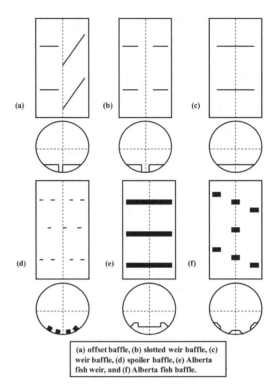

(a) offset baffle, (b) slotted weir baffle, (c) weir baffle, (d) spoiler baffle, (e) Alberta fish weir, and (f) Alberta fish baffle.

FIGURE 4.9. Different types of baffle configurations for culvert retrofits (after Ead 2002).

Baffles have been shown to improve passage conditions not only for strong-swimming fish like trout and salmon but also for weaker swimmers. Spoiler baffles designed to improve fish passage success for common jolly-tail (*Galaxias maculates*) and spotted galaxias (*G. truttaceus*) increased passage for both species to about 80 percent in the most complex baffle configuration as compared to 13.5 percent where no baffles existed (MacDonald and Davies 2007).

If the passage barrier is caused by an excessive outlet drop, a useful approach might be the use of rock or other types of weirs placed downstream of the structure. Weirs create a "backwatering" effect that increases the water depth within the structure while slowing the water velocity.

Retrofitting approaches must be used with caution as they can create additional problems, and in some instances they may contribute to failure of a structure (Blank 2008). Baffles can catch sediment and woody debris, thus reducing the ability of the structure to properly convey water. If enough debris and sediment clog the structure, water may overtop the road or begin to flow along the sides of the structure, eventually causing complete failure or collapse of the road.

Roads that bisect wetlands can cause significant mortality for amphibians and reptiles. U.S. Highway 27 in Florida created conditions that resulted in the direct mortality of large numbers of turtles attempting to cross the highway. Drift fencing to direct the turtles to a highway culvert resulted in significant reductions in direct mortality of turtles attempting passage. Data from research monitoring indicated a reduction in turtle mortality, comparing pre- to post-fence construction, from 11.9/kilometer/day (7.4/mile/day) to 0.9/kilometer/day (0.6/mile/day) (Aresco 2005).

Fishways

Fishways are another option for improving fish passage at barriers (Clay 1995). Fishways work in a way similar to baffling systems. Using specially designed baffles and flow-control structures (such as orifices or vertical slots), they slow water flow, increase its depth, and create passageways through which fish can move to get past otherwise impassable structures. They were originally designed to facilitate passage for anadromous fish around large dams in river systems like the Columbia River in the Pacific Northwest (Clay 1995).

A particularly interesting type of fishway has been used to help freshwater eels (*Anguilla anguilla* and *A. rostrata*) migrate up river systems in the eastern United States and parts of Europe (Clay 1995). These unique fishways use synthetic grass, bristles, and branches to provide a pathway for migrating eels. Eel locomotion includes both swimming and crawling modes. Thus eel fishways take advantage of the eels' unique physical abilities to help them pass over what would otherwise be migration barriers.

Roads and Alterations to Hydrologic Connectivity

Roads can alter hydrologic processes, including the natural surface and subsurface flow paths that transmit water, sediment, and nutrients to and through watersheds, wetlands, streams, and rivers (Trombulak and Frissell 2000, Forman et al. 2003, Wemple and Jones 2003, Coe 2004).

Some effects of roads on hydrology and geomorphology include channelization of streams and intermittent channels, changes to flood dynamics, and changes to groundwater–surface water interactions (National Academy of Sciences 2005). Debris and sediment movement across the landscape are affected by roads (Jones et al. 2000). Chemical characteristics of water flows can also be altered by roads due both to altered hydrology and geomorphology and to runoff from the road surface (Forman et al. 2003).

Research shows that hillslope runoff processes in the Pacific Northwest are dominated by subsurface stormflow (Coe 2004). Road construction often uses cutting and filling of parts of the landscape to establish proper topography for road building. Road cuts can intercept the subsurface stormflow from upslope areas (Jones et al. 2000). The intercepted flow is then redistributed by the local geometry and structure of the road in the vicinity.

Groundwater–surface water interactions through the hyporheic zone (the area under a stream or river where water is actively exchanged between surface water and groundwater) are important for maintaining various types of aquatic processes and habitat (Giller and Malmqvist 1998). For instance, groundwater exchange with surface water can regulate stream temperatures. Groundwater temperatures do not fluctuate as much as surface water temperatures and typically maintain more constant temperatures throughout the year. Groundwater exchange with streams during low-flow and hot periods of the year, such as late summer, can provide much needed cooler water temperatures for cold-water aquatic species like salmonids (Edwards 1998). Bull trout also seek habitat with groundwater upwelling or downwelling for spawning sites (Baxter and Hauer 2000). Increased channel confinement led to decreased hyporheic exchange in the Willamette River in Oregon, regardless of whether the confinement was natural or anthropogenic (Fernald et al. 2001).

Conclusions

Although great strides have been made to improve aquatic connectivity across road systems, considerable challenges face our aquatic habitats, especially when future population growth and associated development pressures are considered. Additionally, there is still a need to invest in new research to help guide and understand the often complex interactions of roads and waterways.

Most departments of transportation across the United States are actively implementing strategies and design protocols that provide aquatic species passage through the road–stream crossing interface. As more attention is paid to the problems presented by barrier culverts, more ecologically friendly road crossings are being designed and constructed than ever before. Even so, solutions to these problems are not complete. Monitoring of new crossing structures that are constructed with the latest techniques is required to evaluate that design protocols and construction methods do provide long-term ecological connectivity in all types of aquatic habitats.

There is some resistance to the stream simulation approaches for passage structure design and construction. This is likely because the initial cost for these structures is greater than that for smaller structures designed only for passing water, even if long-term maintenance is likely to cost less and the chance for complete failure of a crossing designed to accommodate stream function is remote. Unfortunately, data are limited that validate the short-term and long-term cost of various types of road–stream crossings. A cost–benefit study of the range of passage structures across a variety of different types of aquatic environments is needed. Such a study should quantify initial costs for design, permitting, and construction as well as long-term costs associated with operation and maintenance over the life of the structure. In addition, the study should account for ecological benefits provided by larger, ecologically sensitive crossings.

As stated in the beginning of this chapter, most of the research related to aquatic passage is centered on fish species, with the bulk of that work targeting anadromous salmonids. There is still a need to investigate the effect of road–stream crossings and roads on other aquatic species and all types of aquatic habitats.

With regard to fish passage, efforts to perform fish locomotion and behavior studies across a range of life stages and species should continue. These studies should focus on volitional swimming methods rather than traditional methods that relied more on forced swimming. Also, efforts to understand fish behavior in relation to passage structures should continue.

Given the large number of barrier crossings (estimated to be in the tens of thousands to hundreds of thousands in the United States alone); more efficient and cost-effective solutions should be explored. Perhaps there are more creative and cost-effective materials that could be developed into full spanning passage structures.

More research is needed to understand the effects of roads on all hydrologic processes, such as nutrient fluxes, in a variety of geographic settings. In addition, effective approaches to providing hydrologic connectivity for all forms of surface and subsurface flows are limited and not consistently tested and validated.

The goal for the future should be a landscape with aquatically connected habitat for all native aquatic species. Of course, there are situations where connectivity is not desired, as in the case of native species that are threatened by interactions with non-native species. How can we get there? First, all land management agencies across the world must increase their efforts to inventory and assess existing road–stream crossings and other human-made barriers to aquatic connectivity. Next, a regionally based prioritization strategy that places each crossing within a watershed context should be employed to organize and effectively guide restoration efforts. Using the prioritization strategy, problem crossings should be identified then replaced or retrofitted, depending on site-specific considerations, economics, and other factors previously discussed. Once the crossings are restored, effective monitoring should be done on a watershed scale to further our understanding about which techniques are most effective in which aquatic habitats. Finally, land management agencies need to continue to transition their policies toward a more proactive environmental stewardship role.

Acknowledgments

I would like to thank Shane Hendrickson of the U.S. Fish and Wildlife Service for providing the Northern Region Inventory Data. Many thanks are given to Tom McMahon and Joel Cahoon, professors at Montana State University, for their continued positive insight and guidance.

PART II

Ecologically Effective Transportation Plans and Projects

The first part of the book described how the effects of traffic and roads can influence landscapes and local natural environments, but transportation systems also affect local communities and the overall quality of life. Since 1969, the National Environmental Policy Act has required a coordinated review process for any newly proposed project with input from other agencies, such as state and federal land and wildlife management agencies, and the public. Public involvement in transportation project planning has increased across North America as changes in legislation have created more opportunities for environmental stakeholders and interested individuals to engage in the review process. Further, traditional mitigation for transportation impacts has tended to be site-specific, with little consideration of how the project affects the surrounding ecosystem. These issues are exemplified in this part with a chapter covering progressive transportation planning by an interagency group determined to streamline the environmental review process while focusing on ecosystem-based priorities. This part of the book also explores public participation from both transportation agency and public interest perspectives, to stimulate better understanding and communication between transportation agencies and public interest groups for mutually beneficial highway mitigation objectives and goals.

Chapter 5

An Eco-Logical Approach to Transportation Project Delivery in Montana

AMANDA HARDY AND DEB WAMBACH

Safe and efficient transportation is a component of strong economies and a high quality of life, but transportation infrastructure can have negative consequences for fish and wildlife populations (see chapter 1, Evink 2002, Forman et al. 2003, National Academy of Sciences 2005, Clevenger and Wierzchowski 2006). While efficient transportation systems are necessary, these systems should avoid, minimize, and compensate adverse impacts to natural resources. In the United States, the environmental review process for infrastructure project planning and delivery requires reasonable efforts to avoid detrimental impacts to human and natural communities, as well as historic and cultural sites. Nevertheless, unavoidable impacts occur, and responsible parties are legally obligated to offset negative environmental effects to meet regulatory requirements (e.g., Section 7 of the Endangered Species Act, Section 404 of the Clean Water Act). Upon satisfying the terms of the myriad of regulations (see Brown 2006 for a complete list of regulations that must be addressed in the environmental review process), necessary permits are issued and construction is allowed to proceed.

The environmental review process for complex transportation projects is often the most time-consuming part of project delivery (Evink 2002, Government Accountability Office 2003). Traditionally, impact assessment occurs on a project-by-project basis, and the task of developing appropriate measures to mitigate adverse impacts can require significant time and effort, sometimes imposing unpredictable and costly delays. Further, efforts to mitigate impacts are commonly directed toward the affected resources at the project site. While this approach may satisfy regulatory requirements, it is questionable whether the ecological integrity of the disturbed area and adjacent habitats can be fully restored. This approach often overlooks other conservation opportunities in the affected region that might offer a better return for the mitigation investments. Recognizing such shortcomings of the environmental review process, Congress incorporated provisions into the last two transportation bills (the 1998 Transportation Equity Act of the Twenty-first Century and the 2005 Safe, Accountable, Flexible and Efficient Transportation Equity Act: A Legacy for Users) to improve environmental stewardship and expedite the environmental review process for transportation projects (Government Accountability Office 2008).

In response to this challenge, an interagency team compiled guidance and examples for streamlining environmental reviews while more effectively protecting natural resources and ecosystem processes (Brown 2006). Entitled, "Eco-Logical: An Ecosystem Approach to Developing Infrastructure Projects," this document encourages federal and state agencies to strategically collaborate to target ecosystem-based mitigation for regional conservation priorities, early in the project planning and review process. By fulfilling regulatory obligations in advance of final design and construction, this approach has potential to reduce costly delays in project delivery while increasing the cost-effectiveness of mitigation efforts by focusing on prioritized conservation initiatives. The product of extensive discussions between infrastructure-development and regulatory agencies, Eco-Logical has earned executive-level reassurances that a flexible ecosystem approach for environmental reviews can be implemented under existing legal mandates.

An interagency group in Montana adaptively applied the Eco-Logical ideas to create and pilot the Integrated Transportation and Ecosystem Enhancements for Montana (ITEEM) process. This chapter examines the events that motivated Montana agencies to rethink the environmental review process and how the Eco-Logical approach influenced the development of the ITEEM process. We summarize the ITEEM process and describe how the process is being debuted in a pilot study effort. Finally, we synthesize Montana's experiences through the development and initial application of the process thus far, offering insights to other entities that may be embarking on their own path to streamline environmental reviews while improving environmental stewardship.

Historical and Political Setting for the Eco-Logical Approach

In 2002, Executive Order 13274, entitled "Environmental Stewardship and Transportation Infrastructure Project Reviews," directed agencies to streamline environmental reviews while enhancing environmental stewardship for transportation infrastructure projects. Valid concerns were raised that the environmental review process would be compromised in the effort to speed up project delivery. To address these concerns and find creative solutions to the Executive Order's charge, an interagency team of federal regulatory and infrastructure development agencies (including Bureau of Land Management, U.S. Environmental Protection Agency, Federal Highway Administration, National Oceanic and Atmospheric Administration, Fisheries Service, National Park Service, U.S. Army Corps of Engineers, U.S. Department of Agriculture, U.S. Forest Service, U.S. Fish and Wildlife Service, Volpe National Transportation Systems Center, the Knik Arm Bridge and Toll Authority) as well as departments of transportation from several states (North Carolina Department of Transportation, Vermont Agency of Transportation, Washington Department of Transportation), was formed to explore integrated planning approaches to improve stewardship and reduce project delivery timelines. Building on related initiatives that encourage collaborative and balanced conservation approaches to address environmental reviews and mitigation efforts (Office of Environmental Protection 1995, Executive Order 13,352 2004) and the "Enlibra Principles" (Western Governors' Association 1999), the group developed the Eco-Logical guidance promoting ecosystem-based mitigation strategies to improve the environmental review process (Brown 2006).

Eco-Logical defines ecosystem-based mitigation as the practice of coordinating advanced mitigation of infrastructure project impacts by preserving, enhancing, and creating habitat and ecosystem functions where such actions are most needed and where such contributions have been determined to be the most beneficial to regional conservation efforts (Brown 2006). When ecosystem-based mitigation is accomplished early in the planning of infrastructure projects, agencies capitalize on meaningful conservation priorities and opportunities that may be vanishing or becoming prohibitively expensive over time. This increases the cost-effectiveness and ecological benefits of mitigation investments. Advanced mitigation planning can be targeted to fulfill environmental regulatory requirements that avoid costly permitting delays while making important contributions to regional conservation initiatives. The ecosystem approach balances transportation project delivery and ecosystem conservation objectives, reflected in the following goals defined in Eco-Logical:

- *Conservation*: Protection of larger scale, multiresource ecosystems
- *Connectivity*: Reduced habitat fragmentation
- *Predictability*: Knowledge that commitments made by all agencies will be honored—that the planning and conservation agreements, results, and outcomes will occur as negotiated
- *Transparency*: Better public and stakeholder involvement at all key stages in order to establish credibility, build trust, and streamline infrastructure planning and development

To implement an ecosystem approach, Eco-Logical outlines three components that build upon each other through an adaptive feedback loop: integrated planning, mitigation options, and performance measurement (fig. 5.1), briefly described in the next three sections.

Integrated Planning

Establishing regional ecosystem conservation priorities is the basis for ecosystem-based mitigation. Integrated planning is pivotal in determining these priorities. Eco-Logical offers an eight step approach for integrated planning. Briefly, these steps involve developing collaborative partnerships, synthesizing information to identify regional conservation concerns and opportunities, considering how anticipated project impacts might be offset by identified conservation opportunities, and prioritizing opportunities

FIGURE 5.1. Components of an ecosystem approach, as outlined in Eco-Logical (Brown 2006).

that satisfy legal mandates. A consensus-based list of opportunities to offset impacts is then incorporated into the National Environmental Policy Act (NEPA) planning and permitting efforts for the project(s) in question. The final step of integrated planning evaluates how recommended mitigation options that advanced to the NEPA process were incorporated into the final project to address the regional conservation priorities.

Mitigation Options

Eco-Logical describes and offers examples of different mitigation approaches, including project-specific mitigation, multiple-project mitigation, ecosystem-based mitigation, off-site and/or out-of-kind mitigation, as well as mitigation banking, in-lieu-fee mitigation, and conservation banking techniques (Brown 2006). Benefits and drawbacks of each type of mitigation approach are explored in Eco-Logical, as are issues of accountability in ecosystem-based mitigation and conservation banking.

Performance Measurement

The final component outlined in Eco-Logical occurs as infrastructure projects are completed and collaborators assess if desired outcomes were achieved. Based on this evaluation, adaptations to improve the next cycle of Eco-Logical steps are documented. The success of the ecosystem approach depends on adapting priorities, acknowledging successes and failures, and searching for solutions to problematic aspects of the process from one cycle to the next.

The concepts behind the three components just described, along with case studies examples and the executive-level endorsements of Eco-Logical's approach, played important roles in Montana's efforts to create their own ecosystem-based approach to transportation project delivery. Even prior to the release of Eco-Logical, however, there was momentum to create a new approach to transportation project delivery in Montana, prompted by other influential factors.

Rationale for Applying an Eco-Logical Approach in Montana

In 2002, multiple projects planned for the U.S. Highway 93 (U.S. 93) corridor in northwest Montana were identified as high-priority projects under

Executive Order 13274. Legitimate concerns were raised regarding balancing environmental stewardship and expedited environmental reviews along the 460.2 kilometer (286 mile) corridor that traverses important wildlife habitats in the mountains and valleys of western Montana. Stakeholders involved in the U.S. 93 reconstruction (not including projects on the Flathead Indian Reservation, see chapter 8 this volume) began the process of developing a new, defendable approach to increasing the efficiency of the review process for multiple projects while embracing environmental stewardship approaches in the region.

To achieve this goal, an Interagency review team (the review team) was formed with upper-level managers from the Montana Department of Transportation (MDT); the Federal Highway Administration, including Federal Lands Highways (FHWA); the Montana Department of Fish, Wildlife, and Parks; the Montana Department of Environmental Quality; the Montana Department of Natural Resources and Conservation; the Confederated Salish and Kootenai Tribes; the US Environmental Protection Agency; the U.S. Army Corps of Engineers (the Corps); the U.S. Fish and Wildlife Service; and the U.S. Forest Service (USFS). The review team designated representatives from their respective agencies to form a working group to explore more efficient and effective environmental mitigation methods to decrease review times, while upholding important environmental protections in the U.S. 93 corridor.

Despite acknowledging the inefficiencies and inadequacies of the current environmental review system, the working group was understandably tentative about possible legal ramifications of revamping the existing review process. The working group's momentum slowed as unanswered questions and lack of direction overshadowed the task of finding a plausible path toward streamlining the environmental review process for the U.S. 93 projects. In the meantime, standard planning and compliance processes for the U.S. 93 projects were already under way, such that pursuing a new approach for these projects at that point would be counterproductive in terms of shortening the environmental review timeline.

While little new ground was broken to streamline the U.S. 93 environmental review process, the agencies recognized that they had initiated important discussions to improve the environmental review process. The review team asked the working group to redirect their efforts toward the Montana Highway 83 (MT 83) corridor, where two future highway reconstruction projects were in the earliest stages of planning, an important consideration when applying an ecosystem approach. By then, the Eco-Logical document was garnering buy-in from leaders of infrastructure and regulatory agencies alike. The FHWA Montana Division Office staff involved in developing the Eco-Logical document recognized that the working group might be able to apply the ideas in Eco-Logical to make headway on issues that had previously hindered progress. Per the review team's acceptance of the ideas in Eco-Logical, this guidance document was adopted by the working group as a foundational resource offering creative approaches to addressing difficult procedural, legal, and environmental issues associated with planning, environmental review and project delivery.

With review team oversight, a committed working group, a focal region with transportation projects in the earliest stages of planning, and guidance and executive-level endorsements offered in Eco-Logical, momentum and direction materialized out of a period of admitted ambiguity. The final component that helped move the process development phase forward was a project coordinator, hired by FHWA to orchestrate the group's efforts and document the resulting discussions and agreements. Over the course of a year, the project coordinator and working group explored an array of potential approaches, discarded dead-end ideas, and eventually arrived at a common vision and methods for testing an ecosystem approach in Montana. The outcome of their diligent efforts was the first version of the Integrated Transportation and Ecosystem Enhancements for Montana (ITEEM) process (Hardy 2008).

Integrated Transportation and Ecosystem Enhancements for Montana

The ITEEM goals and desired outcomes, roles and responsibilities, dispute resolution process, and tasks to apply the ITEEM process are presented in the following sections. The text here is a simplified version of the document used to guide the pilot study (Hardy 2008), which is discussed later in the chapter.

Desired Outcome and Goals

The desired outcome of the ITEEM process is to balance environmental and transportation values by streamlining transportation program delivery while applying more effective ecosystem conservation. Schedule, cost, safety, quality, public input, regulatory requirements, ecological concerns, and other factors are considered equally with no single factor dominating as the top priority. Specific ITEEM goals expand upon the Eco-Logical goals as follows:

- *Conservation*: Protection of larger-scale, multiresource ecosystems
- *Connectivity*: Enhanced or restored habitat connectivity and reduced habitat fragmentation
- *Early involvement*: Early identification of transportation and ecological issues and opportunities (*issues* refers to potential impacts or concerns associated with the transportation projects under review or regional conservation initiatives; *opportunities* refers to potential options to concurrently address mitigation requirements and conservation priorities)
- *Cost efficiency*: Making the best use of transportation program funding by focusing mitigation efforts where they would be most effective
- *Cooperation*: Finding solutions acceptable to all participating agencies
- *Predictability*: Knowledge that commitments made early in the planning process by all agencies will be honored—that the planning and conservation agreements, results, and outcomes will occur as agreed
- *Transparency*: Better stakeholder involvement to establish credibility, build trust, and streamline infrastructure planning and development

Roles and Responsibilities

Stakeholders share responsibility of finding solutions that meet both transportation and ecosystem conservation goals. One representative and an alternate from each participating agency will commit to serving in the ITEEM oversight group. Individual representatives in the oversight group serve as their agency's point of contact, representing their agency's interests and responsibilities. The oversight group will strive for consensus as they negotiate to optimize ecological conservation opportunities while reducing project development time and increasing the predictability of program delivery. It is the responsibility of the oversight group to identify issues and opportunities, prioritize opportunities to improve long-term cost-efficiency of mitigation efforts, apply programmatic approaches or establish best management practices as appropriate, document recommendations and establish work groups dedicated to implementing recommendations, establish measures of success to evaluate and adapt the process to better meet goals and objectives, and ensure open and ongoing communication between agency representatives and various stakeholders such as non-government organizations (NGOs) and people affected or interested in the project and conservation efforts in the region.

Dispute Resolution

In the event that consensus may not be achieved at any point in the ITEEM process, the dispute resolution process establishes a 2-week timeframe for issues to be resolved via the oversight group, during which time the conflicting parties will focus on resolving the disagreement via open discussion. If the unresolved issue is not critical to the process, the parties with contrasting points of view can respectfully "agree to disagree" with no further implications; these disagreements will be documented for the record. If the issues must be resolved for the process to effectively move forward and parties are unable to come to a solution within two weeks, the issue will be elevated to the review team for upper-level managers to make a final decision.

ITEEM Process Tasks

The ITEEM process consists of six tasks that may be adapted in the future based on evaluation of the process. Each task is briefly described here.

TASK 1: ESTABLISH REGIONAL BOUNDARIES

The oversight group will determine the region where multiple transportation projects in the early stages of planning are programmed to be delivered in the future (e.g., 5–20 years) and where conservation issues and opportunities need to considered in the planning process. This region may be determined based on jurisdictional boundaries, but the region could be delineated by ecologically relevant features such as watershed boundaries if striving for an ecosystem approach.

TASK 2: COMPILE AND PREPARE INFORMATION

With the assistance of a facilitator, participating agencies will be responsible for providing the best available data in a timely manner to identify issues and opportunities relevant to the programmed projects and regional conservation interests. Additionally, the facilitator will hold a public open house to obtain relevant input and information from other stakeholders, including the public and NGOs, ensuring transparency by accommodating public involvement early in the process. Examples of the type of information and data to be collected may include the following (see Brown 2006 and Hardy 2008 for an extensive list of information that may be important to consider in the process): landownership, planned developments, and projected land use change; traffic data and projections; conservation easements; state wildlife conservation plan with fish and wildlife species ranges and critical habitat designations; road kill locations and numbers; habitat connectivity models; wetland locations; water quality impaired streams and local watershed management groups' efforts; culvert locations and fish passage data; and other regional collaborative conservation efforts.

The facilitator will organize and present a comprehensive list of the compiled data and sources to the oversight group, who will determine the final set of information that will be referenced at a collaborative workshop, the next step of the process. The facilitator will document justification for retaining or rejecting data and summarize the final set of information that will be referenced at the workshop; this memorandum will be distributed to participating agencies, allowing them to prepare statements regarding issues and opportunities to be discussed further at the workshop.

TASK 3: WORKSHOP

The facilitator will organize a workshop for stakeholders to collaboratively discuss possible impacts of the proposed transportation projects and regional conservation opportunities that could be tapped to offset negative effects. Workshop participation will be limited to the oversight group, a few agency technical staff, and representatives from local governments and NGOs with regional expertise and an understanding of the ITEEM process. The workshop may occupy two to three days and should be conducted within the identified region to foster a sense of place and facilitate timely field review of potential impact and mitigation sites of interest.

Workshop participants will refer to the information compiled during the previous task to identify issues and opportunities in the focal region at a coarse scale. The facilitator will organize a field review to further discuss and ground-truth (i.e., fact-check in the field) the information to better hone recommendations. Potential opportunities determined to be unfeasible due to physical, social, or land use constraints will be eliminated through consensus. Practical options deemed worthy of further consideration after field review will be documented by the facilitator and may include the following types of details:

- Location, methods, and schedule for implementation of mitigation
- How potential mitigation option(s) offset impacts of the proposed projects and address regional conservation interests, regulatory statutes, and streamlining of transportation project delivery
- Identification of other areas or impacts that could relinquish substantial mitigation improvements in trade for focusing efforts and limited funding on particular conservation opportunities
- Opportunities to leverage funds for collaborative conservation initiatives (i.e., if mitigation can contribute to ongoing regional conservation efforts)
- Workshop attendees' preliminary comments on the identified opportunities

Documented opportunities will be prioritized and a final list of recommendations will be established via consensus among participants. Depending on the complexity of any given recommendation (e.g., recommendations that require coordination between several agencies, conservation easements, land swaps, etc.), work groups may be identified to further detail an implementation plan, including responsible parties, estimated costs, and necessary memorandums or agreements. The final list of recommended opportunities, as well the rationale behind culling other options, will be documented by the facilitator.

Finally, workshop participants will establish measures of success to evaluate and adapt the process for future application. Three factions should be considered when establishing measures of success and performance standards. First, the ITEEM process itself should be evaluated; this may include assessing different facets of the process such as data assimilation, identification of issues and opportunities, workshop field review and prioritization approaches, and agency involvement. Second, infrastructure projects themselves should be appraised in terms of how mitigation recommendations were incorporated into the NEPA process, how permitting proceeds, and overall time to project delivery. Finally, ecological benefits realized through the process should be evaluated in terms of contributions to regional collaborative conservation priorities and initiatives. The participating agencies will determine appropriate and achievable measures for variables related to desired outcomes. Because it will take years to see many of the outcomes, the oversight group should plan periodic follow-up meetings to evaluate measures of success (see task 6) until the commitments are satisfied completely.

TASK 4: DRAFT WORKSHOP REPORT

The facilitator will compile documented recommendations and measures of success established during the workshop into a draft report. Participating agencies will have 45 days to review and comment on the draft report. After revision and agency approvals, the report will be made available for public comment for 30 days. A summary of public and agency comments will be included in the final ITEEM report as an appendix, along with a summary of options not adopted.

TASK 5: FINALIZE ITEEM REPORT

The facilitator will finalize the workshop report, including a signatory page to document agency concurrence. The final report will be referred to as the projects are moved through the NEPA review process; ultimately, recommendations put forth in the ITEEM report should expedite the process of finalizing mitigation plans in these planning documents. Oversight group representatives will continue to serve as their agency's contact for further correspondence regarding the ITEEM report and recommendations.

TASK 6: EVALUATE AND ADAPT THE ITEEM PROCESS

The oversight group will meet periodically (e.g., semiannually) to revisit the final report, discuss progress and outstanding issues, and update measures of success. If changes to existing recommendations are deemed necessary, the oversight group will find a reasonable approach that all agencies can support. Additionally, the oversight group will compile necessary inputs for evaluating measures of success and will document progress, outstanding issues, and suggestions to adapt the process in the future. These periodic meetings will take place until all commitments are fulfilled, which may take many years. Once agencies agree that the commitments documented in the final report have been met, the oversight group will have a final meeting to document lessons learned and recommendations to improve the ITEEM process, resulting in an addendum to the final report, completing a single cycle of the ITEEM process.

Pilot Study: Testing the ITEEM Process

The agencies of Montana took a big leap by committing resources to create the ITEEM process as outlined earlier. Upper-level managers demonstrated their ongoing commitment by allocating additional agency resources to implement the new approach in a pilot study. We recount how the pilot study has been carried out thus far, summarizing efforts taken to prepare for the workshop and how the workshop itself unfolded.

The pilot study began in June 2007 with the review team and MDT identifying two transportation projects that would be the focus of the pilot study. The two projects were reconstruction/rehabilitation projects for sections of the MT 83 corridor between Seeley, Montana, and the Clearwater Divide (fig. 5.2). The study area included a 15-mile corridor straddling the road where the two projects would occur and a larger region encompassing conservation interests beyond the road corridor. Issues and impacts related to the highway projects would be assessed within the 24.1-kilometer (15-mile) corridor. Opportunities to mitigate aquatic resource impacts would be considered across the entire Clearwater drainage plus the section of the Blackfoot drainage in Missoula County. Mitigation to offset terrestrial impacts would be considered across a larger region stretching from the junctions of MT 83 and MT 35 at the north end, and MT 83 and MT 200 at the south end, and extending from the crests of the Mission and Swan Mountain ranges to the west and east, respectively (see fig. 5.2).

Several characteristics of the MT 83 study area offered an excellent testing ground for the pilot study. First, these projects were in their earliest stages of the planning process, lending an opportunity for the pilot study to potentially influence and streamline the environmental review and permitting processes. Second, by focusing on two highway projects, the pilot study would explore the feasibility of addressing mitigation needs for "batches" of projects rather than using the traditional project-by-project environmental review process. The MT 83 study area encompasses important habitats for several federally listed endangered or threatened wildlife species such as grizzly bears (*Ursus arctos horribilis*), Canada lynx (*Lynx canadensis*), bull trout (*Salvelinus confluentus*), and a number of sensitive plant and animal species as well as big game species. The presence of these species meant that the process would have to specifically address Section 7 of the Endangered Species Act as the projects could impact the listed species' critical habitats and habitat connectivity in the region. Additionally, land use and management in this region are overseen by numerous entities, requiring

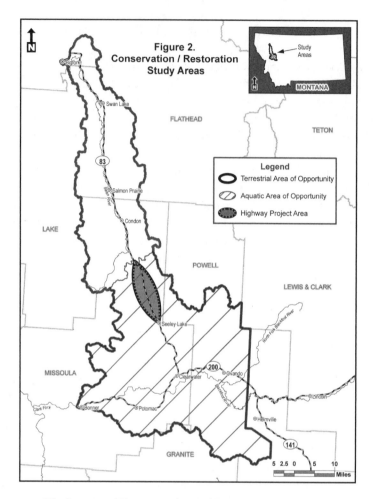

FIGURE 5.2. The Integrated Transportation and Ecosystem Enhancements for Montana (ITEEM) Pilot Study regional boundaries encompassing two Montana Highway 83 projects (highway project area) slated for future reconstruction and the larger area where terrestrial and aquatic ecological conservation opportunities were considered as potential targets for mitigation efforts to offset unavoidable impacts associated with the two highway reconstruction projects.

buy-in from numerous local stakeholders. Further, several watershed-based conservation initiatives and agency management plans were already under way in the region, ensuring that the process would need to consider collaborative partnerships directed at targeting mitigation to address established regional conservation goals.

Four levels of agency involvement participated in the pilot study: the review team (upper-level agency managers), oversight group (midlevel agency managers granted decision-making authority for pilot study implementation), working group (agency representatives most involved in accomplishing the steps of the process), and technical staff (agency representatives assisting with compiling relevant agency documents and information). Reporting to the review team, oversight group members were "the voice" of their respective agencies to make decisions and enter into tentative agreements on behalf of their agency. Working group members served as the primary point of contact for their agency in the day-to-day tasks of the process and ultimately were the true shepherds of the pilot study. There was overlap between the working group and the oversight group members; in many cases the same person served in both roles.

Formal invitations to participate in the pilot study were also extended from MDT to the commissioners of each of the four counties in the study area region. The invitation emphasized the importance of local buy-in to the process, and incorporating future planning, zoning, or development projects on the region as these actions could have notable effects on the identification and prioritization of ecological conservation opportunities.

Upon establishing agency representatives committed to the pilot study and defining the specific study area, MDT contracted a consultant to serve as a facilitator in February 2008 to compile information, facilitate the public involvement process and agency workshop, and document the evolution of the ITEEM process for one year as it is applied in this pilot study. The facilitator's multidisciplinary team, consisting of a project manager, environmental scientists, a geographic information system analyst, a logistics coordinator, and two professional facilitators, embraced the underlying principals of the Eco-Logical approach and demonstrated an understanding of the objectives of the process and goals of the pilot study. A kickoff meeting was held with the working group and the consultant to introduce the consultant's project team, determine the list of stakeholders to invite to participate in the process, and identify appropriate contacts for obtaining data. Format and protocols for data acquisition, pilot study timeline, a review of the dispute resolution process, and refinement of the regional boundaries for the pilot study were also addressed at this meeting.

Following this meeting, MDT and the consultant submitted a letter to working group members and agency technical staff requesting a list of the best available data that each agency wanted considered during the process, including relevant research studies, reports, point data, maps, and geospatial data layers. The letter also asked each agency to prepare a summary of their initial concerns and issues relating to the highway corridor and regional natural resources, along with conservation partnership opportunities in the pilot study region.

To complete the request for relevant information, the consultant advertised a public open house to catalyze public involvement and transparency of the process. Attendance at the open house was sparse, but attendees provided important contacts with several local NGOs that are active in the regional communities and already pursuing endeavors with goals comparable to those of the ITEEM process. Participation by these groups would prove invaluable with respect to local knowledge of issues and opportunities and would later provide a promising avenue for implementation of the pilot study outcomes at the local level.

The consultant spent most of the summer of 2008 compiling and summarizing the data and information for consideration in the process along with the issues and opportunities identified thus far. The working group and consultant met to select a comprehensive yet manageable and relevant subset of information and data detailing critical issues and opportunities that would be discussed at the workshop, and agencies presented their initial list of issues and potential opportunities that they would be advancing to the workshop for discussion. After this meeting, the consultant summarized the pilot study progress to date for the oversight group, including the initial list of identified issues, opportunities, and compiled information and data that would be advanced to the workshop.

The workshop, facilitated by the consultant, took place in Seeley, Montana, over three days in late October 2008 and was attended by the oversight group members and agency technical staff, as well as several local government and NGO representatives. The FHWA representative opened the workshop introductions, an overview of the ITEEM process and the pilot study's progress thus far. Five NGO representatives presented their organizations' respective missions and conservation initiatives within the greater study area. The MDT representative summarized the highway project reconstruction objectives and development process and timeline. The consultant then provided an overview of the compiled data and maps that would be used to inform decision making during the workshop.

The afternoon of the first day of the workshop was spent reviewing the compiled fine-scale data used to identify issues and planning considerations in the 24.1-kilometer (15-mile) highway corridor, and broader-scale information related to potential conservation partnership opportunities throughout the greater study area. Discussions focused on regional ecological resources of interest, including wildlife habitat linkages, grizzly habitats, lynx habitats, big game habitats, other sensitive species and species under special management status, bull trout and westslope cutthroat trout habitats, wetlands, and recreational sites. Within the 24.1-kilometer (15-mile) highway corridor, identified issues fell into six main categories: wildlife permeability, wildlife mortality, aquatic organism passage at stream crossings, water quality, wetland impacts, and adjacent land use and development (as it may affect the long-term efficacy of some mitigation investments in the highway corridor). Beyond the highway corridor, issues and opportunities within the greater study area included wildlife–human interactions (e.g., due to increasing human development and habitat fragmentation/loss, bear conflicts in residential areas), acquiring conservation easements on private lands identified as important wildlife habitat or movement corridors, watershed management and fish passage restoration, and land-use practices and development pressures on private lands. These identified issues would drive the effort to find appropriate opportunities to address associated impacts during the remainder of the workshop.

On the second day of the workshop, participants traveled the highway corridor, stopping at locations where opportunities and issues had been identified, such as stream crossings and areas where wildlife linkage zones intercept the highway. Interactions among participants in the field provided a better understanding of each others' concerns and interests, with significant payoffs realized on the following day when the group would collectively select a final list of recommendations that would be considered in the NEPA reviews for the projects in question.

The last day of the workshop was dedicated to honing the list of issues and opportunities, dropping those unsuitable for further consideration, and prioritizing those remaining for action and implementation. It was suggested that agencies consider establishing a "restoration fund" to augment conservation efforts already under way. Agencies could contribute to the fund in advance of proposed projects while proponents of restoration and conservation projects could apply for monies to address conservation priorities. The securing of conservation easements on other private lands of ecological importance was also proposed. The Nature Conservancy, Trust for Public Land, and Plum Creek had been discussing transactions pertaining to a three-phased purchase of over 121,406 hectares (300,000 acres) in the region over the next two years with the intent to transfer management of these lands to a mix of federal, state, and private ownership. This complex transaction could take years to see completion, and with these negotiations in their infancy, it was not possible to identify distinct conservation needs that could be addressed on these lands, should they be acquired. The feasibility of advanced investments of this nature requires further investigation into funding mechanisms and to determine if such funds need to be directed toward particular resource management objectives to fulfill permit obligations that the mitigation efforts are intended to address; the agencies are exploring these ideas further.

The group identified the purchase of parcels for wetland restoration as a possibility for interagency partnerships to leverage mitigation monies. The MDT had already established a wetland mitigation program in the region, but additional opportunities for compensatory mitigation were identified, such as adding additional wetlands to an existing wetland reserve or to properties adjacent to the highway corridor with potential wetland restoration opportunities that were noted during the field review. The MDT committed to look into the feasibility of wetland mitigation on these properties in an effort to establish additional wetland mitigation credits in the watershed. It was agreed that while wetland purchase and restoration may not be an immediate need within this particular watershed, if an ideal or important wetland project presented itself, the group would consider it a valid opportunity to collaborate with other stakeholders.

Research directed at understanding specific wildlife movements within the highway corridor was suggested as a potential mitigation outlet. Such research would need to consider the goals of increasing permeability of the highway to carnivore movements and reducing animal–vehicle collisions, particularly with ungulate species. Research of this nature could help identify and prioritize locations where crossing structures and exclusion fencing could most effectively intercept and accommodate wildlife movement across the road corridor, with the potential of incorporating such infrastructure into the future highway reconstruction project. Similar to the ideas already suggested, the group recognized that this would require further investigation regarding funding mechanisms and assurances of the validity of applying research to address regulations that mitigation intends to address.

While the foregoing ideas generated more questions than explicit recommendations, the group was able to identify distinct mitigation recommendations specific to highway project planning and design considerations. Mitigation opportunities suggested for the highway corridor region included fish and wildlife passages in combination with exclusion fencing to guide animals under or over the roadway. Roadside vegetation management to facilitate at-grade wildlife crossings in selected areas could be implemented to reduce animal–vehicle collisions. The group also recommended that wildlife warning signage or measures such as animal detection systems that warn drivers of crossing wildlife could be installed to reduce animal–vehicle collisions and increase safety where road alignments provide increased sight distances. For other areas of the highway corridor, a curvilinear highway design that complemented the unique and wild character of the corridor was suggested, along with the minimization of the construction footprint to reduce impacts to habitats near the highway. The inclusion of permanent erosion control facilities such as sediment basins was considered to reduce roadside animal attractants (e.g., to deicing chemicals; see chapter 3) and improve water quality. Potential conservation opportunities to compensate for project impacts across the broader study area included agency partnerships with local grass-roots efforts to facilitate private land acquisition and restoration (particularly in association with the recently initiated Montana Legacy Project, an effort aimed at conserving important forestland currently owned by Plum Creek Timber Company in northwestern Montana). Other opportunities included public education and outreach programs regarding "living with wildlife," and cooperative efforts to open up large stream reaches to fish passage and wildlife movement corridors across private and public lands. These recommendations would be advanced to the NEPA environmental review process in hopes of shortening the time required to incorporate mitigation into project planning efforts.

The workshop concluded with an exploration of the successes and challenges uncovered during the pilot study to date. Everyone agreed that the pilot study had been a worthwhile endeavor, recognizing that the process is iterative and will be improved over time as the strengths and weaknesses of the approach materialize through implementation. The discussion turned to the transition of continued oversight and management through completion of the pilot study. It was proposed that management of the process after the workshop would shift from MDT to another entity capable of stewarding the commitments through implementation within the Seeley-Swan region over the long-term. The group agreed that, if possible, a local government or rural initiative organization would be the ideal entity for carrying the conservation efforts that emerge from the pilot study to fruition, while MDT and FHWA will be responsible for incorporating design considerations along the highway corridor into the NEPA process and scoping for the future highway projects where the ITEEM process may be applied.

The workshop accomplished many of the tasks described in the original ITEEM process, but several tasks remained after the workshop was finished. The group committed to further developing measures of success, identifying areas for process improvement, following through on action items generated from workshop discussions, presenting findings and recommendations to the review team for decision making and documentation of agreements and commitments, and ultimately transferring the MT 83 ITEEM process to a local entity for implementation stewardship.

Following the development of measures of success, the consultant prepared a draft ITEEM MT 83 Pilot Study Final Report and circulated it for review and comment from the working group and oversight group members. The document included a recount of the milestones and tasks as executed through the pilot study, issues and concerns identified as planning considerations for the transportation project development, prioritized conservation partnership opportunities for the study area, and a summation of the established measures of success. A list of considerations and opportunities not advanced by the group, along with the rationale behind those decisions, was included in that document.

After agency review and comment, the revised draft final report will be made available for public comment. Public comments will be addressed by the appropriate cooperating agency and incorporated into the final report. The document will be finalized, including a list of workgroups formed to develop and execute action items and a schedule of future oversight group meetings intended to monitor and discuss the pilot study's progress as agreements and mitigation actions are fulfilled. Once finalized, the MT 83 ITEEM process pilot study commitments will be transferred to an appropriate local entity committed to manage the agreed-upon conservation actions with the assurances of long-term participation and support from the cooperating agencies.

Lessons Learned: Developing the ITEEM Process

Over the years of creating and piloting an ecosystem approach for transportation project reviews, road blocks were encountered. Acknowledging and addressing these setbacks ultimately improved the process, generating lessons learned along the way. We share these lessons to help other groups avoid pitfalls that may commonly be encountered in complex endeavors of this nature.

To start, while Eco-Logical was useful in the effort to develop the ITEEM process, this guidance document did not function as a cookbook with tested recipes guaranteed for success. Rather, examples therein helped participants understand the concepts of an ecosystem approach and how elements of this approach had been applied in other case studies. Further, and perhaps more importantly, the Eco-Logical endorsements by agency executives at the federal level provided Montana's upper-level agency managers reassurances that more creative and flexible approaches can be used while satisfying legal statutes of the environmental permitting processes. The review team's leadership was essential, and having the working group agency representatives in direct communication with upper-level management and decision-making authority helped the group advance through difficult decisions at various stages of developing the process.

As the working group developed the process, it was helpful to focus on specific future highway projects. Representatives were better able to obtain relevant (rather than hypothetical) feedback from their agency colleagues by referring to the specific region where these projects would occur, particularly in regard to compiling regional data and information and the agency's initial list of issues and opportunities. Discussion of appropriate projects to focus on while developing the process also highlighted the necessity to work with projects in the earliest stages of development to successfully reduce project planning time and to capitalize on planning advanced mitigation to address regional conservation priorities before they disappear.

The importance of creating an environment of understanding, respect, and cooperation to work through challenging issues cannot be underestimated when working collaboratively to find a common vision among groups with different missions and interests. While agency representatives in the workshop had worked together for years, they had not previously been motivated to find the greater good for all when applying the traditional project-by-project method for environmental reviews. The traditional approach to project planning did not require agencies to understand each others' interests and missions, resulting in misconceptions that initially hampered true collaboration. The agencies had to develop genuine team camaraderie to find approaches that satisfied not only their particular regulatory mandates but also the interests of the other agencies. Working group representatives were asked to share professional, educational, and personal histories and interests with each other, revealing commonalities that hadn't been discovered despite years of working together. Taking time for representatives to share their agency's mission and management plans further increased mutual understanding between agencies. One-on-one interviews between the project coordinator and representatives illuminated the history of the working relationships among the agencies, allowing for more responsive and strategic facilitation.

As the working group committed to numerous meetings to maintain momentum while developing the ITEEM process, conference calls and video conferencing were used at times to reduce travel costs, but face-to-face meetings supported team-building and helped maintain the group's momentum and accountability. A computer and projector were used to collectively view, comment on, and revise interim products at the meetings, thus reducing the number of individual iterations necessary to finalize the language outlining the process. Agendas were essential to keeping the group on track and also served to document progress made on assigned action items, increasing the accountability of group members.

While developing the ITEEM approach, the Working Group held a 2-day meeting in the study region of interest to examine how an on-site workshop might be incorporated into the process. Participants were better able to focus and engage with each other as they set aside other work demands. Intermingling in smaller groups, participants engaged in more effective exchanges that may not have emerged in more formal settings. The group seemed to relax more with each other as casual interactions occurred during interstitial periods and at meals, further building team camaraderie. This experience solidified the need to incorporate a multiday workshop as a component of the process.

Along the same lines as conducting the field visit, it was useful to carry out proposed steps to reveal how each component of the process might unfold when implemented and what resources might be necessary. With agency technical staff assistance, spatial data layers were compiled in a GIS and displayed to demonstrate how this information could be used to make informed, collaborative decisions. The exercise prompted useful comments that may not have emerged had the group opted simply to envision how this step in the process might occur. For example, by viewing the digitized data, participants realized that the process should accommodate important information that may not be available in digital spatial data layers. Collectively viewing the projected data on a screen catalyzed group discussion, but the group also concluded that having complementary hardcopy paper maps could facilitate easier documentation of issues and ideas on the maps themselves. Further, while this trial only incorporated a handful of spatial data layers from different sources, agency technical staff had to put significant effort into preparing the data (e.g., converting all layers to the same geographic projection) for efficient, comprehensive viewing. Based on this experience, it was clear that this step would require the technical skills and added expense of a paid consultant.

Beyond logistics, this exercise also revealed how strategic conservation investment trade-offs could be collaboratively identified. For example, when the group looked at locations of culverts determined to be barriers to

aquatic organism passage, it was apparent that improving just a few particular culverts to pass fish could effectively open aquatic connectivity for an
entire drainage while the same level of effort at other stream locations
would do relatively less good for regional aquatic connectivity (see chapter
4). This reinforced the importance of collectively viewing and discussing
relevant information as a group and helped root the group's understanding
and faith in the developing and evolving process.

By vetting different aspects of the proposed process, the group was also
able to drop some ideas from further consideration. Discussions about a
credit/debit system to quantify project impacts related to terrestrial mitigation opportunities dominated several meetings, but the group determined
that this task alone could require significantly more time and effort to develop. The group resolved simply to negotiate trade-offs, with the potential
of adopting more formal procedures such as the analytical hierarchy process
approach to guide decision making (Saaty 1980) or other environmentally
sensitive adaptive planning approaches that have been applied elsewhere
(Theobald et al. 2000, Beier et al. 2006, Hilty et al. 2006, Noss and Daly
2006).

The effort to develop the process was demanding but provided the necessary road map for the agencies to apply the process in a pilot study. In
summary, the most important aspects of developing the process included
upper-level management involvement and support, building trust and camaraderie among agencies, group facilitation and accountability, and using
trial-and-error to explore and find feasible approaches palatable to the
agencies that would ultimately carry out the ITEEM process. Lessons
learned as the agencies created the ITEEM process would ultimately be
useful in directing the pilot study.

Lessons Learned: ITEEM Pilot Study

The pilot study was initiated after the ITEEM process was drafted and approved by the collaborating agency leaders. However, even with an agreed-
upon and well-documented process in place, interpretation of the process
was not always congruent between stakeholders for a variety of reasons. In
some cases this was due to agencies appointing new representatives that
had not been involved in the development of the process during the previous year. Given that the pilot study venture diverges so significantly from
the traditional environmental review process, many newcomers struggled
to grasp the overarching purpose of the pilot study and their roles in the

process. Smooth integration of these new players into the process was labored, resulting in delayed task completion and insufficient detail with regard to documenting their agency's issues and opportunities, and feelings of frustration and confusion.

Several approaches to address intraagency coordination challenges were suggested. Inviting all the various levels of agency representation (e.g., upper-level managers, technical staff) to the initial kickoff meeting could improve intraagency understanding of roles, responsibilities, and goals and objectives of the process. Conducting interviews with each agency's team of representatives may flesh out misconceptions and concerns early in the process. Formalizing each agency's internal commitments to the process might unify and affirm their team's efforts; participating agencies could document their approach to internally addressing staff turnover and communication issues affecting their agency's ability to contribute effectively to the ITEEM process. To improve interagency communications, a Web site was created early in the process for the purposes of facilitating information transfer and monitoring those milestones reached and those yet to be completed. Unfortunately, it was underutilized and rarely updated. More careful consideration of the Web site function and design prior to the next cycle will likely enhance its usefulness as a communication tool used to increase the efficiency of the group.

Much of the pilot study prior to the workshop revolved around compiling relevant data that would be used to guide good decision making. The initial call for such information yielded hundreds of sources of data and maps that had to be filtered down to a manageable set of sources that comprehensively addressed issues and opportunities relative to future planning and management activities for the region. Having a consultant proved essential to preparing this information for an effective workshop.

The first day of the workshop was dedicated to introductions, reviewing how the goals of the pilot study relate to stakeholders' interests, and reinforcing the tenets of the ecosystem approach. Discussions of the intent and extent of the pilot study revealed various interpretations of the ITEEM process. Group discussion on the first day helped clarify many of these discrepancies, and the less formal field trip interactions on the second day of the workshop helped solidify the group's understanding of differing perspectives and increased the sense of shared commitment to the goals of the pilot study. The field trip allowed the group to better understand MDT's approach to highway design features, including features incorporated as a matter of standard practice, other features that could readily be incorporated into highway projects, and features that may not be feasible to incorporate into highway design plans given the need to balance other factors of highway design. The field trip allowed the group to see some of the techniques that could be incorporated into highway design to benefit wildlife. In the advent of future workshops, it was suggested that the field trip be held the first day of the workshop, rather than the second, since it provided the group with clarity regarding the issues of scale and practicability.

Because many collaborative conservation initiatives led by local government and NGOs were already under way within the pilot study region, the prioritization of new conservation opportunities seemed to be less critical than understanding how mitigation partnerships stemming from the pilot study might augment these ongoing projects. Additionally the fruition of the Montana Legacy Project seemed to dwarf the pilot study and emphasized the need for the process to adaptively integrate with these established conservation efforts. Newly acquired timberlands associated with the Montana Legacy Project would fall under management of state and federal resource agencies tasked with habitat restoration and long-term sustainability. Regional practitioners and conservation efforts would benefit most from integration of reliable but flexible mitigation commitments stemming from the ITEEM process, such as the restoration fund concept. In terms of timing, the MT 83 ITEEM pilot study lagged behind these multiple large-scale conservation initiatives. While providing opportunities to direct mitigation toward established regional conservation goals, the lag in timing made it challenging for the process to specifically pinpoint more than a handful of prioritized shovel-ready mitigation projects given that the transactions associated with leveraging mitigation funding required further investigation to ensure legal and financial considerations were met.

The ITEEM process was created to integrate ecosystem-based mitigation and streamline transportation project development. The pilot study was initiated under those assumptions but evolved into two distinct yet interrelated arenas of consideration. Issues and concerns pertaining to the future highway projects on MT 83 emerged as the easier, more discrete concept to address. The group's confusion began when it was discovered that the transportation projects were not clearly defined at this early stage in their conception and planning process and would not likely be programmed for a decade or more. Thus focusing on mitigation opportunities for future potential impacts without sideboards, budgets, or a project scope resulted in collective frustration and loss of momentum. The larger-scale concept of identifying and prioritizing conservation opportunities for implementation through interagency partnerships was soon lost to the misunderstandings resulting from surreal highway project impacts and imaginary budgets. While participants were generally on board with the intent of the pilot study, the difficulty in exploring each of these scales separately (e.g., highway corridor versus the more extensive region around the corridor), while understanding their ecological connectedness and how permitting regulations could be met at both scales, required reiterative explanations and ultimately slowed progress. This challenge was not fully revealed until the workshop, nor was the confusion completely overcome by all of the participants at the workshop.

Given present-day risks to ecological integrity and the current conservation initiatives under way in the region, the group agreed that focusing more on collaborative partnerships and less on the planning considerations directed specifically at future highway projects would result in greater realized gains for the resources within the region and help to leverage investments. The group also concurred, however, that early and ongoing coordination with regard to the transportation planning process is essential in fostering a better understanding of MDT's business process, building trust, and streamlining the project development by increasing predictability and efficiency.

The conclusions reached and recommendations made through the pilot study will form the foundation of an interagency collaboration that is committed to work toward streamlining the transportation planning and project development processes along the MT 83 corridor, while conserving the unique roadside culture and diverse biological resources in the area. The lessons learned from the pilot study, coupled with evaluation of the success criteria, will be applied to improve the next invocation of the ITEEM process.

Conclusions

The path to developing the ITEEM process was not always straight given the pioneering nature of revamping the long-standing tradition of environmental review processes combined with diverse missions and interests of the respective agencies. By far, the greatest success of the pilot study was realized in renewed relationships and trust shared among the agency participants. Long-term success of the ITEEM process will be truly realized as interagency partnerships are formed to achieve meaningful conservation projects in advance of threats and impacts. While the recommendation-making phases of the pilot study were concluded in summer 2009 for the MT 83 region, implementation of the agreements will endure for years, and the ITEEM process will be adapted to be applied in another cycle focused on other regions of ecological importance facing necessary planned infrastructure development.

Acknowledgments

We thank Janice Brown, Ted Burch, Paul Garrett, Craig Genzlinger, Carl James, Jim Walther, Jean Riley, Bonnie Gundrum, Pat Basting, Scott Jackson, Todd Tillinger, Steve Potts, Jeff Ryan, Glenn Phillips, Steve Knapp, T. O. Smith, Jim Claar, and Fred Bower for participating in the ITEEM process. We also thank Jeff Berglund and Michelle Arthur, for their contributions. Lastly, we thank Kevin Crooks, Roger Dodds, Dave Willey, Bonnie Gundrum, and Paul Sturm for their helpful reviews of this chapter.

Chapter 6

Improving Conservationists' Participation

PATRICIA A. WHITE

As the issue of conflicts between wildlife and transportation has garnered more attention, the burgeoning science of road ecology has spawned action in agencies, academia, and legislatures. Conflicts between the needs of wild-life and transportation have become the subject of many professional conferences, academic publications, promising research, and best practices. A growing cadre of dedicated people in academia, transportation, and resource agencies is making progress toward raising the ecological standards of our transportation infrastructure. While much is left to learn, enough information, technology, policy, and people exist to turn the corner on this issue.

The task of protecting and restoring core habitat and corridors across the landscape is a daunting one, requiring a full cast of characters: local, state, and federal transportation agencies; local, state, and federal natural resource agencies; elected officials; community leaders; landowners; academia; conservation organizations; and concerned citizens. No one sector can do it alone. Academics can conduct research and provide data, but practitioners must utilize the information. Natural resource managers can create sound plans to protect vital habitat, only to see it compromised by sprawling development. Legislators can change policies, but agencies need to implement them. Even with advances in policy and practice, regulators and

127

action agencies are unlikely to fully realize the potential without oversight, input, and support from conservationists. Everyone adds a piece to the puzzle, and the conservation community can be the glue that holds it all together.

Conservation advocates within nonprofit, nongovernmental organizations bring to the table several advantages and skills to employ in concert with partners within transportation and natural resource agencies. Specifically, conservationists can contribute through media and public outreach, scientific data collection and analyses, lobbying decision makers, and bringing armies of volunteers.

Media and Public Outreach

The general public knows little about the conflict between wildlife and transportation. While millions of people are involved in wildlife–vehicle collisions, very few people understand the full scope of ecological effects of roads upon wildlife. Even fewer are aware of methods to reduce these impacts or understand their own ability to participate in the process. In order to make progress with road ecology, the general driving and taxpaying public must be made aware of not only the problem, but the solutions.

A 2006 study by the University of Denver found four major barriers to effective citizen participation in wildlife sensitive transportation projects:

1. Lack of awareness or only minimal awareness of wildlife and transportation issues
2. Public apathy or a lack of citizen interest in wildlife and transportation issues
3. Ineffective citizen participation techniques and processes
4. Transportation agencies' poor communication with citizens

Conservation advocates are in a prime position to educate the public on the conflict between wildlife and transportation (fig. 6.1). Many conservation organizations have communications experts on staff, skilled at crafting messages for the public ear. Using media and public outreach, conservationists can teach people about the impacts of transportation on wildlife and the variety of solutions. Addressing the problem of wildlife–vehicle collisions and improving habitat connectivity across roads is in the best interest of the motoring public and a valid use of taxpayer dollars. Conservation-

FIGURE 6.1. Many conservation organizations have communications experts on staff, skilled at crafting messages for the public ear. Using media and public outreach, conservationists can inform people about the impacts of transportation on wildlife and the variety of solutions.

ists can also help educate drivers on the need to reduce their speed and increase alertness when driving in areas containing wildlife.

Examples

The I-90 Wildlife Bridges Coalition educated elementary students across the state of Washington about issues surrounding wildlife and roads with a specific focus on the I-90 Snoqualmie Pass East Project. They asked children to express their thoughts through drawings that show how we can collaborate to benefit both animals and people in the I-90 project. Coalition director Charlie Raines and Washington's secretary of transportation Doug Mac-Donald selected the winning drawings.

The Southern Rockies Ecosystem Project spearheaded an education and outreach campaign in Colorado that focused on the human safety issue, while drawing attention to the plight of wildlife on highways. The Colorado Wildlife on the Move campaign urged drivers to watch for wildlife on Colorado highways, especially during times when animals are migrating. The Southern Rockies Ecosystem Project held a media conference with Colorado State Patrol and other partners that reached millions through television, radio, and newspaper coverage. Campaign posters and driver tip sheets are displayed in rest stops, tourist information centers, rental car offices, and other locations across the state.

The Red Wolf Coalition and Defenders of Wildlife partnered with the North Carolina Department of Transportation to install informational kiosks at rest stops to educate visitors about the ecological and economic benefits of red wolf restoration.

Lobbying

Elected officials, from town mayors to state legislators to congressional representatives can influence decisions regarding transportation and impacts on wildlife. Lawmakers at all levels are involved with allocating transportation funds, setting priorities, and even project selection. They can also have a great influence on how state and federal agencies reflect our values through natural resource conservation. Because of the great deal of money involved with transportation, elected officials meet with a great number of special interests who are intently interested in the allocation of transportation funds as well as the prioritization of projects.

Lobbying is defined as the attempt by any individual to influence the passage or defeat of legislation by communicating with lawmakers or their staff. While it has earned a bad reputation of late, lobbying is a vital and important activity. Lobbying is an expression of fundamental rights such as freedom of speech, association, and petition and facilitates the exchange of important information and ideas between the government and private parties.

There are two kinds of lobbying. Direct lobbying occurs when individuals state their position on specific legislation to legislators or other government employees who participate in the formulation of legislation, or if a group urges its members to do so. Grassroots lobbying occurs when individuals or groups state their position on specific legislation to the general public and ask the public to contact legislators through a call to action.

Antilobbying laws prevent federal agency employees from lobbying any federal, state, or local government official with respect to any pending or proposed legislation, resolution, appropriation, or measure. Federal employees may not communicate support or opposition to any pending legislation. They cannot ask members of the public to contact their elected representatives, provide support for lobbying activities of private organizations, or prepare materials for dissemination. Nonprofits on the other hand, can do this kind of lobbying. The federal guidelines that regulate nonprofit advocacy and lobbying incorporate significant flexibility for nonprofit organizations to participate in the public policy process.

Nonprofit conservation organizations can lobby for legislation that promotes the use of measures such as wildlife crossings and the restoration of habitat connectivity. Advocates can also appeal to their expansive memberships and the general public to contact elected officials to support particular legislation through letter writing, phone call, and e-mail campaigns.

> Much of the social change in America had its origin in the nonprofit sector. Nonprofit lobbying is the right thing to do. It is about empowering individuals to make their collective voices heard on a wide range of human concerns.
> —Bob Smucker, Founder, Center for Lobbying in the Public Interest

Examples

In May 2006, voters in Pima County, Arizona, voted to pass a sales tax increase to fund their Regional Transportation Authority's $2.1 billion regional transportation plan. The plan was developed with input from a diverse, 35-member citizens advisory committee that included several members from the nongovernmental organization community. The plan included several highway and transit projects but also set aside $45 million for a "Critical Landscape Linkages" category that will fund wildlife crossing structures and amenities in transportation projects. The crossings are a component of realizing the vision set forth in the Sonoran Desert Conservation Plan, a multiyear, multiagency, multispecies plan to conserve lands and resources in Pima County (see chapter 15). Crossings will complement land acquisitions purchased with a 2004 open space bond, with more planned in the future.

In 2005, the Washington state legislature passed a transportation bill that included $387 million for the Snoqualmie Pass East I-90 Project, which includes several wildlife passages. Members from both sides of the aisle worked to pass this bill and made sure that I-90 remained on the project list. The package was challenged by an initiative to repeal the gas tax, thus threatening the transportation bill that would fund the wildlife passages. But conservation advocates countered with a public education campaign in support of the gas tax and the bill. In the fall of 2005, the citizens of Washington voted to keep the gas tax and supported the Snoqualmie Pass wildlife passages.

Volunteers

Wildlife connectivity measures come in all shapes and sizes, from wide vegetated overpasses to small culverts. Some projects are small scale and low tech, requiring very little expertise. Many successful measures don't require any construction at all. Even for larger projects like overpasses, there are several pre- and post-construction activities that can be outsourced. For these tasks, transportation agencies could save money by using unpaid volunteers.

Many conservation organizations have large memberships from a few hundred to several hundred thousand. They can use their extensive connections with members and the general public to recruit volunteers for small projects such as the following:

- Clearing vegetation near potential crossing locations
- Managing roadside vegetation
- Collecting pre- and post-construction data
- Repairing fencing
- Clearing debris from culverts

Examples

The Miistakis Institute in Calgary, British Columbia, took the citizen science concept to the Web with their "Road Watch in the Pass" project (see chapter 14). Drivers who use Highway 3 through Crowsnest Pass are encouraged to report sightings of wildlife (dead or alive) on a special Web site. Users log in and fill out a simple report on the species, location, and status. Data collected are analyzed and provided to planners, managers, and decision makers in the municipality of Crowsnest Pass and beyond.

Other programs actively engage volunteers in the field. Defenders of Wildlife started the Wildlife Volunteer Corps program to provide their members with hands-on opportunities to benefit wildlife in their own communities. A Corps project in partnership with the Washington State Department of Transportation put volunteers to work, for example, removing invasive plants and the planting of native trees to improve habitat in preparation of a proposed wildlife underpass site in the northern Cascades. Similarly, a Vermont-based project engaged members in searching a wetland for an elusive snake species. The volunteers helped to catch, measure, and photograph a variety of reptiles to assist in habitat conservation efforts. Another project in Washington State involved partnering with Defenders of Wildlife, Conservation Northwest, I-90 Wildlife Bridges Coalition, and Wilderness Awareness School to conduct citizen monitoring of wildlife presence in habitat near proposed wildlife crossing structures. Volunteers installed cameras and routinely returned to check the footage; many images have been captured and significant data were collected as a result.

Obstacles and Opportunities: How Conservationists Can Be More Effective Participants

Together, all the sectors involved have made great strides in advancing the new science of road ecology. Much of the data, technology, policy, best practices, and expertise are in place to protect and restore habitat cores and corridors across our landscape. However, many challenges remain from a lack of public participation to a need for more coordination among transportation and resource agencies. Environmental protection laws often fail to protect vital habitat or fund essential conservation measures. And even our best efforts can be trumped by outdated design standards, rampant development, and political ambitions. Fortunately, members of conservation groups possess the skills and the spirit to meet many of these challenges. Indeed, conservationists are uniquely positioned to address various challenges and help bring together all constituents, policy, implementation, and pieces to make habitat connectivity a reality.

Challenge 1: Transportation Planning Process Is Confusing and Lacks Transparency

By the time construction workers in hard hats and orange vests arrive, the project is reaching the end result of a process that may have taken from five to 10 years, and even decades in some instances. Even before a road project gets to the environmental review stage—where most conservationists have historically become involved—a significant amount of time and money have already been invested. Conservationists need not oppose every new road or road expansion project but should focus on keeping the impacts of highways minimal in their region's most sensitive natural areas and push to make existing roads more compatible with wildlife.

During planning, conservationists have the opportunity to voice concerns early enough to alleviate or reduce many impacts. By the time a poorly designed plan gets to the project stage, usually all that can be done is to minimize and mitigate the negative effects. The sooner conservationists and natural resource agencies become involved and engage with transportation agencies, the greater the opportunity to ensure the project addressed concerns for wildlife and connectivity. Therefore, it is imperative that conservationists and natural resource agencies get involved with planning at the system level, not the project level. An educated conservationist can positively influence wildlife and habitat conservation by encouraging the early adoption of wildlife considerations in transportation planning.

Conservationists are not generally familiar or comfortable with transportation planning, yet this planning guides decisions about where infrastructure is built or expanded. Decisions made today will influence the location, direction, and shape of the development that happens tomorrow and hence the location, types, and quality of habitat that are influenced. Unfortunately, the transportation planning process can be complicated and a bit overwhelming. Because planning is comprehensive and continuing, conservationists must be actively engaged and diligently track several simultaneous processes, plans, and products.

The basic steps in the transportation planning process are as follows:

1. Define the problem, scope, area, issues
2. Set goals, objectives, and criteria
3. Collect data
4. Develop alternatives and scenarios
5. Model—forecast future travel behavior
6. Evaluate alternatives
7. Select a preferred plan
8. Implement the plan through projects

Unfortunately, the planning process is not the decision-making process. If done well, it can provide a framework for informed decision making, but ultimately those elected or appointed to make decisions will make the call. Many transportation planners have instances where good plans were shelved and ill-advised proposals slipped into the process by means of an earmark or other political maneuvering.

Long-range and short-range transportation plans can be very different with vastly different processes and purposes. They both may have opportunities for public input, but what happens in between remains a mystery to many not familiar with the processes. In theory, the short-range plan is supposed to reflect the long-range plan. However, many changes can occur between the time of the detailed short-range plan and implementation of the long-range plan. This means it is important for conservationists to track all the planning activities in their region and to stay engaged with transportation planners and with land and wildlife management agency personnel.

Following are two very different examples of the project selection or programming process at the state level:

1. The Texas Department of Transportation describes its project selection process in five steps: identify needs, build a proposal (funding), begin planning, project development, and construction. In Texas, public involvement does not generally occur until project development, long after project selection, which rests with the commission and local officials.

2. Arizona Department of Transportation district engineers meet with metropolitan planning organizations once a year to develop a list of candidate projects for submission to a selection committee. Projects go through scoping (not environmental review scoping) to flesh out the project details such as traffic, safety considerations, and cost. Using a set annual budget, projects are selected up to that budget amount. The State Transportation Board conducts three public hearings on the draft 5-year construction program. The list of projects is culled from that five-year program, including the federally funded projects, local projects, Federal Lands Highway Program (FLHP) and Bureau of Indian Affairs projects. In Arizona, public involvement generally occurs early in the transportation planning process.

Conservationists can download or request copies of their state and local transportation plans. Working with conservation planners in resource agencies, conservationists can note where and how any upcoming transportation projects or activities will impact the areas of interest, attend public meetings, submit comments, provide information, and offer to make a presentation on the impacts and solutions. Conservationists can also volunteer to serve on citizen focus groups or advisory committees, helping to ensure that wildlife and habitat connectivity issues are brought into the transportation planning process.

Conservation advocates can work with transportation planners to map out the planning-to-project process in their respective region and share it with other concerned citizens. They can also host a citizen training session and invite a transportation planner to discuss the transportation planning process, the plans themselves, and how everyone can more effectively be involved.

Challenge 2: Lack of Public Participation in the Transportation Planning Process

Our transportation planning process isn't perfect, but over time has become more open and accessible to the public. Similar to exercising their right to vote, citizens can exercise the right to participate in planning the best communities possible—but few actually do. Perhaps many people are skeptical about their ability to influence the outcome of transportation plans or projects. Perhaps transportation plans are too abstract and the planning process simply incomprehensible for some citizens. Whatever the reason, without adequate public participation, many transportation plans are made, and ultimately highways are built, with very little input from citizens.

Conservationists can get involved themselves, and they can also get their membership and the general public involved. Citizens can contact their state and local/regional transportation planning divisions and ask to be added to mailing lists to receive newsletters, updates, and other information. Given information on specific public involvement opportunities, citizens can attend public meetings or hearings regarding transportation plans. Advocates can also encourage other concerned citizens to attend these meetings in order to express concerns about the existing, ongoing, and potential impacts of the transportation project on wildlife.

During the planning process, there are numerous instances in which information should be made available to the public for comment. Conservation groups can help citizens draft written comments during public comment periods for plans and recommend solutions that benefit wildlife and connectivity.

Challenge 3: The Transportation Planning Process Fails to Coordinate with Natural Resource Agencies and Inadequately Addresses Wildlife Concerns

Historically there has been a lack of coordination among state transportation and natural resource agencies. In fact, at times the agencies' various missions seem to have been working at odds with one another.

Acknowledging that conservation is much more cost-effective than endangered species recovery, Congress established a program to assist state fish and wildlife agencies in conserving nongame and nonlisted wildlife species through "wildlife diversity programs." The 2002 Department of Interior Appropriations bill included language creating the State and Tribal Wildlife Grants Program, which provides new, dedicated funding for cost-effective, proactive conservation efforts intended to prevent wildlife from declining to the point of becoming endangered. State fish and wildlife agencies receive federal appropriations according to a formula based upon the state's size and population. Projects include the restoration of degraded habitat, removal of invasive vegetation, reintroduction of native species, partnerships with private landowners, research, and monitoring.

As a condition of receiving the grants, Congress charged state fish and wildlife agencies with completing a State Wildlife Action Plan by October 1, 2005. The U.S. Fish and Wildlife Service reviewed each action plan, and state wildlife agencies are required to revisit and update them at least every 10 years to ensure conservation success over the long term. The action plans not only address "species of greatest conservation need" but also the "full array of wildlife and wildlife issues," and they establish a plan of action for conservation priorities with limited funding. To "keep common species common," all plans are based on targeting resources to prevent wildlife from declining to the point of endangerment. Ideally, each action plan will create a strategic vision for conserving the state's wildlife, not just a plan for the fish and wildlife agency.

Congress identified eight essential elements the action plans must contain in order to ensure nationwide consistency:

1. Information on the distribution and abundance of species of wildlife (including low and declining populations) that are indicative of the diversity and health of the state's wildlife
2. Descriptions, including locations and relative conditions, of key

habitats and community types essential to conservation of species identified in (1)

3. Descriptions of problems that may adversely affect species identified in (1) or their habitats, and priority research and survey efforts relevant to restoration and conservation of these species and habitats

4. Descriptions of needed conservation actions and priorities

5. Proposed plans for monitoring species and their habitats, for monitoring the effectiveness of conservation actions and for adapting these conservation actions to respond appropriately to new information or changing conditions

6. Descriptions of procedures to review the action plan at intervals not to exceed 10 years

7. Plans for coordinating, to the extent feasible, the development, implementation, review, and revision of the action plan with federal, state, and local agencies and First Nations that manage or affect significant land and water areas within the state

8. Broad public participation

The practical effect of this new planning requirement was to take advantage of the many disparate, ad hoc, and unrelated conservation planning initiatives, combining them under one all-inclusive, sanctioned, and funded program. The scale is ambitious, yet manageable and fits easily into an existing administrative framework. State Wildlife Action Plans are intended to remain dynamic, serving as the home base for prioritizing conservation efforts in each state and coordinating the roles and contributions of all agencies and conservation partners. Implementation of strategy goals and objectives is aided through continued federal funding, matched by additional sources. In theory, the strategies represent the future of wildlife conservation and management. Collectively, they will create—for the first time—a nationwide approach to wildlife conservation. If each action plan is indeed a strategic vision for conserving the state's wildlife, implementation will require more than the state fish and wildlife agency. For the conservation strategies to be successful, all sectors must embrace the goals, engage in the process, and accept responsibility for their own roles and contributions. State Wildlife Action Plans provide a means for the conservation community to become involved with both transportation agencies and state fish and wildlife management agencies to help interconnect transportation plans with the needs of wildlife.

Serendipitously, the 2005 transportation bill (Safe, Accountable, Flexible, Efficient Transportation Equity Act; SAFETEA-LU) contained a small, unassuming, but very powerful provision that could ultimately protect millions of acres of habitat by changing the way long-range transportation planning is done. For the first time, wildlife conservation will be among the very first things considered, rather than the last. SAFETEA-LU requires each metropolitan planning organization and state transportation agency to consult with federal, state, tribal, and local land-use management, natural resources, wildlife, environmental protection, conservation, and historic protection agencies while developing long-range transportation plans. Each consultation will include a comparison of the transportation plan with conservation maps or inventories of natural and historic resources such as the state wildlife action plans. Each plan will also include a discussion of potential environmental mitigation activities—and potential areas to carry out these activities—that may have the greatest potential to restore and maintain the environmental functions affected by the plan.

In light of this new requirement, the state wildlife action plans are now hardwired into transportation planning and can demonstrate their full value and utility. Beyond their conservation value, the action plans have great potential to aid state transportation agencies in streamlining project delivery. Use of habitat mapping data in the action plans can provide an effective system to identify transportation projects that could have major impacts on wildlife. Early detection of such problems can help avoid costly delays later in the life of projects. Early planning for conservation can also provide a good opportunity to explore mitigation options and identify the best remaining sites for acquisition and restoration (see also chapter 5). Often, by the time a road project develops through the planning, review, and design process, many of the opportunities for high quality and affordable mitigation have been lost. As an added bonus, the transportation agency can adopt a proactive approach to conservation and become a full partner in implementing the action plan for the entire state.

Conservationists can get involved with their state wildlife action plan by requesting or downloading the plan and reading it thoroughly. Once they are familiar with their state's plan, they can work with staff from the wildlife agency to be more effectively involved in the implementation of the plan as it relates to transportation projects. Specifically, conservationists who are also familiar with transportation planning can assist wildlife agency personnel with transportation planning consultations. Public participation in consultations is not required, but many conservation organizations have expertise and sophisticated information, input, data, and resources to contribute, earning a place at the table.

Examples

Maine's Beginning with Habitat is a public–private partnership that combats sprawl by providing communities with practical tools to incorporate natural resource conservation into local land use planning. The initiative brings together crucial wildlife and habitat data into customized geographic information system maps and makes the information accessible to local decision makers, including planning boards, regional planning commissions, community conservation commissions, and land trusts. Beginning with Habitat resource materials, including a road ecology primer, *Conserving Wildlife On and Around Maine's Roads*, are distributed via public presentations and technical assistance. Collaborating with state transportation officials and educating local communities are critical to advancing road ecology. Founded in 2001, Beginning with Habitat is guided by a seven-member steering committee that consists of Maine Audubon, Maine Natural Areas Program, Maine Department of Inland Fisheries and Wildlife, Maine State Planning Office, Maine Coast Heritage Trust, the Maine Chapter of The Nature Conservancy, and the U.S. Fish and Wildlife Service. The program received an Environmental Merit Award from the Environmental Protection Agency and has been recommended by the Association of Fish and Wildlife Agencies for use in all 50 states.

American Wildlands has developed two geographic information system models to locate the highest priority areas for mitigating highways with crossing structures, fencing, or other measures in local landscapes. To prioritize work, habitat cores and corridors from American Wildland's regional Corridors of Life model are overlaid with Statewide Transportation Improvement Plan projects. State transportation departments rely on American Wildland's methodology to justify expenditures of federal appropriations for wildlife mitigation. To date, they have improved five different highway projects in Idaho, Wyoming, and Montana, resulting in the commitment to construct seven wildlife underpasses and two bridges for fish passage in the region. So far, this includes more than $2.7 million for wildlife mitigation and $2.2 million in private land conservation adjacent to highway mitigation.

Challenge 4: Conservation Advocates Have a Contentious History with Transportation Agencies

Conservation advocates may be accustomed to working with government agencies, but all too often as combatants rather than as allies. The history of

America's transportation infrastructure is riddled with controversy. Several early interstate routes were drawn up without regard for impacts to local communities. Because construction of the national highway system was considered in the national interest, building routes trumped local concerns. Low income urban neighborhoods and rural areas were often targeted as prime areas for new highway corridors. The "freeway revolts" of the 1960s brought resident activists and community leaders together to stand up against many proposed routes, leaving a legacy of strife between transportation agencies and advocates.

Today, where high biodiversity and sprawling human activities intersect, the transportation sector often finds itself at odds with the conservation community. Roads may be built to encourage growth or in response to growth pressure. In either case, many conservation organizations have chosen to devote a great deal of time, resources, and effort in trying to stop or significantly alter transportation projects that threaten natural resources.

However, over time the relationship between conservation advocates and transportation agencies is changing. Transportation agencies at the state and federal level are now employing more biologists than ever before. In fact, the discipline of road ecology was borne of conservationist champions within transportation agencies. They worked in concert with academics, natural resource professionals, and advocates, seeking better alternatives to reducing impacts on wildlife and habitat. Perhaps more than others, the issue of wildlife conservation and transportation lends itself to reaching across those physical, political, and ideological boundaries and work collaboratively with government agencies, at the local, state, and federal level. If conservationists have quarreled with their state and federal agencies in the past, building communication and relationships may be more difficult. But conservation advocates can work to build positive relationships with their transportation and natural resource agencies.

One way to begin improving the relationship is to work with the maintenance and operations division of a transportation agency. Road maintenance divisions provide the necessary services to ensure that road infrastructure is in good working order and conditions are safe for the motoring public. While not always recognized, maintenance professionals can be a conservationist's greatest allies. By prolonging the life of our existing infrastructure, they reduce the need to continuously build new highways that may ultimately end up consuming and fragmenting natural areas and essential wildlife habitat.

Maintenance and operations can offer opportunities to reduce the impacts of highways on wildlife and improve habitat quality through voluntary stewardship actions. People who maintain highways are public servants with a natural sense of stewardship. They offer a tremendous, largely un-tapped capacity for improving the environment as part of their daily work. Sometimes small changes in maintenance practices can make a big differ-ence. Conservationists can meet with maintenance personnel to discuss partnership opportunities for wildlife conservation. The following is a brief list of project ideas:

- Host an information exchange to learn more about maintenance practices and brainstorm ideas for making improvements for wildlife.
- Share information on best maintenance practices.
- Volunteer to help with roadside vegetation management like inva-sive species removal and native species plantings.
- Publicly recognize positive efforts to improve environmental stewardship.

Case Study

To educate staff from all departments and levels, the Vermont Agency of Transportation conducts annual habitat connectivity training with Keeping Track, a nonprofit conservation organization based in northern New En-gland (see also chapter 12). Participants are introduced to the habitats and needs of various native species, from moose (*Alces alces*) and black bear (*Ur-sus americanus*) to wood turtles (*Glyptemys* spp.) and salamanders. Exposure to wildlife allows staff to see their work in another context and empowers them to reduce the impacts roads have on wildlife and habitat.

Far beyond my expectations, each department brought to the program a great diversity of personnel, from planners to engineers and executives down to junior staff. Not all of them were card-carrying natural resource enthusiasts when they began. Keeping Track director, Susan Morse.

Inspired by Vermont's success, New Hampshire and Maine recently began their own Keeping Track programs.

Challenge 5: Protecting or Restoring Habitat Connectivity Is not Required under Current Environmental Laws

State wildlife and federal land management agencies are limited to manag-ing habitat they own, such as refuges, forests, and parks. However, public lands make up only a fraction of the habitat necessary to sustain wildlife populations. There is a great deal of essential habitat in private ownership. But without legal protection or financial incentives, private landowners have little motivation to protect or manage their land for ecological value. Under current law, the only types of privately owned habitat that are protected are wetlands (Clean Water Act) and designated critical habitat (Endangered Species Act). Upland habitat cores and corridors—even those areas that are essential to wildlife—are not protected.

Conservation advocates can lobby for increased protection of essential core habitats and corridors. In February 2007, the Western Governors' Association (WGA) unanimously approved a policy resolution, *Protecting Wildlife Migration Corridors and Crucial Wildlife Habitat in the West*. This resolution describes the importance of wildlife corridors and crucial habitat and asks the western states, in partnership with important stakeholders, to identify key wildlife corridors and crucial wildlife habitats in the West and make recommendations on needed policy options and tools for preserving those landscapes. To implement the resolution, WGA launched the *WGA Wildlife Corridors Initiative* (Western Governors' Association 2008), a multi-state and collaborative effort that included six separate working groups (science, oil and gas, energy, climate change, land use, transportation), each of which was charged with developing findings and recommendations for the WGA. The *WGA Wildlife Corridors Initiative* report was approved by the governors during the WGA Annual Meeting in Jackson, Wyoming, on June 29, 2008, with the understanding and condition that implementation of the report will be coordinated and overseen by the WGA through the newly formed Western Wildlife Habitat Council.

Conservationists worked together with transportation professionals and academia in developing recommendations for the WGA initiative. The transportation chapter described improvements to the West's economic vitality, quality of life, and ecological legacy through four specific action items that encompass practice, policy, and fiscal initiatives:

1. *Make the preservation of wildlife corridors and crucial habitat priorities* for transportation planning, design, and construction.
2. *Integrate conservation and transportation coordination, planning, and implementation* across jurisdictions.
3. *Manage and coordinate data information systems* and methodology to increase efficiency and reduce redundancy.
4. *Establish long-term capacity* to staff and fund these initiatives.

The final product provides the conservation community and all road ecology proponents with a solid platform to carry forward not only in the western states but nationwide. Specifically, the following action items lend themselves to conservationist involvement:

- Western governors should consider instructing their respective state department of transportation and fish and wildlife agencies to *conduct an economic analysis* of transportation plans, activities, and structures that may impact state wildlife resources. Such an analysis would inform states about infrastructure improvements that would protect wildlife corridors and crucial habitats, improve public safety, emphasize economic benefits, and evaluate related budgetary considerations. Such an assessment would also provide an analysis of initial financial investment and long-term cost-saving benefits. Also, inclusion of a public outreach component that explains the results of the (cost savings) analysis would provide additional program incentive, justification, and support.
- Western governors should consider directing their state department of transportation and fish and wildlife agency to *develop cooperative, large-scale mitigation plans* with local, state, tribal, and federal agencies to protect and/or restore wildlife corridors and crucial habitats under the intent of SAFETEA-LU Section 6001 planning efforts, the state wildlife action plans, and other applicable laws.
- Western governors should consider directing their respective state department of transportation and fish and wildlife agency to integrate information about wildlife corridors and crucial habitat early in the transportation planning process through *training, guidance,* and specific methods for regional transportation plan development as well as project development for safety and design considerations.
- Western governors should *conduct an assessment of interjurisdictional data compatibility* for use in transportation planning and implementation. The decision support system recommended by the WGA Science Committee's recommendation should integrate such data as road-kill locations; existing infrastructure, such as bridges, culverts, fencing, and the like; and identification of where structural wildlife crossing improvements have already been made.
- The WGA should work with federal, tribal, state, and local transportation agencies to ensure that wildlife corridors and crucial habitat data are integrated into short-range statewide transportation improvement programs, long-range transportation plans, and regional plans. A *transportation geographic information system analysis identify-*

ing wildlife corridors and crucial habitats would greatly facilitate the use of visual tools when planning upcoming transportation projects.

Conservationists can also support efforts to provide financial incentives to private landowners for protecting important habitats. Much of the habitat needed to conserve biodiversity is on private land. Many landowners voluntarily undertake conservation efforts, and many more might if they get appropriate encouragement and assistance through government incentive programs. Although most traditional government incentive programs have been made available to broad classes of landowners, more specific targeting of incentive programs to address specific habitat priorities may be critical in the future, given the limited financial resources available for biodiversity conservation.

Incentive programs for habitat conservation on private lands fall into one or more of these categories:

1. *Property tax benefits*: Many states assess farm- and forestlands at reduced levels for property tax purposes. To maintain this lower assessment, landowners must manage their properties in ways that support these farm and forest uses. Conservation programs with property tax benefits similarly assess lands at reduced levels for property taxes, allowing landowners to participate in conservation practices without forgoing the reduced tax rates. Programs with property tax benefits have a localized financial impact on county governments and special districts with a local tax base.

2. *Income tax credits*: These incentive programs provide a means for landowners to reduce their state income tax burden with a tax credit for part or all of the costs of a conservation practice. Because such programs have a statewide financial impact, they are appropriate to accomplish conservation objectives with statewide benefits, rather than just local benefits.

3. *Regulatory streamlining*: A landowner can enter into a stewardship agreement with a participating agency, when the landowner's conservation efforts exceed those required by law. The landowner, in return, may receive regulatory certainty, expedited permit processing, or higher priority access to other programs.

4. *Direct funding*: Various state and federal agencies (and private sources) provide direct contributions to private landowners or landowner organizations to improve water quality, protect, restore, and enhance fish and wildlife habitat, and improve land management

practices. These include grants, purchase of conservation easements, cost sharing, and reimbursement of expenses.

5. *Technical assistance* (including education and conservation planning): Landowners may need assistance with identifying and understanding relevant programs, understanding regulations, developing conservation plans, applying for permits or programs, or designing specific conservation elements. Assistance is available through a wide array of government and nongovernment sources.

Challenge 6: Protecting or Restoring Habitat Connectivity Receives No Dedicated Funding

Congress authorized $286 billion total in the last highway bill and some believe that is not enough. The cost of road building is continually rising, sometimes dramatically outpacing other sectors of construction and land development. The United States invests enormous sums in our transportation systems—significantly more than is spent on natural resource and land management. Transportation law provides dedicated funding for several different categories, such as the interstate system, maintenance, research, and others. But no dedicated funding source is available for protecting or restoring habitat connectivity, making it difficult for conservation advocates and champions to get wildlife crossings built into transportation projects.

First, conservation advocates can promote more effective use of existing funding for wildlife habitat connectivity. There are several available funding sources that are underutilized.

Transportation Enhancements

The Transportation Enhancements (TE) program sets aside 10 percent of all Surface Transportation Program dollars for community-based projects that expand travel choices and enhance the transportation experience by improving the cultural, historic, aesthetic, and environmental aspects of our transportation infrastructure. TE is a federal aid reimbursement program, whereby the federal government pays 80 percent of the project cost and the project sponsor pays the nonfederal match of 20 percent. While TE uses federal funding, state transportation agencies retain most of the responsibility for implementing the program, and each state does so in its own way.

Each state devises its own application, selection process, and selection criteria, but they all have some characteristics in common, such as eligibility, advisory committees, project implementation, innovative financing, and streamlined project development. To qualify for consideration, projects do not have to be associated with a specific highway project, but they must be within the acceptable categories and must relate to surface transportation.

In 1998, Congress included Activity 11, known in law as "environmental mitigation to address water pollution due to highway runoff or reduce vehicle-caused wildlife mortality while maintaining habitat connectivity." The provision provides communities with funding to decrease the negative impacts of roads on the natural environment—including water pollution and habitat fragmentation. To reduce water pollution from stormwater runoff, TE funds can be used for pollution studies, soil erosion control, or river cleanups. To address wildlife passage and habitat connectivity, TE funds can be used for crossing structures and monitoring and data collection on habitat fragmentation and vehicle-caused wildlife mortality.

From 1998 through 2006, state transportation agencies programmed just $53 million for Activity 11 projects, most of which went to stormwater projects. Only $11.5 million was programmed to "reduce vehicle-caused wildlife mortality while maintaining habitat connectivity." Of the 23,000 TE projects, only 71 have been related to wildlife habitat connectivity. Just 20 states have implemented wildlife-related TE projects, averaging $161,971 per project. Conservation advocates and natural resource managers are missing a golden opportunity. Since 1998, $8.1 billion has been authorized for all TE projects. If each of the twelve categories received equal portions, that would mean $675 million for wildlife, more than $61 million per year. Conservationists, largely unaware of the TE program and the benefits for wildlife, have submitted few applications. Even after 10 years of TE Activity 11, conservationists' lack of participation has not only lost millions for wildlife, but state TE programs and selection committees still have very little experience with wildlife and habitat connectivity issues.

The Federal Lands Highway Program

The Federal Lands Highway Program (FLHP) is an adjunct to the Federal-Aid Highway Program, created in 1982 to fund a coordinated roads program for transportation needs of federal and Indian lands that are not the responsibility of a state or local government. Often referred to as "the DOT for federal lands," FLHP's purpose includes the following:

- Ensure effective and efficient funding and administration for a coordinated program of public roads and bridges serving federal and First Nation lands
- Provide needed transportation access for Native Americans
- Protect and enhance our nation's resources.

FLHP funds are distributed to each category, where project selection is delegated to users (federal land management agencies, tribes, and states) according to 3-year transportation improvement plans. Roads owned by the Bureau of Land Management, the Bureau of Reclamation, and the U.S. Army Corps of Engineers and other Department of Defense agencies do not receive dedicated funding and have to compete for funds under a discretionary category. FLHP funds are 100 percent eligible for wildlife mitigation measures. SAFETEA-LU provided $4.5 billion for the FLHP through 2009, which is eligible for wildlife mitigation measures on highways within or serving our public lands system.

Safety

Because wildlife–vehicle collisions are now more widely recognized as a serious safety hazard for the traveling public, safety funding can be used to build wildlife crossings or any other mitigation measure. SAFETEA-LU clarified the eligibility of safety funds with a provision in the Highway Safety Improvement Program. "The addition or retrofitting of structures or other measures to eliminate or reduce accidents involving vehicles and wildlife" is now considered a highway safety improvement project and therefore eligible for safety funding.

Bridge Construction

Along with constant maintenance and upkeep of highways, transportation agencies are fastidiously checking and rechecking all the bridges and culverts in all states. They keep records of the conditions and schedule them for maintenance, restoration, and full reconstruction when necessary. Bridge reconstructions are an excellent time to rethink the opportunities for better aquatic and terrestrial passage under the bridge.

Flexible Funding

A one-size-fits-all approach to funding would never work for our vastly different states. A hallmark of the Transportation Equity Act (TEA) bills is the permission for state transportation agencies to "flex" dollars from one pot to another, based on their own needs and priorities. Because one state might prioritize public transportation more than another, roughly 75 cents of every federal highway dollar can be used for transit investments such as bus, rail, or streetcar systems. Highway dollars can be flexed for fix-it-first or pedestrian and bicycle safety initiatives. Unfortunately, only a handful of states have taken advantage of the flexibility. According to the Surface Transportation Policy Partnership, 87 percent of flexible funds given to state transportation agencies in the 1990s went to highway and bridge projects. Most of that flexible spending (82 percent) happened in just five states (New York, California, Pennsylvania, Oregon, and Virginia). But, in theory, the transportation bill allows state and local governments, transit operators, and metropolitan planning organizations to build a multimodal transportation system to meet their unique needs.

Second, conservation advocates can lobby for a dedicated source of funding for protecting and restoring habitat connectivity. In early 2009, the Western Environmental Law Center spearheaded a 1% for Wildlife campaign to encourage 1 percent of all transportation dollars allocated be set aside for wildlife crossings. These dollars can come from existing programs, such as Public Lands Discretionary, TE, and the Forest Highways program. For example, if Colorado receives $500 million in transportation dollars, $5 million would be available to provide safe passage for wildlife in Colorado. The 1% for Wildlife campaign can create thousands of jobs and stimulate the transportation industry while also restoring habitat connections for wildlife in the face of climate change.

Examples

Conservationists teamed up with the Colorado Department of Transportation and others to explore building a wildlife bridge just west of Vail Pass on I-70. The location was recognized as a high-priority habitat linkage for a diversity of species by an interagency group called A Landscape Level Inventory of Valued Ecosystem Components. When finished, the bridge will reconnect critical wildlife habitat fragmented by the interstate and restore one of the last remaining forested connections for wildlife moving north–south through the heart of the Rocky Mountains. In 2005, Congress appropriated $500,000 through FLHP's Public Lands Highway Discretionary Program to conduct preliminary studies and planning, and additional funds are expected. The project brings highway dollars into the state without bringing more highways and because it is funded under the PLHD program, no match is required from the Colorado Department of Transportation or local governments.

In 2006, Defenders of Wildlife's Florida office applied for a TE project to improve a small bridge on U.S. 41 in the Big Cypress National Preserve for wildlife passage. Despite lowered speed limits, seven Florida panthers had been killed within 4 kilometers (2.5 miles) of the bridge. Florida Department of Transportation checked their records and discovered that the bridge was already scheduled for reconstruction. As a result, they will use bridge replacement funds for the project, supplemented with $425,000 in TE funds for preconstruction monitoring and design.

Challenge 7: Highway Design Standards Fail to Recognize the Importance of Habitat Connectivity

The American Association of State Highway and Transportation Officials began publishing highway design standards in the 1930s, now commonly known as the "Green Book." The official title is *A Policy on the Geometric Design of Highways and Streets*, and it is considered the definitive reference for highway design. The Federal Highway Administration has formally adopted parts of the Green Book as the national standard for roads in the National Highway System, which includes interstates and some primary routes.

However, in pursuit of standardized design, other cultural, aesthetic, and environmental values are overlooked. The "wider, flatter, straighter" formula is not always appropriate in wildlife areas. Conservationists can work with AASHTO to update the Green Book to include modern, wildlife-friendly design standards.

Example

The Arizona Wildlife Linkages Workgroup is a collaborative effort between public- and private-sector organizations to address habitat fragmentation through a comprehensive, systematic approach (see chapter 15 and box 15.1). Workgroup partners, including conservation nongovernmental organizations, conducted a statewide assessment to identify blocks of protected habitat, the potential wildlife corridors between them, and the factors threatening to disrupt these linkage zones. After four successful workshops and many hours spent coordinating, meeting, mapping, and writing, the Workgroup presented their initial findings, methodology, and recommendations in December 2006—a product that is intended to evolve and ultimately be used as a planning instrument.

Challenge 8: Politics Can Trump Good, Sound, and Balanced Decision Making, Leaving Wildlife Conservation and Connectivity Behind

Lawmakers run for office for many different reasons, and some of those reasons might be at odds with the needs of wildlife. At the local level, many city councils are populated by business leaders and large landowners who prioritize development and growth over conservation. They are more likely to work for policy and projects that work in favor of their business interests over the short term rather than natural resources over the long term.

At the state and national level, elected officials are under pressure to deliver money and jobs to their constituencies during their terms in order to be reelected. Too often, they look to transportation appropriations and high-priced highway projects as a major source of economic growth. However, highway construction jobs are temporary, but the highways are permanent. New highways may be unnecessary or inappropriate in many locations. Because of the long-term consequences of additional highway lanes, they should never be used for short-term political gain.

Of all the players on the stage of road ecology, conservationists and citizen advocates are in the best position to hold politicians accountable for their actions. Most government agency staff are prohibited from actively influencing legislation or communicating their support/opposition on legislation to the public. Conservationists can actively inform politicians on the impacts of highways and lobby for positive measures for wildlife, using all the tools in the conservationists toolbox:

- Legislators spend a good deal of time at their respective capitals, but they return to their home districts on weekends, holidays, and during district work periods. Conservationists can take those opportunities to meet in person with lawmakers and/or their staff to discuss solutions for addressing the impacts of transportation on wildlife.
- In addition to or in lieu of personal meetings, conservationists can touch base with lawmakers via phone calls, letters, and e-mails. To

reinforce their position, they can also recruit members and supporters to make contact with elected officials, expressing public support for habitat connectivity measures.

- Legislators often hold town meetings or listening sessions during home visits to solicit constituents' input. Conservationists can attend these events, bringing several like-minded citizens, and use the forum to discuss the benefits of road ecology solutions.
- Conservationists can invite lawmakers to attend their own events (meetings, workshops, conferences) to discuss wildlife and transportation.
- If conservation organizations have particular sites or projects of interest, they can work with transportation and resource agency staff to organize a field trip and invite elected officials. By visiting the site, decision makers can gain a greater understanding of the issue and see the impacts firsthand.

Conservationists can lobby lawmakers for positive changes, and they can reward elected officials for making better decisions that benefit wildlife. Advocacy groups can use their mailing lists, Web sites, and the press to publicly recognize positive actions by lawmakers.

Challenge 9: The Public Is Uninformed about the Conflict between Wildlife and Transportation

When it comes to pressing national issues, the conflict between wildlife and transportation falls pretty far down the list. And without significant public interest and input, making the large and lasting changes in policy and practice can be more difficult. In fact, the debate is often framed in terms of a false dichotomy: wildlife habitat must be sacrificed for public safety and traffic mobility. To the extent that the debate is carried out in the press, the media for the most part have done a poor job of covering the issue of wildlife and transportation. However, with a wider public discussion of transportation's role in climate change and the need to protect wildlife from the impacts of climate change, the media are more likely to cover the issue. In fact, the press has begun to pay more attention to this issue. Over the last few years, papers in major cities have covered the story, bringing regional and national attention (e.g., Mapes 2008, Robbins 2008). Conservation advocates can use communication skills to promote more press coverage. At the local level, regional organizations can contact local journalists to tell the story. At the national level, conservation organizations can encourage the large national and international outlets to cover the issue.

Recommendations for Future Transportation Policy

Currently, wildlife habitat protection and restoration measures carried out by transportation agencies are often inconsistent or temporary due to the lack of dedicated resources and staff at the federal, state, and local level. Many road ecology champions employed at state or federal agencies are lost to private practice or retirement, leaving gaps in experience and enthusiasm. And at the first sign of budget reductions, environmental measures are cut as if they were unnecessary luxuries. Implementing road ecology measures over the long term will require creating, capturing, or redirecting additional capacity. Conservationists should lobby key decision makers with the following policy recommendations:

- *Create a priority funding status* for projects that either do not impact or actually restore and enhance crucial habitat and wildlife corridors. Transportation projects can take decades from planning to project delivery. Between the time a project is planned and built, new information can come to light, making the project less desirable. Transportation agencies could conduct periodic audits of project lists and prioritize according to importance. Special funding incentives such as increased federal match could be provided for projects that address current needs but do not impact and perhaps even restore crucial habitat and wildlife corridors.
- *Supplement maintenance funding* for states that commit to a Fix It First policy. "Fix It First" means maintaining and improving the existing infrastructure rather than exhausting transportation budgets on expensive, major new construction projects. When transportation agencies fix it first, they invest in improving existing communities and reduce the pressure to build more highways that further fragment remaining crucial habitat areas and wildlife corridors. Fix It First principles would also allow the possibility to upgrade existing infrastructure to be wildlife friendly, permeable, and safer.
- *Standardize wildlife–vehicle collision data collection* and support the sharing and analysis within and among states or provinces. A recent study by the Western Transportation Institute found that, while many transportation agencies collect information on wildlife–vehicle

collisions, the quality varies tremendously among states. Without accurate information about the animal and the location of the accident, authorities cannot estimate the magnitude of the problem or identify locations for corrective measures like wildlife crossings.

- *Integrate measures to assist wildlife responses to climate change* in transportation planning. Transportation planners should work with natural resource agencies to inventory critical wildlife movement corridors in light of climate change projections to determine whether, when, and where existing or planned highways might impact wildlife movements in response to climate change.

- Make the federal to state funding ratio for transit projects equal to that of highway projects. Most highway projects receive 80 percent federal funding while new transit projects receive just 50 percent federal funding. This discrepancy sets up a perverse incentive for states to continue to rely on additional highways rather than transit to meet transportation needs because they initially cost the state less out of pocket. If transportation needs are always met with more highways, more habitat will be lost.

- Increase the State Planning and Research (SP&R) set aside from 2 percent to 2.5 percent. States are required to set aside just 2 percent of the apportionments they receive from the Interstate Maintenance, National Highway System, Surface Transportation, Highway Bridge, Congestion Mitigation and Air Quality Improvement, and Equity Bonus programs for state planning and research activities. Of this amount, states must allocate 25 percent for research, development, and technology. In order for planners to take advantage of opportunities to protect and restore crucial habitat and wildlife corridors from the landscape level, planning funding must be increased.

Conclusions

Protecting and restoring essential core habitat for wildlife and corridors is a big job and will take considerable investments from all sectors from transportation and natural resource professionals, elected officials, landowners, academia, and communities. In the world of wildlife and transportation, conservation advocates often have the resources and skills to bring all the partners together.

PART III

Effective Partnerships

Calls for new solutions to the conflicts between roads, fisheries, and wildlife are increasingly heard from the public, agency biologists, the transportation community, and decision makers. Federal and state/provincial transportation agencies have recognized that ecosystem approaches and early stakeholder involvement are needed in identifying issues and areas of concern if their projects are to be environmentally sustainable and streamlined. Therefore, partnering and collaborative approaches are essential to developing ecosystem and habitat conservation initiatives.

The third part of the book presents a series of case studies from a variety of partnerships occurring across North America. These examples highlight the successful implementation of ecological and engineering solutions on the ground. They also serve to illuminate the cooperative efforts emerging as a result of transportation agencies and nongovernmental organizations finding common ground. Each case study explains the development of the partnerships, the rationale for the projects, the critical factors that made the project happen, the outcomes of the projects, and the lessons learned from each project. These examples illustrate highly innovative and productive partnerships resulting in institutional changes in project planning, design, and implementation, with the common goal of ecosystem conservation.

155

Chapter 7

The Banff Wildlife Crossings Project: An International Public–Private Partnership

ADAM T. FORD, ANTHONY P. CLEVENGER,
AND KATHY RETTIE

The Trans-Canada Highway (TCH) in the Bow Valley of Banff National Park (hereafter referred to as Banff) was newly built in the early 1950s, and like many scenic, low-volume, two-lane highways, it probably had little impact on wildlife at that time. In the last 50 years the highway has transformed into a major commercial thoroughfare and become Canada's economic lifeline, connecting goods and people between the Atlantic and Pacific coasts. Since the mid-1970s, collisions between vehicles and large ungulates on the TCH have been a major concern of Parks Canada (Flygare 1978, Holroyd 1979, Damas and Smith 1982). In the last 15 years, the TCH was recognized not only as an important source of mortality but also as a potential barrier for large mammal movement in the mountain parks and the substantially larger Central Rocky Mountain ecosystem (Banff–Bow Valley Study 1996).

This chapter describes how the implementation of various science and policy measures has created one of the most intensely mitigated and studied stretches of highway in the world. We begin with a description of the ecological processes disrupted by the TCH and then discuss adaptive management approaches established to address these impacts, and finally, end the discussion with an overview of the various institutional innovations that were developed to make this project succeed over the long term.

Geographical, Historical, Political, and Social Setting

Established in 1885, Banff is Canada's oldest national park and was created to provide tourist access to a natural hot spring, although the boundaries and size of the park have changed several times since then. Banff has always attracted a "wilderness"-focused luxury tourism industry with access to remote but generous accommodation. Currently, Banff sustains about 4 million visitors per year that recreate in both front- and backcountry facilities. The towns of Banff and Lake Louise, with a combined population of 15,000, exist to provide accommodation for tourists and to house tourism operators and spin-off industries. In addition to tourists and the two townsites, Canada's busiest transportation corridor bisects the park. The TCH averages 25,000 vehicles per day during the summer while the adjacent Canadian Pacific Railway averages thirty trains per day (Parks Canada, unpublished data). Banff and neighboring Yoho National Park are the only protected areas in North America bisected by a major four-lane highway. Recognizing the vital role of the TCH in Canada's national transportation network and the level of human use within Banff, Parks Canada has emphasized mitigating the deleterious effects of the roadway on wildlife while improving motorist safety.

Roadway and Environmental Issues

Roads are known to affect wildlife populations by increasing mortality, creating a barrier to movement, removing habitat, and facilitating the spread of invasive species (Forman et al. 2003, see chapter 1). The most obvious effect of roads on wildlife is mortality due to wildlife–vehicle collisions (WVCs). WVCs are a concern to human safety, property damage, and insurance costs (Conover et al. 1995, Huijser et al. 2007), as well as to wildlife conservation. In some cases, rates of WVCs may be high enough to cause wildlife populations to decline. For example, the 1990 elk population in Banff was estimated at 800 individuals and was predicted to fall to fewer than 175 individuals by 2010, largely due to WVCs along the TCH (Woods 1990). From 1981 to 1996, 48 percent of all ungulate mortality and 65 percent of carnivore mortality in Banff was road related (Shury 1996). Mortality from WVCs can affect population growth rates if mortality is greater than recruitment from immigration and birth. Also, if few animals can safely cross the road then rates of gene flow between subpopulations divided by roads may be low. This could cause genetic structuring to occur, which may in turn jeopardize population viability (Gerlach and Musolf 2000, McRae et al. 2005, Epps et al. 2005, see also chapter 1).

In addition to mortality, the barrier effect of roads can impede animal movement through avoidance behavior. Road avoidance has been documented in songbirds (Cassady St Clair 2003), small mammals (Adams and Geis 1983, Ford and Fahrig 2008), ungulates (Rowland et al. 2000), and large carnivores (Mace et al. 1996, Kerley et al. 2002). Many organisms need to move daily to locate forage, and they may also move intraannually to breeding areas or to seek seasonally available resources (Mattson et al. 1991, Wright et al. 1998, Berger 2004). Many species also move interannually as part of a juvenile dispersal process (Sutherland et al. 2000). Disruption of these movements by roads can have negative consequences for individual survival and population persistence (Carroll 2006). Genetic connectivity across the TCH in Banff was being mediated by male grizzly bears, but demographic connectivity was being disrupted because female movement across the road was limited (Gibeau 2000, Proctor 2003).

The effects of roads on wildlife in general, and of the TCH in Banff specifically, are to reduce wildlife population viability through increasing mortality and disrupting animal movement across the highway. Attempts to minimize these effects must therefore focus on WVC reduction while at the same time ensuring wildlife can access food, shelter, and mates across the landscape and throughout the year. Achieving these goals requires cooperative efforts from a suite of disciplines, including civil engineering, environmental design, transportation planning, and biological sciences.

Rationale for the TCH Twinning, 1980–1997

In 1978, the federal government proposed to expand the width of the TCH in Banff from two to four lanes, a process known as twinning the highway (McGuire and Morrall 2000). The TCH twinning project has proceeded in a series of phases, beginning with Phase I in 1979 and continuing through the current day on Phase IIIB (fig. 7.1).

Phase I: Kilometer 0 to 13

The 1979 Federal Environmental Assessment and Review Process for Phase I identified that WVCs were a major concern for human safety and wildlife conservation values. For example, 45 percent of the 780 ungulate

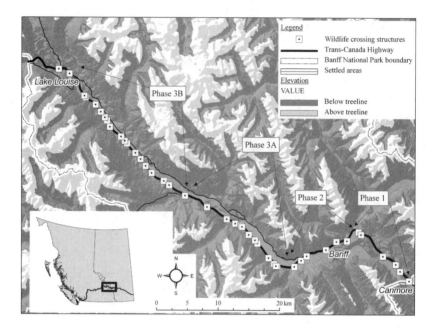

FIGURE 7.1. Map of the study area showing Trans-Canada Highway mitigation and twinning phases from the eastern boundary of Banff National Park to the British Columbia–Alberta border.

road mortalities between 1964 and 1979 within Banff occurred along this 13 kilometer (8.1 mile) stretch and in a single year (1978) over 110 elk–vehicle collisions were recorded (FEARO 1979). Given the high frequency of WVCs in this area, the role of the highway right-of-way as grazing habitat for large ungulates, and the predicted declines in moose (*Alces alces*) and elk populations as a result of road mortality, it was recommended that mitigation focus on altering ungulate movement patterns to minimize the probability of WVCs. WVCs involving carnivore species and small mammals were raised during the review but as "lesser concerns" (FEARO 1979, 26). A wildlife exclusion fence (2.4 meters [7.9 feet] high) was installed along both sides of the highway to reduce animal access to the highway right-of-way (fig. 7.2), and six wildlife underpasses were installed to allow animals to safely cross beneath the road.

Phase II: Kilometer 13 to 27

A Phase II environmental review was completed shortly after construction began on Phase I. During Phase I public meetings, Parks Canada identified

FIGURE 7.2. Elk grazing near the wildlife-exclusion fence along Phase I of the Trans-Canada Highway, Banff, Alberta. The bottom portion of this fence is not buried.

that TCH twinning to 27 kilometers (16.8 miles) from the East Gate was warranted, though traffic levels and terrain differences in these two stretches were different enough to merit a separate environmental review process (FEARO 1982). Highway aesthetics were less of an issue in the environmental assessment for Phase II, and mitigation measures proceeded seamlessly from Phase I. Some members of the environmental assessment panel argued that monitoring of Phase I continue longer to allow for a more rigorous assessment of management actions. However, construction began in October 1984, with fencing and twinning complete by September 1987 (FEARO 1982).

Again, unburied fencing and four additional wildlife underpasses were constructed along this stretch following the design of the Phase I crossing structures. Once Phase I and II were completed in September 1987 there were a total of ten wildlife underpasses on 27 kilometers (16.8 miles) of the TCH.

Phase IIIA: Kilometer 27 to 48

In the mid-1980s the federal government submitted a proposal to further twin the TCH to Castle Junction (see fig. 7.1). With the advent of ecological-integrity based management in the national parks system (Rettie et al. 2009), Phase IIIA was the start of a new era in highway mitigation. Unlike the environmental assessments from previous phases, the review was conducted independent of Parks Canada, and recommendations indicated that large carnivore conservation should be a priority in the implementation of proposed mitigation measures (Parks Canada 1995).

With the growing loss of montane habitat to human development in the Bow Valley (Banff–Bow Valley Study 1996) the footprint of the fenced TCH was of increasing concern. The Phase IIIA fence was designed to have the smallest footprint possible and was situated at the minimum distance from the road as possible (the "clear zone"). The fence was buried with a 1 meter (3.3 feet)-deep section of chain-link fence material as carnivores were found to dig under the unburied fence on Phase I and II (Bunyan 1990).

Initially wildlife underpasses similar to those on Phase I and II were planned for construction on Phase IIIA. Many internal meetings discussing the need to provide movement across the TCH for large carnivores (primarily grizzly bears, wolves, and cougars) resulted in two 50-meter (164 feet)-wide overpasses being constructed along with ten wildlife underpasses. Mitigation measures on this phase were completed in November 1997. By that time twenty-three wildlife crossing structures were built on 45 kilometers (28 miles) of TCH.

Monitoring Performance

Generally, there has been a lack of indicators or criteria developed prior to the construction of mitigation measures to adequately assess how well crossing structures ultimately perform in meeting land management and transportation objectives. Performance can be measured in terms of WVC reduction and rates of passage at the wildlife crossings and their effects at the individual and population level (Clevenger 2005). Management within Banff has evaluated mitigation performance through long-term monitoring of both mitigated and unmitigated highway sections.

Wildlife–Vehicle Collisions

WVCs along the TCH in Banff have received considerable attention over the past 25 years (Flygare 1978, Damas and Smith 1982, Clevenger et al.

2002a, Hebblewhite et al. 2003). With a large pool of trained staff (e.g., park wardens and research personnel), Parks Canada has been able to track the numerical and spatial distribution of WVC for a variety of large mammal species along hundreds of kilometers of highways in the mountain parks (Flygare 1978, Clevenger et al. 2002a, Gunson et al. 2009). Parks Canada staff regularly travel the TCH within Banff so WVC data are gathered opportunistically on a daily basis. This information is stored in a central database and each record includes the date, coordinate and descriptive location, species, number of individuals, and physiological information from necropsies. During the 1980s the focus of this effort was directed at ungulate, specifically elk, mortality as well as large carnivores (Woods 1990). Shorter-term monitoring of small mammal, amphibian, and bird WVCs has also occurred as part of specific research projects. Furthermore, Parks Canada wardens have a fence intrusion reporting system whereby observations of large animals inside the highway fence are recorded in a central database containing information on the location, species, and timing of the event.

As each phase of twinning and mitigation was completed, researchers discovered replicable patterns in WVC reduction and fence intrusions. Wildlife fencing dramatically improved motorist safety and reduced WVC occurrence. Fencing reduced WVCs by over 90 percent for ungulates and 86 percent for all large mammals (Woods 1990, Clevenger et al. 2002a). Furthermore, fence intrusions were 83 percent lower on highway sections with buried fence aprons compared to those with unburied fence sections (Clevenger et al. 2002a). These results indicate that highway mitigation measures were effective at reducing WVCs within Banff.

Wildlife Crossing Structures

Determining the rate of crossings by various species over time was one of the key tasks that the crossing structure monitoring was designed to address. The primary method for determining the frequency of crossing structure use was based on sand track pads at the crossings (see Clevenger and Waltho 2000, 2005 for details). For the past 12 years, track pads were visited every two days during the summer and every four days during the winter. At each visit to a track pad researchers recorded the species, direction of movement, and number of individuals for all mammals coyote-sized and larger.

In recent years, track pad checks have been supplemented with motion-sensitive cameras to monitor species use of the crossing structures. These cameras (i.e., Reconyx Inc., Holmen, WI) also provide information on time, animal behavior, and ambient temperature during each crossing event. Cameras are a more reliable, cost effective, and less invasive means of monitoring crossing structure use than tracking alone (Ford et al. 2009).

Since 1996, when systematic monitoring began, over 190,000 crossing events have been recorded by eleven species of large mammals, including moose, bighorn sheep, deer (*Odocoileus* spp.), lynx, cougar, coyote, wolf, grizzly and black bear, and wolverine (Clevenger et al. 2009). When these data were analyzed it was found that human activity at or adjacent to the crossings was the most important factor negatively affecting wildlife use (Clevenger and Waltho 2000). Also species had specific preferences for crossing structure designs: cougars and black bears preferred smaller crossing structures; whereas grizzly bears, wolves, and ungulates preferred larger and more open structures (fig. 7.3, Clevenger and Waltho 2005).

Monitoring the Phase IIIA crossings after their construction strongly suggested that time was required for animals to adapt to the new structures (fig. 7.4). These results have significant implications for design of monitoring programs aimed at performance evaluations. Recommendations for future crossing structure designs based on only two years' monitoring data would be substantially different (i.e., larger and more costly crossing structures would have been recommended) than based on more than 5 years of data.

Phase IIIB Recommendations

Results of monitoring and research of Phase I, II, and IIIA mitigation measures were used to guide the planning and design of mitigation on Phase IIIB. This adaptive management approach was sought by Parks Canada to streamline planning by obtaining recommendations based on credible science. The location of future wildlife crossings was based on modeling empirically based movement data, simulating movements of five large mammal species (wolf, grizzly bear, black bear, elk, moose) and validated by independent data in the study area (see chapter 2, Clevenger and Wierzchowski 2006). The locations of the wildlife crossing structures were prioritized based on the key habitat linkages (primary, secondary, tertiary) identified in the model (Clevenger et al. 2002b). Design specifications of the crossing structures were derived from research identifying what attributes facilitated passage of large mammals (Clevenger and Waltho 2000, 2005).

FIGURE 7.3. Images of different wildlife crossing structure designs along the Trans-Canada Highway in Banff National Park, Alberta (clockwise from top left): (a) Bull elk at Morrison Coulee, a 4-meter-diameter culvert; (b) Five-Mile Bridge at the Bow River; (c) Wolverine Overpass with Wolverine Underpass (4 meters × 7 meter metal elliptical culvert) in the foreground; (d) Wolverine Creek underpass, with riparian walkway.

Final TCH Twinning in Banff

Phase IIIB: Kilometer 48 to 83

Building on the lessons of previous phases and research, the final stretch of TCH twinning in Banff is currently under construction. The upper Bow Valley (Phase IIIB) is characterized by an assemblage of large mammal species with low population densities and sensitivity to human disturbance (wolverine, lynx, grizzly bear, moose) compared to the typical fauna of the

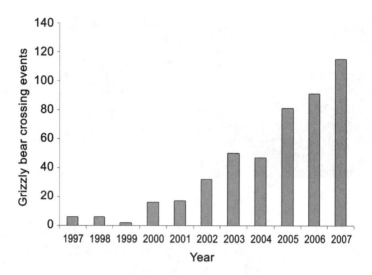

FIGURE 7.4. Grizzly bear use of wildlife crossings structures along Phase I, II, and IIIA from November 1996 to December 2007, as measured by track pad records.

middle and lower Bow Valley (cougar, wolves, black bear, elk, deer) where previous phases were constructed. Consequently, to restore connectivity on Phase IIIB, mitigation plans were altered from previous phases. A mitigation strategy was recommended that incorporated a greater number of high-quality (i.e., large underpasses and overpasses) wildlife crossing structures and at more frequent intervals (roughly 1.5 kilometers [0.9 mile] between structures) than in the past (Clevenger et al. 2002a). For example, the 70-meter (230-feet) width proposed for overpasses on Phase IIIB was seen as a suitable guideline for the sensitive species living in the upper Bow Valley. The 20-meter (66-feet) increase in width relative to previous mitigation designs was a rational next step for the assessment of crossing structure performance. However, Parks Canada failed to follow this design recommendation for "primary" wildlife crossing structures and scaled down the overpasses to 60 meters (197 feet) wide (T. McGuire, pers. comm.).

Maintaining Support for Monitoring and Research

From 1996 to 2002 funding for the research and monitoring of the TCH mitigation was provided entirely by Parks Canada's Highway Service Centre. The research project responded with a rigorous, 5-year study (Clevenger et al. 2002a); however, once this initial phase of monitoring was complete, Parks Canada scaled back funding to maintain only basic monitoring of the crossing structures, with little in the way of support for a continued research program and personnel. Other funding sources were secured and run through the Western Transportation Institute, and Parks Canada's senior managers agreed to allow continued monitoring and research (Banff Wildlife Crossings Project [BWCP], Clevenger et al. 2009).

An International Public–Private Partnership

Today, the BWCP research, education, and outreach efforts are a combined endeavor of a public–private partnership including a federal agency, a university institute, and private foundations. Parks Canada manages the construction and maintenance of the wildlife crossings and assures that they conform to the national park's commitment to protect, as a first priority, its natural heritage. With a dedicated road ecology program, the Western Transportation Institute at Montana State University leads the scientific research, education, and outreach activities of the BWCP.

The BWCP's monitoring and wildlife research, coupled with outreach and education, aim to inform the transportation community, wildlife managers, and public of the environmental and societal benefits of Banff's highway infrastructure investments. This will allow other communities to develop sustainable transportation practices that provide ecological connectivity across their transportation corridors as well as maintain motorist safety. In this light, the specific objectives of the BWCP were divided in four areas:

1. *Partnership*: Maintain and continue the international public–private partnership in conservation science and management of transportation systems in natural and working landscapes. Garner support and interest in the BWCP from Canadian foundations in addition to support from U.S. foundations.
2. *Science*: (a) Conduct research measuring gene flow of grizzly and black bears using wildlife crossings of the TCH and model population viability; (b) continue monitoring and research of wildlife crossings use by a variety of wildlife species, including newly constructed Phase IIIB crossings; (c) based on these results, develop science-based guidelines for designing effective wildlife mitigation for transportation projects.

3. *Technology transfer and education*: Present the research findings in major international journals, books, and conferences on transportation and ecology. Provide greater professional understanding and knowledge of measures to reduce highway impacts on wildlife and fisheries through training courses. Support university graduate students active in the project and use information for university classes, courses, and symposia.

4. *Community awareness*: Provide greater understanding and general public awareness of the BWCP and research findings to allow leveraging for similar mitigation for highways across the Yellowstone to Yukon bioregion and North America via field trips, workshops, media coverage, and educational venues (i.e., schools, universities, museums).

Accomplishments

Between 2005 and 2009 the BWCP advanced research on the conservation value of the Banff wildlife crossing structures, continuing long-term monitoring of wildlife use of the crossings, transferring technology learned from Banff to transportation practitioners, and communicating science to the public (Clevenger et al. 2009). Over 10,000 DNA bear hair samples were collected from individuals using the crossings and from the population in the surrounding landscape. The preliminary genetic data indicate not only that the crossings allow for sufficient movement of individuals across the TCH (Clevenger et al. 2009), but that bear movement is not sex-biased; both sexes mix freely across the TCH (Clevenger and Sawaya 2010). These preliminary data suggest that the twenty-four crossing structures are functional from a genetic and demographic connectivity standpoint (Proctor et al. 2005).

Many presentations were made to transportation practitioners and wildlife scientists describing the research and implications for highway management. The BWCP organized training courses and workshops for Canadian transportation engineers. In the 4-year period, nine peer-reviewed publications (including one book) were produced, along with five more manuscript submissions, one being accepted for publication.

Since 1996, 180 school presentations were made to over 5,000 students and teachers in the Bow Valley, Columbia Valley, Crowsnest Pass, and Calgary areas. Two wildlife crossing art contests were held in conjunction with the presentations. Wildlife crossing exhibits were held at Banff's

Whyte Museum of the Canadian Rockies (which subsequently won the Banff Heritage Tourism Award), the Calgary Zoo, and Parks Canada's information centers. Numerous articles about the BWCP appeared in local and international media, including the *Globe and Mail* and *New York Times*.

Lessons Learned

Evaluating the success of highway mitigation is dependent on the specific objectives established at the planning stages. In this case, WVC reduction and the maintenance of wildlife movement across the TCH for large mammals were important conservation goals for Parks Canada that mitigation was able to achieve. However, the success of the mitigation, monitoring, and research efforts requires a broad view of "audience" when disseminating results. Key to garnering support for mitigation is to first dispel notions that the structures were not used or only used by some common species and were a waste of taxpayers' dollars.

Conservation Lessons

Over time, questions used to evaluate the effectiveness of mitigation have evolved:

- Did we reduce WVCs? (Phase I era)
- Are ungulates using the underpasses? (Phase I and II eras)
- Do the wildlife crossings provide population-level benefits for large mammals? (Phase IIIA era)
- Can genetic and demographic connectivity be restored across the TCH for low-population-density species such as wolverines, lynx, and grizzly bears? (Phase IIIB era)

At least three innovative management approaches enabled researchers to effectively evaluate the success of highway mitigation. First, management took a long-term approach to monitoring WVC and crossing structure use. This approach required not only patience and planning but also consistent funding and logistical resources to conduct the work. Had monitoring ceased within the first two years, the normal duration for most crossing structure monitoring studies prior to 1996, conclusions about mitigation performance would be inaccurate. Long-term monitoring has enabled researchers to rigorously evaluate trends in WVC frequency and distribution, as well as species use, adaptation, and response to the wildlife crossings.

The second key management innovation was to build a diversity of crossing structures. By diversifying the design of these structures, engineers and wildlife researchers were able to determine the most cost-effective means of ensuring wildlife movement across the highway. Full experimental control is often elusive in evaluating highway mitigation because of the permanent nature of the structures and the cost of their construction. Nonetheless, management now has the information needed to tailor mitigation toward focal species, in a variety of terrain conditions and with a suite of associated costs.

Lastly, evaluating the success of highway mitigation from a WVC reduction and landscape connectivity view must address several levels of biological organization. Namely, mitigation effectiveness needs to be evaluated against criteria such as the preservation of genetic, species-population, and community-ecosystem functions and processes (Noss 1990, Clevenger 2005). From the top down, mitigation has ensured that wide-ranging species are able to move at scales that invariably overlap with the TCH. Monitoring has shown that all species common to the Bow Valley ecosystem are using the crossing structures, and more recently, we are assessing rates of gene flow across the TCH.

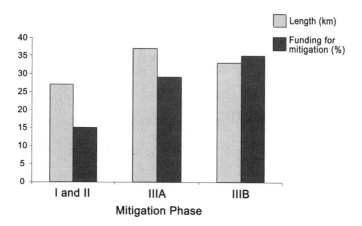

FIGURE 7.5. Proportion of total project expenditures directed at mitigation (fencing and wildlife crossings) for different phases of the Trans-Canada Highway twinning in Banff National Park, Alberta.

Conclusions

Funding for highway mitigation draws upon taxpayer dollars allotted for the twinning of the TCH (fig. 7.5). To many taxpayers, it is more valuable to have extra miles of four-lane highway than fewer lanes with wildlife mitigation. It is essential that Canadians understand the purpose and effectiveness of wildlife crossing structures and how people and wildlife are afforded safe passage because of them.

Without focused public education and outreach programs, rumors continued to circulate among the public and Parks Canada staff in the early days of the BWCP. Key tools for informing and engaging the public are the

FIGURE 7.6. Images from remote cameras documenting wildlife use of crossing structures on the Trans-Canada Highway, Banff, Alberta. Clockwise from top left: Cougar using a culvert underpass; moose stepping over bear-hair-snagging wire on a wildlife overpass; wolf passing at a concrete box-culvert underpass; bull elk and wolf encounter at an open-span bridge underpass.

scientific data recorded from the track pads, but the most unequivocal evidence of success has been produced from still and video cameras, along with the results of the genetic analysis. Indeed, the most compelling educational resources used by the BWCP are the photographic images (fig. 7.6) and video footage (http://www.blogs.westerntransportationinstitute.org/road ecology/archive/2009/01/13/what-s-new-wildlife-research-vidoes.aspx) of wildlife using the crossing structures.

Scientific research has an important role in communicating science to peers and management. However, scientific research must be effectively communicated to the public and particularly local communities and decision makers to gain political support for road mitigation measures (Jacobson et al. 2006).

Chapter 8

Reconstruction of U.S. Highway 93: Collaboration between Three Governments

DALE M. BECKER AND PATRICK B. BASTING

Ancestors of the members of the Confederated Salish and Kootenai Tribes (CS&KTs) have lived in the northern Rocky Mountain region since time immemorial. The day-to-day existence of Salish, Kootenai, and Pend d'Oreille people who inhabited the area was tied inextricably to the natural resources of this area (Fahey 1974, Malouf 1998, Smith 1998). Abundant wildlife and natural resources provided for subsistence, cultural, and spiritual needs of the people. Tribal peoples' lives were intertwined with those of the native animals and plants, and their activities, movements, lifestyles, and well-being depended upon those resources. Wildlife played, and continues to play, dual roles as both natural and cultural resources for the CS&KT, and as they make the Flathead Indian Reservation their homeland today, they continue to care deeply about their aboriginal territory and the animal inhabitants on the reservation and across the broader region. Tribal members also continue to rely heavily upon the wildlife resources, both on and off the Flathead Indian Reservation, for subsistence, cultural, and spiritual needs (Tony Incashola, Salish and Pend d'Oreille Culture Committee Director, pers. comm.).

In the early 1990s, the Montana Department of Transportation (MDT) and the Federal Highway Administration (FHWA) announced a proposal for reconstruction of a 97 kilometer (60 mile) portion of U.S.

Highway 93 that traverses the Flathead Indian Reservation between the communities of Evaro and Polson, Montana. The proposal resulted in consideration of a wide variety of issues and concerns important to Tribal people and their culture, including wildlife and wildlife habitat (Federal Highway Administration and the Montana Department of Transportation 1995, Becker 1996).

Work on the proposal throughout much of the 1990s involved substantial disagreements between the CS&KT government and members of the public and the MDT and other citizens (Marshik et al. 2001). Both the CS&KT and MDT recognized and agreed upon the need for improvements in design and safety, but strong disagreement on the details persisted. The CS&KT expressed substantial concerns about potential deleterious impacts that the reconstructed highway would have upon the tribes' unique culture, including the landscape and its physical, biological, and historical components.

In the mid-1990s, the three governments (MDT, FHWA, and CS&KT) collectively prepared a Final Environmental Impact Statement (FEIS) and a Section 4(f) Evaluation (under the Federal Highways Act) (Federal Highway Administration and the Montana Department of Transportation 1995) that described the proposed project alternatives and social, economic, and environmental impacts. The FEIS evaluated several alternatives ranging from "no action" to construction of a divided four-lane highway. Each alternative, however, included limited proposals for mitigation for cultural and wildlife and wildlife habitat issues that were significant concerns for the CS&KT. The document also deferred decisions on Section 4(f) determination related to wildlife preserves and other significant habitat issues until the governments agreed upon contentious issues such as lane configuration bypass issues for towns, design features, and mitigation features.

In its official response to the FEIS, the CS&KT expressed concerns about the preferred alternative, which encompassed a divided four-lane configuration, as well as other concerns related to cultural, wildlife, and mitigation issues. As a result, the project continued to be a topic of substantial discussion and disagreement between the CS&KT and MDT.

The FHWA subsequently issued a Record of Decision and a later modification (Federal Highway Administration 2001), which selected the existing alignment for improvement throughout the proposed project, as well as other important factors related to design and mitigation of cultural and natural resources impacts. However, without resolution of the CS&KT's concerns and consensus between the CS&KT and MDT, the FHWA, in response to its federal trust responsibility to the CS&KT, was not able to support moving forward on the project. As a result, during the next three years, there were efforts to continue negotiations between the CS&KT and MDT, but those efforts failed to bridge the differences adequately enough to break the stalemate and move the proposed project forward.

In 1999, an effort to resolve the final disagreements on the project was initiated by the three governments. Progress on several contentious issues occurred, and the negotiation process moved forward. The result of the negotiations was an extensive document entitled "Memorandum of Agreement—U.S. 93—Evaro to Polson" (Montana Department of Transportation, Federal Highway Administration and Confederated Salish and Kootenai Tribes 2000), which was signed by representatives of the three governments in December 2000. This document, hereafter referred to as the MOA, set the course for development of highway reconstruction designs for the project, as well as agreed-upon mitigation measures. In the MOA, the governments agreed upon conceptual design and alignment, design guidelines and recommendations, traffic operational and safety analysis report, wildlife mitigation, and design components roadway improvements for 66.6 kilometers (41.4 miles) of the highway. They also agreed to (1) prepare a Supplemental Environmental Impact Statement for the remainder of the project where resolution of differences was not reached; (2) reevaluate the environmental impacts of the preferred conceptual improvements for necessary changes; and (3) continue to work cooperatively to achieve physical construction of improvements to meet the needs of each government and that are in the best interest of the traveling public, the residents of the Flathead Indian Reservation, and the members of the CS&KT. The MOA established a technical design committee to work cooperatively on design and other related issues and a policy oversight group to establish and maintain policy for the environmental and design phases of the project and provide direction to consultants and staff in their joint efforts in performing environmental and design work.

Geographical, Historical, Political, and Social Setting

The Flathead Indian Reservation, located in west central Montana, was created as a permanent homeland for the Salish, Kootenai, and Pend d'Oreille people under the terms of the Treaty of the Hellgate of 1855. Under the treaty, the three tribes relinquished ownership to most of Montana lying west of the Continental Divide, as well as portions of central Montana, eastern Idaho, and eastern Washington, in return for exclusive use of the lands encompassed within the reservation boundaries. Later, allotment of Indian lands, government withdrawals, and the opening of the reservation to settlement resulted in substantial permanent changes to the environment on the reservation. Those changes continue today, and they relate directly to the highway reconstruction project discussed herein.

The CS&KT, as a sovereign nation, have inherent authority to regulate many aspects of the U.S. Highway 93 Evaro to Polson project to the exclusion of state authority. Tribal land is held in trust by the federal government for the benefit of the tribes and is not subject to state powers of eminent domain, so those portions of the highway corridor owned by the tribes cannot be used without tribal consent (FHWA Memorandum revising ROD 1998).

The Flathead Indian Reservation encompasses approximately 505,875 hectares (1.25 million acres) within its exterior boundaries (fig. 8.1). The land base comprises a wide variety of habitats ranging from semiarid shrub–grasslands to diverse wetlands and riparian habitats to subalpine habitats. It consists of four distinct valley complexes bounded by mountain ranges. The primary subject of this discussion is the eastern side of the reservation, where the exiting and proposed alignments of U.S. Highway 93 are located. A dominant feature forming the eastern boundary of the reservation is the Mission Mountain Range, with elevations up to 2,994 meters (9,822 feet) above sea level and the Rattlesnake Range to the south.

The dominant land use of valleys to the west of these mountains is agriculture, predominantly irrigated and dryland farming and livestock production. A significant geological feature is the extensive wetland complex centered near Ninepipe and Kicking Horse Reservoirs. Several small rivers and streams drain into the Flathead River, which bisects the reservation.

The reservation provides a diverse array of habitats for a large number of wildlife species. This fauna includes 309 species of birds, 66 species of mammals, 9 species of amphibians, and 9 species of reptiles (Tribal Wildlife Management Program, unpublished data).

Rationale for the Project

The high rate of population growth and increased tourism throughout western Montana are sources of growth in traffic on U.S. Highway 93. The highway is important to safety, social well-being, and the economy and is the major north–south transportation route in western Montana. It pro-

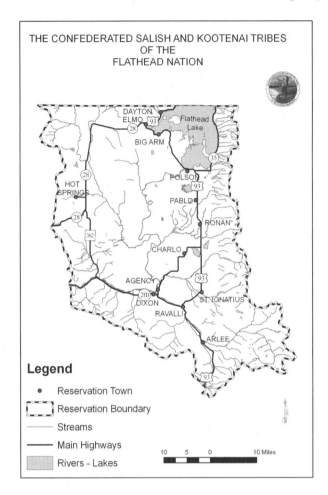

FIGURE 8.1. Map of the Flathead Indian Reservation.

vides interstate, regional, and local access to natural resource–based industries such as agriculture, forestry, mining, tourism, and recreation (Montana Department of Transportation, Federal Highway Administration, and Confederated Salish and Kootenai Tribes 2000). U.S. Highway 93 had various geometric features that did not meet standards for safety and design. The level of service was poor, and it was projected that traffic volume would exceed capacity by the design year 2020. Safety was a major issue since the highway was known as one of the most dangerous roads in Montana. Many local vehicles were adorned with bumper stickers that read "Pray for me. I drive Highway 93."

A safety analysis conducted in the U.S. Highway 93 corridor con-cluded that traffic accidents in the corridor increased by 10 percent between 1995 and 1999 (Montana Department of Transportation, Federal High-way Administration, and Confederated Salish and Kootenai Tribes 2000). Total accidents had increased faster during the study years (3.9 percent per year) than the growth in traffic volumes. The study also found that the cor-ridor experienced 30 fatal accidents and 279 injury accidents during the five-year study period. The proportion of fatal accidents in the corridor (4.8 percent) was much higher than the statewide average for National High-way System routes (1.7 percent). The proportion of nonfatal injury acci-dents in the corridor (44.2 percent) was also much greater than the statewide average for comparable roads (37.1 percent). Thus high accident severity (i.e., high risk of death or injury when an accident occurs) was a major concern in the corridor. Many of the accidents were related to exces-sive speed for existing conditions, head-on and rear-end collisions due to the highway, and many occurred at the numerous access points entering the right-of-way (Montana Department of Transportation, Federal Highway Administration, and Confederated Salish and Kootenai Tribes 2000).

Given its location in rapidly growing western Montana, daily traffic volume ranging from 6,440 to 12,070 vehicles on certain segments of the highway was recorded, with anticipated levels of from 12,490 to 23,420 vehicles per day by 2024 (the end of the project period) (Montana Depart-ment of Transportation, Federal Highway Administration, and Confeder-ated Salish and Kootenai Tribes 2000).

Analysis of the proposed project also spotlighted several substantial and related environmental issues especially given projections of large in-creases in traffic on the roadway during the anticipated 25-year life of the project. Those projections, compiled by the Midwest Research Institute, as part of the MOA, indicated peak traffic range of 864 to 1,264 vehicles per hour in 2000 and a projection of 1,200 to 1,770 vehicles per hour in 2024 (Montana Department of Transportation, Federal Highway Administra-tion, and Confederated Salish and Kootenai Tribes 2000).

The growing volume of traffic was a cause for concern about the frag-mentation of wildlife habitat and the loss of population linkage on the reservation and attendant adverse impacts on a regional scale upon several species, such as grizzly bears (*Ursus arctos horribilis*), black bears (*U. ameri-canus*), deer (*Odocoileus* spp.), elk (*Cervus elaphus*), gray wolves (*Canis lu-pus*), Canada lynx (*Lynx canadensis*), western painted turtles (*Chrysemys picta*) and other species (CS&KT Wildlife Management Program, unpub-lished data, Fowle 1990, Becker 1996, Ruediger et al. 1999). Documentation of wildlife–vehicle collisions was provided by the Tribal Wildlife Management Program (unpublished data) and MDT maintenance records (unpublished data), as well as other sources. One interesting source was a local high school science class project, which recorded annual mortality of approximately 300 to 450 western painted turtles in a 9.7 kilometer (6 mile) section of the highway.

The proposed project also offered potential opportunities to improve several adverse habitat impacts due to past construction. Several riparian areas were severely restricted by the use of small culverts, resulting in seasonal flooding and little, if any potential for use as wildlife passage. Past construction had also bisected diverse wetland habitat complexes, individual wetlands, and the home ranges of numerous resident wildlife species. The project provided opportunities for restoring, constructing, or conserving wetlands, as outlined in the CS&KT's Wetlands Conservation Plan for the Flathead Indian Reservation (Confederated Salish and Kootenai Tribes 2000). This plan established a goal of halting the loss of the remaining wetland and riparian habitats and the decline in wetland and riparian quality on the reservation. The long-term goal of the plan was to increase the acreage of wetlands and riparian areas and improve the quality of the resource. It outlined strategies for avoiding wetlands and for conservation and mitigation of adverse impacts upon wetland and riparian habitats, as well as procedures to address these issues.

Mitigation planning efforts included a thorough assessment of existing wetland and riparian habitats and potential impacts of the project. This plan required that wetland impacts must be avoided and/or minimized where practicable. If these impacts were unavoidable, mitigation had to be undertaken to replace or restore an appropriate amount of wetland habitat. The plan defined the types of mitigation possible and appropriate mitigation, according to CS&KT regulations.

Roadway and Environmental Issues

Wildlife utilize the habitats adjacent to the entire U.S. Highway 93 corridor on the Flathead Indian Reservation, as indicated by observations and by the number of road-killed deer, bears, turtles, small mammals, nongame and game birds, and other species observed within the right-of-way. A number of areas receive higher levels of wildlife use than others and exhibited a corresponding increase in vehicle-related wildlife mortality. These areas included wetlands, riparian areas, forested areas, and certain agricultural lands.

Observations of vehicle-caused wildlife mortality and important known and anticipated wildlife movement areas also provided prospective locations for wildlife crossing structures along the highway route. In addition, an analysis of habitat features, such as vegetative cover and hiding cover, indicated where animals might be expected to attempt to cross the highway. Remote-sensing cameras placed near wildlife trails provided added insight about the numbers and species of animals using the trails (Becker et al. 1993). Collectively, the information provided by these methodologies assisted in indicating where animal use was occurring and the degree of that use.

Critical Factors

Despite the often-contentious nature of the relationship among the three governments and their respective representatives, everyone involved viewed improvement to U.S. Highway 93 as important whether for public safety, cultural, or ecological perspectives. The determination of each of the governments to reach agreement through reasonable compromises and proceed with the reconstruction of the highway was a significant factor in the project. In addition, the mutual respect that developed between the involved staffs of the governments as the negotiations and later the project planning and design phases proceeded was essential to reach the final outcome. That respectful working relationship provided valuable exchanges of ideas, rationale, and knowledge between everyone involved. The diversity of the three governments was an interesting aspect of the process, but the diversity of people, which included administrators, managers, and staff from varied backgrounds (engineers, planners, archeologists, cultural resource specialists, and biologists) also added to the unique nature of the effort. Their combined efforts and knowledge collectively enhanced the project planning efforts and the quality of the project.

Outcomes of the Project

Sites of documented wildlife mortality, especially multiple-recurring mortality, indicate the location of many, if not most, wildlife passage problems. To alleviate these problems to the greatest extent practicable, a number of different design features were included in the reconstruction plans for Highway 93. Collectively, these design features will decrease the amount of wildlife mortality caused by traffic on the highway, as well as mitigate for the habitat loss, degradation, and fragmentation that currently exist.

Instead of maintaining the current configuration or designing a divided four-lane highway, consultation with Jones and Jones, a landscape architect firm from Seattle, Washington, resulted in final designs that incorporated a variety of configurations that included both, as well as other configurations where necessary. Unique variations such as the addition of curvilinear designs, recontouring of adjacent terrain, and the integration of the highway and the surrounding landscape were incorporated. Construction was planned to occur in phases for various segments, accounting for limitations due to anticipated seasonal weather constraints and the need to maintain traffic flow.

Reconstruction plans included provisions for 42 metal pipe culverts or concrete box culverts designed to facilitate wildlife crossing the highway right-of-way (fig. 8.2). Twenty-three of these structures ranging in sizes from 3.7 meters × 6.5 meters (12.1 feet × 21.3 feet) to 5.0 meters × 7.4 meters (16.4 feet × 24.3 feet) in size and structures approximately 1.3 meters × 2 meters (4.3 feet × 6.6 feet) were planned, with most completed as of

FIGURE 8.2. Wildlife crossing structure, with adjacent wildlife jump-out and wildlife fencing.

2008. Seven bridges, ranging from 12 meters (39.4 feet) to 110 meters (361 feet) in length and with a minimum 3.7 meters (12.1 feet) of vertical clearance to facilitate wildlife passage and revegetation are also being constructed across the larger rivers and streams that bisect the highway. One 46 meter (151 feet) to 61 meter (200 feet) wildlife overpass was completed in 2009.

Wildlife crossing structures were designed and placed to ensure maximum opportunities for wildlife passage across the highway right-of-way by all local species. Many concepts for these structures were originally developed elsewhere. Extensive literature reviews of similar structures used elsewhere in North America (Evink et al. 1999, Clevenger and Waltho 2000, Messmer and West 2002, Transportation Research Board 2002b, Forman et al. 2003) and in Europe (Damarad and Bekker 2003) assisted in the development of design concepts. Site visits to examine other similar crossing structures in Florida, Alberta, Arizona, and Wyoming by representatives of the three governments provided conceptual ideas for designs and configurations, but specific locations and local concerns were necessary to complete the designs (Montana Department of Transportation, Confederated Salish and Kootenai Tribes, and the Federal Highway Administration 2000).

Plans for mitigating wildlife mortality included the construction of 26.7 kilometers (16.6 miles) of 2.4-meter (7.9-feet) high page wire fencing with wing fencing at terminal locations to accompany the wildlife crossing structures in areas of the highest wildlife use. Continuous fencing throughout the entire project was not planned due to the excessive costs, difficulty in dealing with the numerous access points to the highway, and the fact that most collisions with larger wildlife species generally occurred at selected locations. The fencing is being placed to encourage wildlife movement toward the crossing structures. In areas where burrowing or digging animals are a concern, a buried apron (see chapter 2) will be added to discourage digging animals from breaching the fencing. In areas where fences span distances of more than 1 kilometer (0.62 mile), openings in the fence are guarded by the installation of wildlife guards (similar to cattle guards), and 2.6-meter (8.5-feet) vertical jump-out structures that lead to openings in the fence will allow for the escape of animals that are caught inside the fencing along the highway right-of-way. Similar structures have been installed elsewhere and have exhibited use and success by larger mammals (Bissonette and Hammer 2000).

Informative signing will take two forms—signs to warn motorists of potential wildlife hazards and signing to inform motorists of wildlife crossings. Warning signs will assist motorists regarding potential wildlife hazards in the right-of-way. Informative signing will also assist motorists in learning about the presence of wildlife crossing structures, as well as the rationale for their construction.

A report by the Western Transportation Institute (Hardy et al. 2007) provided preconstruction baseline information and performance criteria agreed to by the three governments. It also provided a baseline for future comparative monitoring of post-construction use of the wildlife crossing structures. These efforts will be coordinated by representatives of the three governments, academic institutions, and other entities to enhance knowledge derived from the research. A post-construction monitoring plan has been developed, and the three governments have located partial funding to commence the post-construction monitoring program in 2010 when most construction within the corridor is completed. In the interim, CS&KT Wildlife Management Program and Western Transportation Institute staff are monitoring wildlife use of completed wildlife crossing structures (fig. 8.3). Methodology includes use of sand tracking beds and motion-sensitive cameras placed at strategic locations inside some crossing structures and at some wildlife jumpout structures. Data collected during this monitoring effort indicate regular use by a variety of species (figs. 8.4 and 8.5).

Most habitat mitigation associated with the reconstruction of Highway 93 has been associated with anticipated losses of wetland and riparian

FIGURE 8.3. Confederated Salish and Kootenai Tribes wildlife biologist clearing tracking bed during wildlife monitoring.

FIGURE 8.4. White-tailed deer image recorded in a wildlife crossing structure.

FIGURE 8.5. Two mountain lions using a wildlife crossing structure at night.

habitats due to construction activities. Detailed on-site wetland mitigation concepts were developed for several construction project segments through a consensus process that included representatives from the MDT, FHWA, CS&KT, the U.S. Army Corps of Engineers, and other participating resource agencies. Design firms responsible for each project segment used the concepts to produce detailed on-site mitigation plans. Wetland mitigation sites were selected based on several criteria including the following: (1) proximity to planned fish and/or wildlife crossing structures; (2) proximity to other protected lands; (3) position on the landscape likely to support self-sustaining wetlands; and (4) suitability for restoration of forested, shrub, wet meadow, or emergent habitat in order of preference (Montana Department of Transportation 2004). These on-site wetland mitigation efforts involved stream restoration as well as wetland restoration, creation, and enhancement.

The three governments also developed a unique wetland reserve project to appropriately mitigate for any remaining unavoidable wetland impacts. Under directives for wetland mitigation, the CS&KT acquired a tract of drained and degraded wetland habitat, restored the habitat on the site, and then sold the wetland mitigation credits to the MDT. MDT utilized the site to assist in meeting its wetland mitigation obligations under the Federal Clean Water Act administered by the U.S. Army Corps of Engineers and the CS&KT wetland mitigation requirements.

Habitat fragmentation is a constant concern on the Flathead Indian Reservation, and fragmentation due to land uses, highway and road construction, and subdivision activities are major issues (Confederated Salish and Kootenai Tribes 1994, 1996). The CS&KT's concerns related to these issues have a direct bearing on the Highway 93 reconstruction project. As a result, the CS&KT utilize a variety of policies and planning tools such as their Aquatic Lands Conservation Ordinance and Comprehensive Resource Plan to manage human growth pressures, habitat degradation issues, and highway construction impacts and subdivision pressures. Because much of the reservation consists of nontribal lands, the CS&KT work closely with other government agencies at the local, county, state, and federal levels in an attempt to decrease the adverse impacts of activities upon tribal resources.

Lessons Learned

The assemblage of the right people from all three governments was necessary to break the stalemate on this project. Leaders who were able to build personal and working relationships, work together, build trust, and listen and learn from each other so that a greater vision could be achieved were crucial. A dispute resolution process clearly laid out rules, roles, and responsibilities that each party was to abide by.

Other lessons learned were that, despite our best efforts to review plans, some details were occasionally overlooked, such as places where fencing should have been extended, or jump-outs that might have been sited at better locations. The flexibility of technical design committee members working together and maintaining open communications created a cooperative and productive environment to deal with these types of issues.

Another valuable lesson learned was that the original jump-out designs as prepared by design engineers were not what biologists had visualized. After reviewing the jump-outs in the field, MDT and CS&KT worked together to modify the designs so that they would function better, as well as blend into the landscape. The new design also gave the project manager flexibility to direct the contractor yet still allow nuances to fit the situation without requiring change orders.

Open and free communication proved crucial. MDT construction project managers handled new ways of doing business, constant demands on their time, and intense workloads from managing multiple construction projects simultaneously. The CS&KT's construction reviewer and wildlife and wetlands programs staffs worked directly with MDT Environmental Services and construction project managers and contractors to resolve problems, answer questions, and give advice in a timely fashion. Some issues required rapid on-site changes, but by having fundamental trust and commitment to work together and get the job done, everyone worked through these issues to appropriate conclusions. Early coordination between CS&KT and MDT staffs to develop corridor-wide special provisions also greatly aided in reducing project-by-project workload to draft special provisions for contracts and provided consistency and predictability on many issues for contractors and MDT project managers.

CS&KT and MDT coordinated a great deal in the development of project-specific special provisions. An example was that when stream restoration work was to be performed, specific requirements were put in place to assure that experienced contractors and oversight personnel were available to assist MDT during construction.

Ongoing activities occurring elsewhere indicated that the ability and innovation to do something positive for wildlife and habitat in designing a highway was both possible and practical. These activities were certainly preferable to the traditional means of highway planning. As a result, the potential for designing and building a highway that integrates into the landscape and helps solve many of the environmental issues was accomplished.

New Standards

At this point, no new standards have been incorporated as policy by the MDT. This project has, however, led to a new way of considering environmental issues in highway construction. Much of the information from this project has been considered for other projects elsewhere in Montana, and it has also been shared with other state departments of transportation throughout the western states.

Conclusions

The planning effort for the reconstruction of U.S. Highway 93 through the Flathead Indian Reservation has been a long and arduous task that continues toward resolution since the completion of a Supplementary Environmental Impact Statement (Montana Department of Transportation, Confederated Salish and Kootenai Tribes, and Federal Highway Administration 2007) on the last segment of the highway, located in an extensive wetland complex. The insistence of the CS&KT that the highway be designed as a safe, environmentally friendly road instead of other alternatives has set the stage for a new vision for future highway designs.

Chapter 9

Citizens, the Conservation Community, and Key Agency Personnel: Prerequisites for Success

MARK L. WATSON AND KURT A. MENKE

With state and federal wildlife and land management agency budgets continually shrinking, agencies are relying more on involvement by concerned citizens to achieve conservation goals (Lee et al. 2006, New Mexico Department of Game and Fish 2006). In New Mexico, concerned citizen and conservation community volunteers are becoming more fully engaged in transportation planning and project development, particularly with highway projects that may cause adverse effects to wildlife populations and habitat connectivity. In the case of the Tijeras Canyon Safe Passage Project, the involvement of volunteers from local communities and conservation organizations was necessary to encourage the New Mexico Departments of Transportation (NMDOT) and Game and Fish (NMDGF) to seek alternative mitigation strategies to ensure wildlife permeability across Interstate 40 (I-40) in Tijeras Canyon and reduce the potential for wildlife–vehicle collisions.

Within the last six years, volunteer efforts to provide safe passage for wildlife and motorists and improve wildlife habitat connectivity across New Mexico's highways has occurred at several scales. These efforts include recommending specific mitigation activities at the project level (e.g., fencing, animal detection systems) (Tijeras Canyon Safe Passage Coalition 2005, Marron and Associates, Inc. 2007), to statewide and regional-level planning for habitat connectivity (The Wildlands Project 2003). The Tijeras Canyon Safe Passage Project relied on both levels of planning, initiated by volunteers from local communities and conservation organizations.

Implementation of the Tijeras Canyon Safe Passage Project also required development of a relationship of mutual trust between concerned citizens from local communities, conservation organization volunteers, and key state agency personnel from NMDOT and NMDGF. Full commitment to project goals by key agency personnel was necessary to successfully implement the Tijeras Canyon Safe Passage Project. Project proponents, including key agency personnel, shared common values and recognized the importance of increasing human safety while maintaining a wildlife linkage across a high speed, high-volume transportation corridor.

Geographical, Historical, Political and Social Setting

Tijeras Canyon is located approximately 1 kilometer (0.62 miles) east of Albuquerque, New Mexico. The canyon bisects two mountain ranges; the Sandia Mountains to the north and the Manzano Mountains to the south. These two ranges were formed approximately 20 million years ago and are considered by geologists to form a single north–south trending geologic unit referred to as the Sandia–Manzano range. Combined, they are approximately 105 kilometers (65 miles) long, 9.7 to 13 kilometers (6–8 miles) wide, and form the eastern side of the Rio Grande rift (Julyan 2006). Both ranges have peaks over 3,050 meters (10,000 feet) above sea level. Both ranges are part of the Cibola National Forest, and each range contains its own wilderness area: the 15,328 hectare (37,877 acre) Sandia Mountain Wilderness, and the 14,961 hectare (36,970 acre) Manzano Mountain Wilderness, established in 1978 by the Endangered American Wilderness Act (Julyan 2006). Both of these wilderness areas and surrounding national forestlands provide important refugia for native wildlife, plants, and ecological processes. The Western Governors' Association's Wildlife Corridors Initiative (Western Governors' Association 2008) directed western state wildlife agencies to identify "crucial habitats and corridors" for a suite of wildlife species in an effort to protect these areas from further development. As a result, NMDGF identified the Sandia and Manzano Mountain Wilderness Areas and the surrounding Cibola National Forest as both "crucial habitat and corridors" for mule deer (*Odocoileus hemionus*), black bears (*Ursus americanus*), and cougars (*Puma concolor*), with Tijeras Canyon identified as an important "corridor" that connects the two wilderness areas and Cibola National Forest lands.

Elevations along the bottom of Tijeras Canyon vary from 1,707 to 2,134 m (5,600–7,000 feet) above sea level. Tijeras Canyon drains to the west via Tijeras Creek, which joins the Rio Grande a few kilometers to the west in Albuquerque. Although Tijeras Creek becomes an ephemeral wash after exiting the canyon, it contains perennial reaches and springs throughout the canyon that provide critical water sources for wildlife. Historically, Tijeras Creek supplied human-made acequia systems that irrigated agricultural fields for local Native American and Hispanic communities (Quintana and Kayser 1980).

Tijeras Canyon has been used by humans for millennia as a gateway to the Rio Grande valley from the eastern plains (Cordell 1980, Julyan 1998, 2006), and so has been critical for both human and wildlife movements. Today, the village of Carnuel is situated in Tijeras Canyon at the western end, the village of Tijeras at the eastern end, and the remnants of the Cañon de Carnue Land Grant occur throughout the eastern half of the Canyon. The Cañon de Carnue Land Grant was established in 1763 to promote occupation by Spanish settlers to act as a buffer to Albuquerque from repeated attacks by Comanche warriors from the east (Quintana and Kayser 1980). According to the 2000 U.S. Census, approximately 1,000 residents occupy the Tijeras Canyon area.

Roadway and environmental issues

The land north and south of the canyon is largely unroaded habitat of mixed ownership that provides a buffer to the wilderness areas and habitat for many large mammals, including mule deer, black bear, bobcat (*Lynx rufus*), gray fox (*Urocyon cinereoargenteus*), and cougars. Cougars and black bears move long distances for dispersal, and the protection of habitat linkages between suitable habitats is necessary for the persistence of relatively isolated populations of these two species (Parsons 2003). Wildlife populations in the Sandia Mountains are becoming increasingly surrounded by human development. Maintaining effective wildlife corridors between the Sandia and Manzano mountains is especially important for the persistence of wildlife populations in the Sandias (Western Governors' Association 2008, The Wildlands Project 2003).

There are two major parallel high-speed, high-volume highways running through Tijeras Canyon; I-40 and New Mexico Highway 333 (NM 333). Originally NM 333 was U.S. Route 66 and now serves as a frontage road to I-40. I-40 was built in the 1960s and is one of the primary east–west highways for commercial truck traffic in the nation, connecting east

FIGURE 9.1. Tijeras Canyon looking west. (Mark Watson 2008)

and west coasts (fig. 9.1). During May 2007, the average daily traffic count on I-40 in Tijeras Canyon was 18,442 vehicles per day (Marron and Associates, Inc. 2007). Tijeras Creek is the primary water source for wildlife in the immediate canyon area, meandering between and adjacent to the two highways. Wildlife in this relatively dry environment is attracted to the riparian habitat at Tijeras Creek, creating an ecological trap due to the high potential for wildlife–vehicle collisions (Marron and Associates, Inc. 2005). The construction of each of these highways and associated increases in traffic volumes have created a filter barrier to wildlife movement between the Sandia and Manzano mountains, as indicated by increased wildlife mortality (Clevenger et al. 2002a, Forman et al. 2003, The Wildlands Project 2003, Marron and Associates 2005, 2007).

Accident report data for the Tijeras Canyon area suggested that the majority of wildlife–vehicle collisions on I-40 occur with commercial truck traffic, while on NM 333 they occurred primarily with local residents. Mule deer are hit most often, with black bears and cougars being hit less often. Estimating the total number of wildlife–vehicle collisions in Tijeras Canyon is complicated by the destruction of road-killed animals by high traffic volumes on I-40 and local motorists taking carcasses for meat on NM 333.

Rationale for the Project and Key Players

An intolerance of the continuing wildlife mortality on highways as collateral damage, and a growing awareness of the danger of wildlife–vehicle collisions, led to a convergence of events in New Mexico that facilitated the formation of the Tijeras Canyon Safe Passage Project. These events included identification of Tijeras Canyon as a crucial wildlife linkage at local, regional, and continental scales; passage of wildlife conservation legislation; and implementation of a collaborative landscape-level planning workshop. The synergism of these events culminated in 2003 with the formation of the Tijeras Canyon Safe Passage Coalition (Coalition), the key conservation organization responsible for implementation of the Tijeras Canyon Safe Passage Project.

New Mexico Highlands Wildlands Network Vision

The Wildlands Project (TWP) is a 501(c)(3) nonprofit conservation organization founded in 1994. TWP's mission is to restore and protect ecological integrity at a continental scale by establishing a series of wildlife "megalinkages." Each megalinkage represents a large-scale ecologically functional landscape with sufficient habitat connectivity to facilitate wildlife movement. Megalinkages are identified and analyzed within regional planning-level documents called Wildlands Network Designs (WNDs). TWP's work is based on the scientific understanding that providing landscape-level connectivity for large carnivores is crucial to ecosystem function (Terborgh et al. 1999).

TWP has partnered with regional conservation organizations to map and analyze priority areas within each WND. In 2003 TWP released the landscape-scale analysis entitled *New Mexico Highlands Wildlands Network Vision* (The Wildlands Project 2003). The *Wildlands Network Vision* was the first attempt to look at the New Mexico landscape in terms of core wildlife habitat, compatible use areas, and dispersal corridors. A series of spatial analyses identified the portions of the landscape that need to be protected to support healthy ecosystems in New Mexico; however, the corridors were only vaguely identified and were called linkage zones. Concurrently, TWP had worked with other regional conservation organizations to complete adjoining WNDs along the "Spine of the Continent" megalinkage. This megalinkage consists of the Rocky Mountain chain from northern Sonora, Mexico, northward to Alaska. The wildlife linkage zones identified within each WND form conceptual links for maintaining wildlife habitat connectivity at local and regional scales, and each WND is an equally important component for maintaining wildlife habitat connectivity at a continental scale.

To initiate protection of priority lands, TWP identified the most endangered wildlife linkage in each WND, using a consensus-based process among the authors. Endangerment was determined by identifying the level of threats to the linkage from development. Tijeras Canyon was selected for the New Mexico Highlands Wildlife Network Design, and was therefore considered a crucial linkage for wildlife movements at both regional and continental scales within the Spine of the Continent megalinkage.

Wild Friends Pass House Joint Memorial 3

In 1991, attorneys and educators at the University of New Mexico's Center for Wildlife Law (Institute of Public Law, University of New Mexico School of Law), collaborated in the development of a wildlife conservation–based education program called Wild Friends (University of New Mexico Center for Wildlife Law 1992). The pilot project was funded by the W. K. Kellogg Foundation, and the New Mexico state legislature has provided an annual appropriation for the program since 1995. Wild Friends is a hands-on program that integrates civics, wildlife science, and language arts to help students understand the democratic process, develop good citizenship skills, and contribute to wildlife conservation by involving them in public policy projects of their choice. Each year, New Mexico students in grades 4 through 12 (ages 9–18) write laws to improve wildlife conservation in the state. They vote by ballot to select a wildlife conservation–related subject, research the issues and current laws (or lack thereof), and then receive assistance from natural resource attorneys to write a bill or memorial and find a legislative sponsor. Each year during the legislative session, Wild Friends students travel to Santa Fe to lobby and provide testimony for passage of their bill or memorial. Wild Friends has been very successful. Over 10,500 students from Wild Friends have participated, initiating a total of seventeen memorials and bills in the legislature to improve conservation of New Mexico's wildlife and environment.

In January 2003, New Mexico's 44th state legislature passed House Joint Memorial 3 (HJM 3), which was developed by Wild Friends, sponsored by New Mexico Representative Mimi Stewart and signed by Governor Bill Richardson. HJM 3 directs NMDOT and NMDGF to collaborate during highway project planning and implementation to reduce the potential for wildlife–vehicle collisions in New Mexico. HJM 3 directed both agencies to utilize the best available technologies, such as constructing wildlife-proof fencing to direct wildlife to existing underpasses, although no money was provided to implement these recommendations.

Critical Mass Workshop

As a direct result of HJM 3, in 2003, the Critical Mass Workshop was sponsored by the New Mexico Carnivore Working Group, with assistance from the U.S. Fish and Wildlife Service, U.S. Forest Service, and NMDGF. The working group consisted of agency wildlife biologists and conservationists interested in promoting landscape-level habitat connectivity for large carnivores and their prey species. The working group was a state-level offshoot of the Southwest Carnivore Committee, affiliated with Defenders of Wildlife. Members shared a strong interest in enhancing landscape-level habitat connectivity for carnivores across New Mexico.

The workshop was attended by approximately 100 people representing state and federal wildlife and land management agencies, NMDOT personnel, environmental consultants, members of the conservation community, and concerned citizens. Over the course of two days, the group generated a consensus-based prioritized list and map of 30 highway segments in New Mexico identified for further analysis to assess the need for mitigation to reduce wildlife–vehicle collisions. Tijeras Canyon was identified as one of the top four critical highway segments in the state.

The prioritization of highway segments was based on three criteria: (1) large game animal–vehicle accident report data (one year's data only); (2) proximity of highway segments to important wildlife habitats on public land; and (3) the perceived potential for adverse effects of highways and associated traffic to threatened, endangered, and sensitive species. Maps from the New Mexico Highlands Wildlands Network Design were an important part of the process because they were the only statewide maps showing the existing network of core protected areas, compatible use areas, and potential habitat linkages.

From its inception, the Critical Mass Workshop was intended as a preliminary effort to identify important wildlife–vehicle conflict areas, due in part to the limited wildlife–vehicle collision data available for review. More importantly, the Critical Mass Workshop represented the first collaborative effort among state and federal wildlife, land management, and transportation agency biologists and conservation community activists to initiate wildlife safe passage projects in New Mexico.

Therefore, in 2003 Tijeras Canyon was identified by two independent assessments as one of the most endangered wildlife linkages in New Mexico. This classification meant that it was recognized as an important connection for wildlife between two mountain ranges that had been surrounded by highways and suburban development. Furthermore, HJM3 provided critical legislative leverage to support additional follow-up efforts.

Tijeras Canyon Safe Passage Coalition

The formation of the Tijeras Canyon Safe Passage Coalition was a direct outcome of the Critical Mass Workshop and was facilitated by an impending I-40 Tijeras Canyon shoulder-widening and repaving project. A small subset of Critical Mass Workshop participants met in 2004 to determine how to address the wildlife mortality and habitat connectivity issues in the Canyon. The Safe Passage Coalition, which is primarily an unpaid volunteer organization of local concerned citizens, conservation community members, and state and federal wildlife agency personnel, was created during this meeting. Agency members include biologists from the U.S. Fish and Wildlife Service and NMDGF. Local conservation groups supplying members include The Wildlands Project, Animal Protection of New Mexico, the New Mexico Chapter of the Wildlife Federation, and Sierra Club. All members share the common goal of providing safe crossings for wildlife and safer travel for people through Tijeras Canyon. The Safe Passage Coalition quickly grew to include over 100 individual members and two dozen organizations. However, a core group of approximately 10 non-paid volunteers were responsible for the majority of work done to implement the Tijeras Canyon Safe Passage Project.

One of the first key decisions of the Coalition was to request that NMDOT initiate a feasibility study to analyze the need for wildlife–vehicle collision mitigation in association with the impending I-40 repaving project. Primary Safe Passage Coalition concerns that led to this request were (1) rapidly increasing human development and associated increases in traffic volume; (2) the likelihood for increased barrier effect of I-40 from the planned installation of tall median barriers, which would likely decrease successful wildlife crossings and increase wildlife mortalities in Tijeras Canyon; and (3) the threat of climate change to the persistence of local wildlife populations from increased frequency and intensity of drought and wildfire and thus the need for wildlife to migrate northward to cooler habitats.

The NMDOT Environmental Bureau project lead initiated the study (Marron and Associates, Inc. 2005). No other study of its kind had been proposed for a highway project in New Mexico, so the study represented a precedent for NMDOT and for transportation-related wildlife conservation in the state.

Recognizing that the cooperation of the Cañon de Carnue Land Grant was critical for implementing any safe passage project in the canyon, the coalition offered a permanent position on the steering committee to a Land Grant board member. Land Grant members maintain a continuing distrust of outsiders, due in part to the historical context of continual reduction of communal lands. For example, construction of I-40 led to the condemnation of Cañon de Carnue Land Grant lands, effectively splitting the village of Carnuel in half (Quintana and Kayser 1980, 52). Acceptance by the Land Grant of any coalition recommendation for a wildlife safe passage component of the broader I-40 project was essential to move forward. Proposals to increase wildlife habitat connectivity would affect Land Grant members by redirecting wildlife to strategic crossing locations that occurred on Land Grant lands. A proposal to enlarge a concrete box culvert to allow mule deer to travel through (the only concrete box culvert within the project that contained enough overburden to do so) was ultimately rejected by NMDOT, NMDGF, and the Safe Passage Coalition, due to the fact that it would have occurred directly across from an area proposed for a future campground. The enlargement of this culvert to allow mule deer passage was considered by the Land Grant to provide a greater opportunity for cougars and black bears to occur in this area, so therefore was not supported by the Land Grant.

The Feasibility Study

NMDOT contracted an environmental consulting firm to conduct a feasibility study. This firm was recommended by the coalition for its reputation of conducting rigorous wildlife assessments. The study was initiated to determine whether wildlife–vehicle collision mitigation was important for the I-40 project. The study's contractors were requested to do the following: (1) conduct extensive right-of-way surveys of the project area, which included documenting game trails, carcasses, and skeletal remains from wildlife road-kills; (2) analyze multiyear animal–vehicle collision data collected by NMDOT; and (3) survey local citizens for anecdotal reports of wildlife sightings and mortalities within the project area. This survey facilitated local citizen input, which not only helped to clarify wildlife movement patterns and high-risk areas in the canyon but also increased buy-in of the project by local community members.

The study documented a relatively high rate of wildlife–vehicle collisions throughout Tijeras Canyon on both I-40 and NM 333 (Marron and Associates, Inc. 2007). These included several wildlife collision "hotspots" that were likely the result of wildlife movements (primarily mule deer) influenced by local topography, water availability, and other ecological factors. The study recommended mitigation fencing to prevent wildlife from entering I-40 and direct them to existing culverts and underpasses. Escape ramps or jump-outs (see chapter 2) were also recommended to allow wildlife that accessed the I-40 right-of-way to exit safely. Lastly, animal detection systems were proposed at key at-grade crossings on NM 333 (Marron and Associates, Inc. 2005).

The study's recommendations ultimately became the basis for wildlife mitigation components constructed on the ground. These components included approximately 2.4 kilometers (1.5 miles) of 2.4 meter (8 foot) chain-link fence, 3.2 kilometers (2 miles) of electric fencing and electromats (Electrobraid Company, Dartmouth, Nova Scotia, Canada), twelve escape ramps, and two animal detection systems that include motion detection cameras (Econolite Control Products, Inc. Annaheim, California, USA), which alert motorists of impending wildlife crossings at overland crossing sites. The electric fence was installed in mountainous terrain where human contact is unlikely. Chain-link fence was installed near the communities of Edgewood and Carnuel. No new underpasses or culverts were installed during this project.

The Safe Passage Coalition was instrumental in reviewing the study and provided further recommendations to enhance the success of the wildlife safe passage component of the I-40 project. These additions included (1) the installation of angle iron strips in the floor of a large, trapezoidal concrete box culvert that funnels Tijeras Creek under I-40. The angle iron is intended to collect sediment to provide a surface more favorable to wildlife movement than bare cement and will enhance wildlife track monitoring through the culvert; (2) relocating guardrails at a paved underpass under I-40 and scarifying the paved shoulder for similar reasons; and (3) construction of a wildlife access ramp (primarily for mule deer) to facilitate access into the trapezoidal culvert. Before the ramp was constructed, mule deer access into the culvert was precluded by a deep scour pool filled with boulders on the downstream side.

One measure of citizen involvement in this is the 80 pages of comments provided by the coalition on the study, which NMDOT accepted as peer review along with project review by state and federal agencies. This close scrutiny of the study and the ongoing engagement of the coalition in the process provided the NMDOT Tijeras Canyon project lead sufficient leverage with NMDOT decision makers to justify implementation of the study's recommendations. This caused the project design engineer to comment that in his long career he had never seen as much public involvement in a highway project.

When it became clearer that the wildlife–vehicle collision mitigation components recommended by the study would be implemented, the coalition turned its attention to the potential for enhancing wildlife use of three existing bridges on I-40 at Tijeras Creek. The coalition enlisted the assistance of the Friends of the Sandias to remove non-native brush (primarily tree of heaven [*Ailanthus altissima*], Siberian elm [*Ulmus pumila*], and tamarisk [*Tamarix ramosissima*] in two of the three bridges to encourage wildlife passage. This effort set another precedent in that it represented the first time that a volunteer organization's efforts were included in construction plans by NMDOT.

The repaving of the I-40 project was completed in June 2007, and during an on-site dedication ceremony, the project was hailed by Lieutenant Governor Diane Denish as a precedent for New Mexico with regard to enhanced protection for wildlife and motorists. However, construction of the wildlife fencing was not completed until September 2007. Post-treatment monitoring of wildlife–vehicle collisions on I-40 and NM 333 and wildlife use of underpasses and culverts was initiated at that time. The NMDGF and NMDOT project leads jointly developed a monitoring plan whereby annual monitoring reports developed by NMDGF are provided to NMDOT. The reports evaluate success at reducing wildlife–vehicle collisions and maintaining habitat connectivity for a minimum of five years. If success cannot be determined after five years, implementation of Phases II and III should be reconsidered. Phase II proposes to install a large culvert under NM 333 at the Hawkwatch property (the primary overland wildlife crossing with an animal detection system) to prevent wildlife–vehicle collisions at this location. If Phase II is not effective in reducing collisions, Phase III recommends construction of a wildlife overpass. The only location in Tijeras Canyon where a wildlife overpass spanning I-40 and NM 333 could reasonably be constructed is where NM 333 crosses below I-40 east of Deadman's Curve (fig. 9.2). The land on the south side of I-40 is

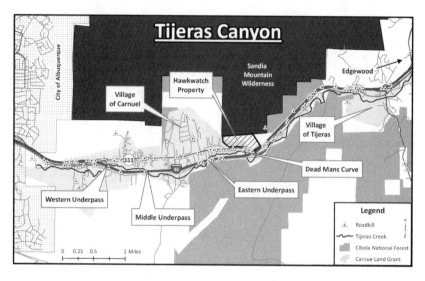

FIGURE 9.2. Map of Tijeras Canyon. (Kurt Menke, Bird's Eye View 2009)

owned by the Cañon de Carnue Land Grant, but a discussion with them about the possibility of installing a wildlife overpass at this location has not taken place. Should a wildlife overpass spanning I-40 and NM 333 be proposed, implementation will require further collaboration among local citizens, including Land Grant members, conservationists, and agency personnel.

Hawkwatch Property

Coalition volunteers were instrumental in protecting a critical piece of the wildlife linkage from development by converting it to public domain that will be managed as wildlife habitat. The 25.5 hectare (63 acre) Hawkwatch property is situated in a narrow side canyon that naturally funnels wildlife down into Tijeras Canyon and Tijeras Creek (fig. 9.3). However, wildlife must first cross NM 333 at a location locally known as Deadman's Curve, due to the sharp curve cut into the mountainside as NM 333 passes beneath I-40. The area directly below the Hawkwatch property was also identified to be a high-risk location for wildlife–vehicle collisions on the I-40 and NM 333 corridor through Tijeras Canyon (Marron and Associates, Inc. 2007). The funneling effect of the canyon in which the Hawkwatch property occurs and the permanent water source in Tijeras Creek were believed

FIGURE 9.3. Hawkwatch property. (Kurt Menke 2006)

to best explain the high incidence of wildlife–vehicle collisions at this location (Marron and Associates, Inc. 2005).

The Hawkwatch property was donated to Hawkwatch International, a nonprofit raptor conservation organization based in Salt Lake City, Utah. Due to a need to generate operating revenue, the property was offered for sale. Although not preferred by Hawkwatch, the sale of the Hawkwatch property to developers was likely imminent. The coalition immediately recognized that protection of the Hawkwatch property was needed for a successful implementation of the Tijeras Canyon Safe Passage Project, due to its key location in the Canyon. The New Mexico Land Conservancy, a member of the coalition, stepped in to facilitate the sale of the property to the City of Albuquerque, which purchased the property for $650,000 as Albuquerque Open Space. This purchase would likely not have been possible without the intervention of an Albuquerque City councilor, who encouraged the city council and mayor to protect the parcel as open space. This parcel became the first Albuquerque Open Space unit to be purchased specifically for its value as a wildlife corridor. A conservation easement was placed on the property, which will be managed for wildlife conservation into perpetuity. The Hawkwatch property enhances wildlife habitat connectivity north of the I-40 corridor, as it abuts another Albuquerque Open Space property, both of which now provide an undeveloped buffer to the Sandia Mountain Wilderness.

Conclusions

Preliminary results after one year of monitoring indicate that wildlife–vehicle collisions on I-40 have been reduced, and that mule deer and other wildlife use the three major underpasses routinely. However, it is not clear at this point if wildlife–vehicle collisions have been reduced on NM 333. We believe additional monitoring may show a decrease in mule deer collisions on NM 333 as resident mule deer learn the locations of crossing structures.

Since completion of the Tijeras Canyon Safe Passage Project, two additional wildlife safe passage fencing projects on I-40 east of Tijeras Canyon were facilitated by an expanded second I-40 corridor feasibility study (Marron and Associates, Inc. 2007). Credit for these additional projects is attributable in large part to the NMDOT Tijeras Canyon project leader, who facilitated both I-40 corridor studies internally within NMDOT. This study was implemented without pressure by the coalition and represents what we hope is a policy decision by NMDOT to proactively consider wildlife mitigation for future highway projects. For NMDOT to consider wildlife needs in all highway projects fulfills one of the primary goals of the coalition.

The first of the two wildlife fencing projects will tie in with the eastern end of the existing wildlife fence at the village of Tijeras and continue eastward approximately 5 kilometers (3.1 miles) to Zuzax. As with the Tijeras Canyon project, the wildlife fencing and escape ramps will be installed in conjunction with an I-40 repaving project.

The second wildlife fencing project on I-40 west of Edgewood is not directly connected with highway improvements. This project represents yet another important precedent for New Mexico, as it will be the first wildlife mitigation project not associated with a highway improvement project. The second feasibility study identified that along the entire 32-kilometer (20-mile) I-40 corridor from Albuquerque to Edgewood, the segment west of Edgewood had the highest rate of deer–vehicle collisions in the entire corridor. The occurrence of this north–south mule deer movement corridor is of ecological interest as it occurs at an ecotone where juniper (*Juniperus monosperma*) woodland and savanna give way to Western Great Plains shortgrass prairie (Marron and Associates, Inc. 2007). Federal transportation safety funds will be used to implement this project. This project would not have been possible without the willingness of the NMDOT Tijeras Canyon project lead to consider innovative funding sources such as the safety funds that will be used for this project.

Some important issues remain to be worked out with the Tijeras Canyon Safe Passage Project that we suspect may be shared with other similar projects across North America. These include (1) the effectiveness of the animal detection system at warning drivers of impending wildlife crossings, (2) continued maintenance of the fencing to prevent wildlife from accessing the highway right-of-way, (3) protection of key parcels of undeveloped land within the corridor, and (4) maintaining public and agency support to keep the safe passage project functional over the long-term.

We believe that the success of the project lies to a large degree in the mutual support and cooperation between coalition activists and key NMDOT and NMDGF personnel. The NMDGF project lead was involved in the development of House Joint Memorial 3, organization and implementation of the Critical Mass Workshop, and the provision of agency support for the Tijeras Canyon Safe Passage Coalition. The NMDOT project lead and project engineers exceeded public expectations by, in part, participating in evening public meetings at the behest of the coalition, independently acquiring novel (for NMDOT) wildlife-related highway funding, and working cooperatively with the coalition to find solutions within the budget. From the coalition's perspective it was invaluable to have access to agency information, opinions, and insights throughout the process.

From an agency perspective, the coalition was able to work on aspects of the project that would be inappropriate for state employees to address. This included lobbying decision makers, reaching out to local communities for support, working with the local media, and guiding and providing support for agency proposals.

The NMDOT and NMDGF project leads were able to leverage public support for the project to gain internal support from decision makers within their respective agencies. The agency project leads were also able to maintain agency support by encouraging positive media coverage that emphasized the win–win goals of reducing wildlife–vehicle collisions for motorists and wildlife.

Project proponents now recognize that long-term protection of the Tijeras Canyon wildlife linkage between the Sandia and Manzano mountains will require more than maintaining the effectiveness of the Tijeras Canyon Safe Passage Project infrastructure. Protection of key tracts of undeveloped land that surround the wildlife passages and connect to national forest lands is necessary to maintain functionality. Because of rapidly increasing human development in the area, the time window for doing so is closing. Also, a long-term commitment by safe passage advocates to stay engaged in and sensitive to local community concerns is necessary.

The success of the coalition in initiating the Tijeras Canyon Safe Passage Project is undeniable. As a result of their work, the thousands of motorists that commute daily through Tijeras Canyon realize that alternatives exist to accepting high rates of wildlife–vehicle collisions, and demands for similar projects are expanding across New Mexico.

Safe passage projects create win–win situations for both wildlife and motorists. We believe that citizen efforts in support of such projects can be highly successful, especially when citizens support state and federal agency personnel to promote science-based solutions. We recognize that the development of public support and creation of citizen and government alliances will always be unique for each safe passage project. However, our experience suggests that having the mutual support of local communities, committed volunteers, and key agency personnel that share common values is required for successful implementation of a project. We urge biologists with state wildlife agencies and transportation departments to take the lead in creating similar coalitions by facilitating connection of interested parties. We sincerely hope that the Tijeras Canyon Safe Passage Project model can be replicated for many additional safe passage projects across North America.

Chapter 10

The I-75 Project:
Lessons from the Florida Panther

DEBORAH JANSEN, KRISTA SHERWOOD,
AND ELIZABETH FLEMING

Prior to 1900, Big Cypress Swamp was a sparsely populated and roadless wetland encompassing 640,000 hectares (1.6 million acres) in south Florida. The first major road to traverse the swamp and impact the natural north to south hydrologic flow was Tamiami Trail (U.S. 41), finished in 1928. As humans discovered the alluring year-round climate, the vast natural resources, and the beauty of the Atlantic Ocean on the east and the Gulf of Mexico on the west, they came to south Florida in droves, and stayed. Coastal roads and towns sprang up. The next major highway across south Florida was completed in 1967 (Duever et al. 1986). Named Everglades Parkway, State Road 84, and more commonly Alligator Alley, it had also been called "Slaughter Alley" as the wildlife that lived there were unable to avoid a new and unfamiliar danger of this road in their environment.

As human settlement in south Florida continued to increase, the need for completion of an interstate highway system to link the east and west coasts became apparent. As a result, the south Florida segment of I-75 was first proposed in the 1968 Federal Highway Act. Of the five alternatives presented in 1971, a 122 kilometer (75.8 mile) road from Ft. Myers on the west coast to Andytown on the east coast was selected (H. W. Lochner, Inc. 1972), 64 kilometers (39.7 miles) of which paralleled the two-lane Alligator Alley (fig. 10.1). Construction along the existing State Road 84 was not

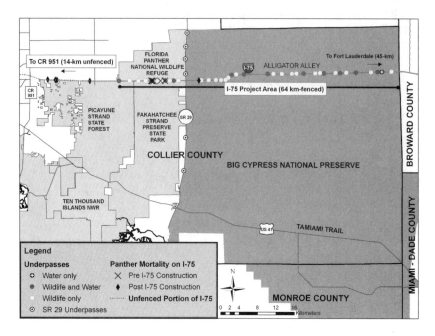

FIGURE 10.1. Interstate 75 project area showing underpass type and panther mortality pre- and post-project completion.

only determined to be more cost effective, but also was thought to provide an opportunity to remedy the already apparent negative hydrologic effects of the only four-year-old Alley.

It is somewhat ironic that one of the original intents of this interstate project was to provide passage for water, not passage for an animal that many people think detests water, a cat. The anticipated transportation and hydrologic restoration needs of south Florida were the motivations for the project at its inception in the late 1960s. The realization that a critically endangered species could benefit from design improvements came more than 10 years later. And it was an additional 10 years, until construction was completed, that enabled the cat, the Florida panther (*Puma concolor coryi*), along with the water to once again more easily traverse the land they shared.

Improvements for terrestrial wildlife (i.e., the underpasses with their associated canal crossings and continuous fencing), were built along a 64 kilometer (40 mile) stretch of Alligator Alley, starting 14 kilometers (8.7 miles) east of State Road 951 (SR 951) in Collier County and east to the

Broward County line (see fig. 10.1). This segment of I-75 is the setting of this chapter and is hereafter referred to as the I-75 project. Here we focus on the Florida panther, how it influenced the project's construction, and how the completed project impacts the panther today.

Geographical, Historical, Political, and Social Setting

Growing awareness of impending development and concern for the hydrologic health of the state of Florida was addressed in the Environmental Land and Water Management Act of 1972, enacted to protect the natural resources and environment by "ensuring a water management system that will reverse the deterioration of water quality and provide optimum utilization of our limited water resources, facilitate orderly and well-planned development, and protect the health, welfare, safety, and quality of life of the residents of this state." With a projected human population increase in south Florida from 2.4 million in 1970 to 4.5 million in 1993 (H. W. Lochner, Inc. 1972), concern next focused on the future of the southern tip of the state. As a result, the Big Cypress Conservation Act was passed in 1973 "for the purpose of conserving and protecting the natural and economic resources as well as the scenic beauty of the Big Cypress Area of Florida." This act established the Big Cypress Area of Critical State Concern with the goal of protecting the 634,500-hectare (1.6 million-acre) Big Cypress watershed and its three key resources: Everglades National Park, the estuarine fisheries, and the freshwater aquifer of south Florida (Department of Administration 1973).

In 1974, 230,000 hectares (568,000 acres) of the Big Cypress watershed came under public ownership when Big Cypress National Preserve (Big Cypress) was established to protect the area's natural resources and recreational values. That same year, 30,400 hectares (75,000 acres) of the Fakahatchee Strand, the largest strand in the watershed, were acquired. Administered by the Florida Department of Environmental Protection, it is today known as Fakahatchee Strand Preserve State Park (Fakahatchee).

The habitat of south Florida (i.e., low-lying, poorly drained sand and limestone flatlands, with a seaward slope of only 5 centimeters [2 inches] every 1.6 kilometers[1 mile]) posed unique highway construction challenges. The elevation of this watershed ranges from mean sea level to 9 meters (29.5 feet), with the terrain along the proposed project corridor at only 3.7 meters (12.1 feet; Bob Sobczak, pers. comm. 2008). The subtropical climate of south Florida is characterized by long, warm summers and short, mild winters. The heavy seasonal rains that begin in June and end in October fill its shallow aquifer. The annual rainfall ranges between 130 and 150 centimeters (51.2–59 inches) and supports the area's major plant communities: cypress swamp, wet prairies, freshwater marshes, pine-palmetto flatwoods, hardwood hammocks, tidal marshes, and mangrove swamp, as well as the wildlife species that inhabit them.

Rationale for the Project

Little mention was made of wildlife crossings in early I-75 expansion correspondence and documents. No mention was made of the Florida panther because, at the time, it was generally believed to be extinct in the wild. Although pumas (colloquial names such as panther, cougar, mountain lion, and catamount are used for the same species, *Puma concolor*) were once found in almost every state, they could not endure the dramatic loss of their prey base nor their deliberate persecution as an unwanted predator during the settling of the eastern United States. Even those panthers that inhabited the roadless and "hostile" environment of the Big Cypress Swamp were subject to organized hunts by humans and their hounds.

The Florida panther was first safeguarded when it was federally listed as an endangered species in 1966 under the Endangered Species Preservation Act, then under the Endangered Species Conservation Act of 1969, and ultimately under the Endangered Species Act of 1973. Though finally protected by law, most thought that help came too late. An occasional panther road-kill or reliable sighting report were the impetus for the World Wildlife Fund to hire Roy McBride, a well-known cat tracker from Texas, to search for panther sign in south Florida. In 1972, McBride confirmed that a few panthers still remained when his hounds treed one in Glades County and he found their tracks and scent markers in the Fakahatchee (Nowak 1973). Subsequent wildlife agency meetings focused their discussions on the feasibility and impacts of removing all remaining panthers from the wild and attempting captive breeding. However, advocates who supported the study of panthers first in the wild prevailed.

The decision to study the wild population resulted in formalizing a panther sighting "clearinghouse" in 1976 (Belden et al. 1991) and a research project involving radio telemetry in 1981, headed by the Florida Game and Fresh Water Fish Commission (now the Florida Fish and Wildlife Conservation Commission). These early efforts revealed that a few, but very few, individuals remained. The loss of even one panther, therefore, was of concern, as emphasized in a 1982 letter from the Commission to the Governor's Office of Planning and Budgeting, which stated, "Because of the very small number of Florida panthers believed to be remaining, the loss of a single animal should be considered significant, and any redesign of roads in known panther habitat should fully accommodate this concern." Where once the importance of wildlife crossings was ranked below aquatic weed control, a littoral shelf in the canal, and recreational access in planning the interstate, landscape connectivity for an endangered species now became paramount. As one of the first examples of implementing wildlife crossings in roadway design in the United States, the I-75 project distinguished Florida as an innovator in protecting wildlife in transportation planning.

Roadway and Environmental Issues

In the 20-year time period between conception and completion of the I-75 project, the predominant issues were hydrologic restoration, wildlife crossings, severance of private property and public recreational access, and inclusion of an interchange at State Road 29 (SR 29), the only major north–south road within the I-75 project area (see fig. 10.1).

The I-75 project was designed to provide hydrologic restoration by placing two new lanes between the existing Alligator Alley and the canal that paralleled the highway. The construction of new water conveyance structures and the upgrade of existing bridges were planned to provide flow under the roadway into a series of discontinuous borrow canals and then sheet-flow into the southern wetlands.

With little documented research for reference, it was unclear how many or how large the underpasses should be to ensure wildlife use. The Commission's recommendations for wildlife passage included a minimum of 25 new bridges measuring 1.8 meters (6 feet) high by 61 meters (200 feet) long and the reconstruction of thirteen existing bridges for both water flow and animal travel (May 22, 1984, letter from Director Brantley to Assistant Secretary Lewis, Florida Department of Transportation [FDOT]). New crossing sites were selected based on known panther and bear travel routes and the presence of vegetative features that provided the cover preferred by panthers (Logan and Evink 1985). In 1985, the Commission reduced the number of new crossings to 23 and revised the dimensions to 2.4 meters (8 feet) high by 30 meters (98 feet) long, for an estimated savings of $5.3 million (April 2, 1985, letter from Director Brantley to FDOT Secretary Pappas).

The underpass design consisted of two 13.1 meter (43 feet)-long open-span bridges separated by a 22.3 meter (73 feet)-wide median that would be open overhead. Two sections of 3.7 meter (12 feet)-high fencing topped by four to six strands of barbed wire installed in the median would prevent wildlife from getting onto the roadway. A continuous 3.4 meter (11 feet)-high chain-link fence made of galvanized steel topped by a 1 meter (3.2 feet) outrigger with three strands of barbed wire was planned for both sides of the 64 kilometer (40 mile) I-75 project to keep wildlife off the highway and funnel them to the crossings. Other specifications included placement of the canal crossings at natural ground elevation with native vegetative cover and appropriate canal slopes to prevent pooling of water.

The loss of entry points to private property was an issue needing resolution, since the construction of an alternative access road system was "neither feasible nor permissible" (April 4, 1985 Memorandum from Secretary Pappas to Governor Graham), mainly due to wetland laws and cost constraints. The habitat loss associated with vegetation removal along the corridor and damage to wetlands was to be mitigated through acquisition of adjacent lands and construction of wildlife crossings. Consequently, several large parcels of land were purchased and came under public ownership. The Arizona–Florida Land Exchange Act of 1988 provided for the acquisition of 59,100 hectares (146,000 acres) adjacent to Big Cypress and the SR 29 corridor owned by the Collier family. It also authorized the transfer of 8,000 hectares (19,768 acres) of Collier family lands to the U.S. Fish and Wildlife Service to create the Ten Thousand Islands National Wildlife Refuge. In 1989, 10,700 hectares (26,440 acres) were purchased that became the Florida Panther National Wildlife Refuge (Evink 2002, see fig. 10.1). Not only did this acquisition provide secure habitat for the panther, but it also protected part of the vast watershed that flows south through the Fakahatchee and into the Ten Thousand Islands estuary.

A number of small, privately owned parcels, mostly hunting camps, in Big Cypress and Fakahatchee were accessed from Alligator Alley. In most cases, these camp owners were provided "severance of access" compensation that allowed them to retain their camps but reach them by another route. Recreationists (i.e., hunters, anglers, hikers, and off-road vehicle enthusiasts) also lost their unlimited access to these public lands. After extensive discussion and compromise, six access sites, three in Collier County within the I-75 project area and three in Broward County, were agreed upon. To assure that wildlife would not get onto the interstate, the I-75 Recreational Access Plan stipulated that gates installed at each access point be "sturdy, require low maintenance, easy to operate, be self-closing/self-latching and aesthetically non-intrusive" (National Park Service 1990).

Although an interchange at SR 29 was in the conceptual design for the I-75 project in 1972, it was removed due to concern that increased traffic on SR 29 would only intensify panther mortality there. Both agencies and environmental organizations, particularly Audubon of Florida and the National Wildlife Federation, debated the issue, but, in 1984, agreed to the interchange with certain stipulations (August 24, 1984 letter from Director Brantley to the Executive Office of the Governor). Foremost was the purchase of the four sections of land intersected by the roads and the prevention of any development within their boundaries. Other endorsement criteria focused on the subsequent impacts of the interchange. On SR 29, corridor tracts were to be acquired to preclude development, the shoulders were to be widened to increase sight distance and reaction time for both wildlife and motorists, the existing panther warning signs and nighttime 45-mph (72-kph) speed limit zones were to be maintained, and three wildlife crossings were to be constructed if two or more panthers were killed in the panther speed zones in a three-year consecutive period.

Construction on Alligator Alley began in 1986 and was phased in ten segments with emphasis on areas where panthers had been killed. The entire I-75 project was completed in 1993 and had an annual average daily traffic volume of 10,330 that increased to 18,700 by 2007 (Mark Schulz, FDOT, pers. comm. 2008). Initially, pamphlets were provided at the local tollbooths informing travelers about the status of the panther and the improvements made for wildlife as part of the interstate construction.

Most of the funding for the project was provided by the Federal Highway Administration (FHWA) and through secured federal toll bonds. Additional funding was provided by FDOT to cover the cost for construction of the new wildlife crossings (Platte 1985). The 64 kilometer (40 mile) I-75 project portion cost $99 million (M. Schulz, FDOT, pers. comm., 2008) with an additional estimated $155 million for the large-acreage acquisitions and severance of access damages (May 21, 1985, letter from Secretary Pappas to FHWA Administrator Barnhart).

Critical Factors

The I-75 project was the result of a growing awareness by the public, natural resource agencies, and environmental advocacy groups that a critically endangered species could benefit from a transportation project designed

with wildlife in mind. Public awareness of the status of the panther intensified as their deaths made front-page news and research began to put the
pieces of panther demographics together. Agencies took notice when, in
1982, Florida's schoolchildren chose the Florida panther as the state animal. Legislators saw the creature they were asked to save when a south
Florida naturalist strolled his captive puma through the halls of the state
capitol. Governor Bob Graham furthered this awareness when he included
hydrologic restoration and panther preservation in his 1983 "Save Our
Everglades" initiative. Hearts nationwide were won over to save the last of
the large cats east of the Mississippi River. Between 1981 and 2008, 164
Florida panthers were radio-collared and monitored by aircraft three times
per week, resulting in more than 80,000 locations. Information on dispersal movements, home range size, and causes of mortality has provided insight into the relationship between panthers and roads. Between 1981 and
2008, 111 panthers were killed on Florida roads. Collisions with vehicles
accounted for 47 percent of all documented panther mortality.

Following the release of female panthers from Texas into south Florida
in 1995 to offset inbreeding depression, the population has grown steadily.
The result today is a threefold increase in their numbers, a decline in deleterious traits, and a reoccupation of lands devoid of panthers for many years
(McBride et al. 2008). Although it was found that road mortality was not
the reason for the low population numbers, the I-75 project was significant
in facilitating panther population expansion.

Outcomes of the Project

Prior to completion of the I-75 project in 1993, two radio-collared and
three uncollared panthers had been struck and killed on Alligator Alley.
Since completion, two panthers have been killed within the 64 kilometer
(40 mile) project area. One was killed on the entrance ramp of the unfenced
SR 29 intersection, and the second one was killed 3 kilometers (1.9 miles)
from the interchange near a humanmade breach in the fencing. Six panthers
have been killed in the unfenced 14 kilometer (8.7 mile) section between
the western end of the project area and Naples. One panther was killed at
the eastern end of the project fencing.

Only two published studies have evaluated the effectiveness of the I-75
project. Foster and Humphrey (1995) examined wildlife use of four underpasses for an average of 10 months (range = 2 to 16 months) during the
construction phase from 1989 to 1991. The width of these underpasses

ranged from 21.2 to 25.8 meters (70 to 85 feet). The distance from the north fence to the south fence averaged 48.5 meters (159 feet), and bridge clearance was 2.1 to 2.4 meters (7 to 8 feet). They monitored radio-collared bobcats (*Lynx rufus*) as surrogates for the less common panther. They found that female bobcats rarely crossed roads, whereas males crossed them frequently, making them more susceptible to highway-related mortality. They also placed cameras beneath the underpasses and recorded ten crossings by two male panthers in addition to other species of wildlife. They concluded that wildlife crossings on I-75 would compensate for the obstruction of the fenced highway by providing panthers a safe means of travel across the highway corridor. They stated that panther–vehicle collisions along I-75 would be prevented if breaches in the fencing for human trespass were eliminated, if the fencing was inspected frequently and promptly repaired, and if a contingency plan was developed to rescue any panther caught within the fencing.

A second study involving two underpasses on I-75 was conducted in conjunction with an evaluation of two SR 29 underpasses (Lotz et al. 1996). The study found that panther use of the underpasses on I-75 had increased since Foster and Humphrey (1995) had conducted their work, and it was predicted that more panther use will probably occur as individuals learn the underpass locations. The study also found that one of the two underpasses evaluated on I-75 could not be monitored for four months due to high water and that standing water was present in the SR 29 wildlife crossings. The authors stated that excessive water depth likely reduces use of the wildlife crossings and, therefore, recommended grade adjustments.

Since little has been published on post-construction monitoring of the I-75 project underpasses, we examined panther home range data to determine to what extent panthers have used the underpasses since their completion. Our sample consisted of ninety-one independent panthers, sixty-one males and thirty females, whose home ranges fell within 1.6 kilometers (1 mile) of I-75. Only five (17 percent) of the females had crossed I-75 and had done so an average of two times (range = 1 to 3 times). Figure 10.2 depicts four females, one who crossed two times and three who never crossed. In contrast, thirty-two (52 percent) of the males with home ranges within 1.6 kilometers (1 mile) of I-75 crossed the road. Four "dispersers" crossed only once. The remaining twenty-eight male panthers averaged fifty-five crossings (range = 2 to 243 times).

We field-checked the current status of 35 of the 37 wildlife underpasses within the 64 kilometer (40 mile) fenced portion of I-75. We did not check the two existing roadways, County Road 839 and Levee-28, that the

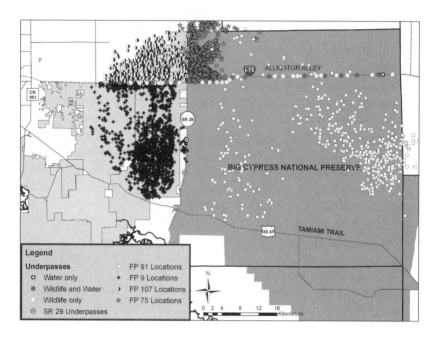

FIGURE 10.2. Sample of radiotelemetry locations for four female panthers with home ranges within 1.6 kilometers of I-75.

interstate crosses because no modifications were necessary for wildlife passage. We surveyed the underpasses in October 2008 and again in April 2009, as representative samples of the area's wet and dry seasons. Based on our analysis, we classified each underpass as "wildlife only," "water and wildlife," and "water only." The "wildlife only" classification reflects an underpass designed for minimal, if any, water flow and adequate dry ground for terrestrial wildlife use year-round. The "water and wildlife" underpasses were designed to convey water throughout the year and provide a pathway for terrestrial wildlife to cross the canal on the north side of the interstate and travel under the roadway along one side of the water.

We measured the underpass dimensions and documented cuts in the fencing under the roadway using a global positioning unit with an average error of 4 meters (13 feet). We considered the width to be the distance between the fences beneath the road, which, in most cases, were placed at the base of the slopes. We acknowledge that the actual span size provides added "openness" for wildlife. If we could access the underpass through a breach in the fencing, we recorded water depth, wildlife tracks, invasive plant species, and the density of the vegetation in the median and along the wildlife pathway. Our efforts gave us a cursory look at wildlife use of the underpasses as well as obstacles that may inhibit wildlife use.

Of the 35 underpasses, 23 were built for wildlife use only and 12 for both water and wildlife passage. One additional bridge provided for water flow only. The "wildlife only" underpasses averaged 22 meters (72 feet) in width, whereas those with water averaged 40 meters (131 feet) in width with a wildlife pathway (averaging 9 meters [30 feet]) on one side (fig. 10.3). All underpasses averaged 73 meters (240 feet) in length from fencing to fencing, of which 51 meters (167 feet) was the distance an animal would need to travel to successfully cross the two spans and the open median. The height of the "wildlife only" underpasses averaged 2.3 meters (7.5 feet), whereas the clearance on the pathways of the "water and wildlife" underpasses averaged only 1.6 meters (5.3 feet). The average distance between the underpasses was 1.67 kilometers (1 mile) (range = 0.43 to 5.13 kilometers [0.25-3.2 miles]).

Although the fencing along the I-75 project is continuous, animals have the potential to access the interstate through breaches in the fencing, caused by either a vehicle accident or deliberate cutting by humans to access

FIGURE 10.3. An open-span "wildlife only" underpass within the I-75 project area. (Photo by Ralph Arwood)

the canal for fishing or adjacent lands for recreation. Since the installation of an Impact Detection and Alarm System along portions of the fencing adjacent to canals in 2005, emergency personnel have been able to respond quickly when a vehicle diverts from the roadway and goes through the fencing (P. B. Farradyne, Inc. 2006). Openings or breaks made in the fencing beneath an underpass to access fishing or hunting areas, however, are less obvious.

Inspection of the fencing within the I-75 project is not done on a routine basis. Whenever damaged fencing is reported by the public or right-of-way maintenance crews, the contractor is required to make a temporary repair immediately and a permanent repair within five days (Scott Teets, FDOT, pers. comm., 2008). During our evaluation in October 2008, we found 38 openings in the fencing large enough for human or large animal passage beneath 27 underpasses and documented tracks of deer and bear outside several openings from which they could access the interstate. Subsequently, we reported the locations of these openings to the Florida Department of Transportation. During our April 2009 evaluation, 29 of the 38 breaches had been repaired. It is possible that the other nine were also repaired but were cut again. We found an additional 21 breaches in April.

Human activity in the vicinity of underpasses has varying impacts on wildlife use (Clevenger and Waltho 2000, Gloyne and Clevenger 2001); however, Foster and Humphrey (1995) stated that routine maintenance of the fencing along the I-75 project was an important component of its success.

Although the underpasses were designed to remain dry year-round, we found an average of 14 centimeters (5.5 inches; range = 0 to 35 centimeters [0 to 13.8 inches]) of standing water in 20 of the 23 "wildlife only" as well as inundation of the canal crossings and pathways in the "water and wildlife" underpasses. In April, these areas were dry.

In October, we found wildlife tracks at seven underpasses that we could access and were not too wet for tracking. We found panther tracks at four underpasses. In April, we were able to access 15 underpasses with suitable substrate for tracking. We found panther tracks, two of which were females, at 12 underpasses. We also found sign of white-tailed deer (*Odocoileus virginianus*), American alligator (*Alligator mississippiensis*), raccoon (*Procyon lotor*), opossum (*Didelphis virginianus)*, bobcat, black bear (*Ursus americanus*), marsh rabbit (*Sylvilagus palustris*), domestic dog (*Canis familiarus*), coyote (*Canis latrans*), humans, birds, turtles, and snakes of unknown genus.

During our site visits, we found non-native plant species in most medians. The majority were Brazilian pepper (*Schinus terebinthifolius*) and melaleuca (*Melaleuca quinquenervia*), both of which are serious invasives of south Florida. We also noted that the density of the vegetation on the canal crossings and pathways under the roadway of the "water and wildlife" underpasses likely was both a visual and physical impediment for species such as deer, bear, and possibly panthers (Forman et al. 2003). The predominant vegetation (i.e., willow [*Salix caroliniana*], cattail [*Typha* spp.], and alligator flag [*Thalia geniculata*]) indicates that these areas are inundated by water for a considerable portion of the year, consequently impacting use of the underpass by terrestrial wildlife.

Lessons Learned

Providing safe passage for wildlife along a 64-kilometer (40-mile) roadway in south Florida was the first effort of this magnitude in the world. It was the cooperative spirit among agencies and advocacy groups coupled with public support that made the endeavor a reality. It also exemplified the extent to which U.S. citizens care about the welfare of other species, especially those whose future is in question.

The I-75 project has been very successful in preventing panther deaths on the 64 kilometer (40 mile) stretch of road that traverses over 300,000 hectares (740,000 acres) of public land. The wildlife crossings within the I-75 project are functional for Florida panthers based on the criteria that they have reduced road-kills, maintained habitat connectivity, enabled genetic interchange to continue, and allowed for dispersal and recolonization (Forman et al. 2003). The road, however, is a filter barrier, especially to female panthers, based on our findings that 48 percent of the male and 83 percent of the female radio-collared panthers monitored within 1.6 kilometers (1 mile) of I-75 have not crossed the road. If this were due to the standing water beneath the underpasses, it would be expected that some females would shift their home range use during the dry season. However, this does not appear to be happening. In addition, some females that do not cross the interstate do, in fact, cross the two-lane SR 29 (see fig. 10.2).

Possible factors causing this reluctance to cross include the 51 meter (167 feet) crossing distance, the increased traffic volume and associated noise levels, the periodic water inundation, the lack of visibility due to dense vegetation, and the presence of human activity. Smith (2003) found that carnivores near culverts did not cross 84 percent of the time when the average annual daily traffic volume was greater than 6,900 and that medium-sized mammals crossed only 13.9 percent of the time when the volume was greater than 7,500.

In the 16 years since this project has been completed, much has been learned about making roads safer for wildlife. Key findings include the need to monitor project effectiveness and to maintain the passages for continued wildlife use (Clevenger and Waltho 2003, Smith 2003, Cramer and Bissonette 2005, Cramer and Bissonette 2007, Bissonette and Cramer 2008). Our cursory evaluation of the 35 wildlife underpasses has provided insight into some obstacles to their use by wildlife as well as an opportunity to remedy them.

Our recommendations include routine inspection and repair of the breaches in the fencing, removal of invasive plant species, and mechanical treatment of native plant species that limit visibility for wildlife, especially on the canal crossings and pathways. We also suggest a more formalized study to determine the extent that underpasses are used by other wildlife species and whether use would increase if dry pathways were available year-round.

Florida panthers continue to be killed on roads outside the I-75 project boundaries, with the highest mortality occurring on SR 29, the road on which the controversial interchange was built. Between 1998 and 2008, 20 panthers were killed on a 38 kilometer (24 mile) section of this road that traverses vast tracts of public land (see fig. 10.1). A key stipulation for endorsement of the interchange at SR 29 by agencies and environmental organizations was the construction of wildlife underpasses on that road if panther deaths increased with the subsequent traffic volume increase. In 1991, the Florida Fish and Wildlife Conservation Commission and the United States Fish and Wildlife Service recommended that six underpasses be built at locations on SR 29 where the highest mortality was occurring. Two underpasses were completed in 1995, one in 1997, one in 1998, and the last two were finished in 2007.

Although the initial design was to be similar to the underpasses on I-75, it was agreed in 1992 to build structures of varying widths on an experimental basis to evaluate wildlife usage. All six underpasses crossing the two-lane road were similar in height (2.4 meters [8 feet]) and length (14.6 meters [48 feet]). The first two underpasses on SR 29, prefabricated box-culverts, were the smallest in width (7.3 meters). Each structure cost $110,000 compared to $175,000 for a two-lane-span on I-75 (Lotz et al.1996). The next four underpasses were open-span bridges. The two completed in 1997 and 1998 provided crossing widths for wildlife of 22 and 35 meters (72 and 115 feet), respectively, and cost $200,000 per structure. The two completed in 2007 were 15 meters (49 feet) in width, with a total project cost of $3.8 million for each (Mark Schulz, FDOT, pers. comm., 2008). It is important to note that the cost for construction and materials more than doubled between 1995 and 2007. For instance, the cost of fencing increased from $12 to $26 per linear foot.

The canal crossings at the first four underpasses were built at grade level and provided an unobstructed view of the habitat on the far side, a desired component for carnivores (Beier, 1995, Foster and Humphrey 1995, Forman et al. 2003, Ruediger 2007). The canal crossings built in 2007, however, were elevated about 1.5 meters (5 feet) above the adjacent natural grade as required by South Florida Water Management permitting (Mark Schulz, FDOT, pers. comm., 2008). As a result, the approach to the underpass by wildlife does not provide a clear view of the habitat on the other side. In addition, the structure over the canal was covered with sand, and, two years post-construction, has no vegetative cover (fig. 10.4). A camera-based study of wildlife use of these two underpasses is not yet available

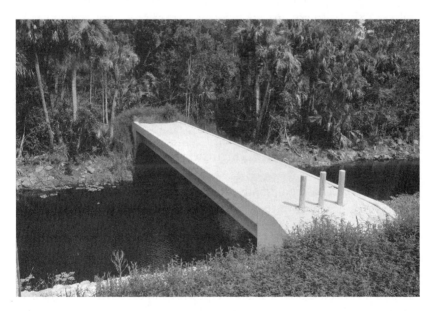

FIGURE 10.4. The elevation above natural grade and the lack of cover on the structures over the canal at the State Road 29 underpasses built in 2007 may impact wildlife use. (Photo by Ralph Arwood)

(E. Serdynski, FDOT, pers. comm., 2009); however, some use by panthers has been documented (Roy McBride, pers. comm., 2009).

Continuous fencing was not installed on SR 29 because of the distance between the underpasses, the existence of private property, and agency concern over severing access to the adjacent canal for anglers. The first two underpasses were 3 kilometers (1.9 miles) apart with 5.2 kilometers (3.2 miles) of fencing on each side. The average length of fencing of the last four underpasses was just over 0.5 kilometer (0.3 mile) north and south of each crossing. The fencing between the last two underpasses was only installed on one side of the road for a length of 2.3 kilometers (1.4 miles).

Panther mortality continues on SR 29 where fencing and underpasses do not exist. In addition, because the fencing is not continuous, two panthers have been trapped and killed on the road between the fencing. Clevenger et al. (2001a) found that wildlife collisions with vehicles were clustered at fence ends. In 2008, an additional 400 meters (1,300 feet) of fencing was added in an area where panther and bear usage and deaths were occurring.

Although the smaller and more cost efficient underpass designs built on SR 29 have proven to be used by panthers, it is apparent that the six underpasses and 12 kilometers (7.5 miles) of fencing in 38 kilometers (23.6 miles) of SR 29 do not provide the same level of protection to panthers as found in the I-75 project. Although steps have been taken to reduce the extent of panther mortality on SR 29 since the construction of an I-75/SR 29 interchange, more underpasses and continuous fencing on that road would provide safe passage for wildlife between two vast areas of public land in south Florida.

Conclusions

The I-75 project has significantly reduced panther mortality along 64 kilometers (40 miles) of an interstate highway that traverses the heart of habitat used by the Florida panther. Through this project, Florida has been recognized as an innovator for large-scale wildlife crossing implementation and ecological consideration in transportation planning. Every effort should be made to replicate this design (i.e., abundant and closely spaced underpasses coupled with continuous fencing), especially in areas with large tracts of public land and for species that require vast expanses of natural habitat. Post-construction evaluation and routine monitoring of such a project are critical to assure that goals continue to be met and obstacles to wildlife use identified and remedied.

Florida continues to recognize the benefits of providing safe passage for wildlife across its roads statewide by incorporating wildlife concerns into its highway planning (Schaefer and Smith 2000, Swanson et al. 2005, Smith et al. 2006). With safe wildlife corridors that connect the remaining natural habitat in Florida, the panther's chances of repopulating its historic range will be greatly improved.

Acknowledgments

We thank Steve Schulze and John Kellam, Big Cypress National Preserve, for their assistance in home range analysis and graphic preparation. We thank the Florida Fish and Wildlife Conservation Commission for providing historical documents on I-75 and panther telemetry data from their study area. We thank Mark Schulz, Florida Department of Transportation, for providing information on project funding and cost details.

Chapter 11

Wildlife Underpasses on U.S. 64 in North Carolina: Integrating Management and Science Objectives

MARK D. JONES, FRANK T. VAN MANEN,
TRAVIS W. WILSON, AND DAVID R. COX

U.S. Highway 64 (U.S. 64) is an important route in eastern North Carolina connecting major population centers and highways, including North Carolina's urban "Piedmont Crescent" and Interstate Highways 40 and 95 with the northern Outer Banks. In 1992, the North Carolina Department of Transportation (NCDOT) initiated planning studies for improvements to U.S. 64 between the towns of Plymouth and Columbia in northeastern North Carolina (Washington and Tyrrell counties). The existing roadway was a two-lane, rural road with a posted speed limit of 55 miles per hour (89 kilometers per hour) and lower speeds in developed areas. Because of increasing traffic volumes from tourism and a push for economic development in this region, NCDOT proposed to improve 45 kilometers (28 miles) of U.S. 64. New routes were proposed for a large portion of the project (fig. 11.1) to improve driver safety and increase speed limits to 70 miles per hour (113 kilometers per hour). The projected cost of these improvements was $195 million.

NCDOT's first involvement with safe wildlife passages began in 1991 when the North Carolina Wildlife Resources Commission (NCWRC) requested that passages for black bears be considered on a new section of Interstate Highway 26 (I-26) in the southern Appalachian Mountains of Madison County, adjacent to the Pisgah National Forest. At that time,

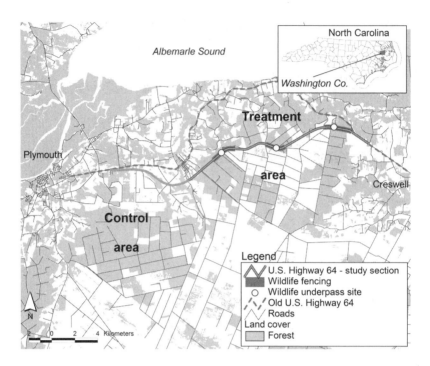

FIGURE II.I. Project area for new section of U.S. Highway 64 with three wildlife underpasses, Washington County, North Carolina. Remote camera and track surveys were conducted during 2006–2007 to document usage of the underpasses, and highway surveys were conducted along the new section of highway (including protected [fenced] and unprotected [unfenced] survey sections) to document wildlife mortalities due to vehicle collisions.

NCWRC and NCDOT staff had limited experience with wildlife passages of any type. Because construction had already started on the I-26 project before these discussions were initiated, NCDOT settled on structures that could be accommodated within the existing roadway design (i.e., two 2.4 meter × 2.4 meter concrete box culverts [47.3 and 42.7 meters in length] and a standard 1.4 meter height right-of-way fence). NCWRC personnel believed these underpasses were too small and did not have an appropriate "openness ratio" for most large mammals. Indeed, a later study of the effectiveness of these passageways concluded that several factors negatively influenced wildlife use, including human use of the passageways, high traffic levels of the interstate highway, small structural design, and lack of appropriate fencing (Jones 2008).

For the last two decades, NCWRC and NCDOT biologists have been opportunistic in approaches to wildlife passage issues by incorporating passages at new locations when possible and designing bridge replacements to enhance permeability for wildlife. Over time, improvements in the effectiveness of wildlife passages were noted that were then applied to new highway projects. For example, extending bridge structures to span a portion of a riparian corridor (usually a minimum of 7.6 meters [25 feet]) and raising the height of bridges to accommodate target species (e.g., minimum height of 2.4 meters [8 feet] for large mammals) were important improvements. NCWRC biologists also worked with NCDOT and their contractors to minimize the installation of riprap along riparian areas because NCWRC biologists believed the permanent rock armor composing riprap may impede the movement of certain wildlife species. Similarly, the installation of culverts was adjusted to better accommodate wildlife passage. By designing certain culverts using low–normal flow cells and floodplain cells, culverts remain dry during normal flow periods providing passage for small to medium-sized wildlife. Previously, most culverts were installed only at the stream level. For larger systems with a high flow capacity, NCDOT sometimes installed three culverts (2.4 meter [8 feet] × 2.4 meter) side by side, thereby widening the channel and allowing water to pass through all three culverts. To improve that system, NCWRC recommended NCDOT utilize two of the culverts (2.4 meter × 2.4 meter) in the existing stream channel with the third culvert at floodplain elevation. Therefore, under normal flow conditions, two culverts have flowing water and the third culvert remains dry. When the water rises during storm events, the third culvert becomes functional. This new design represents a minor and inexpensive change offering improved permeability through the riparian area for both aquatic and terrestrial fauna.

Geographical, Political, and Social Setting

Washington and Tyrrell counties are in the northern coastal plain of North Carolina. Topography is flat with an elevation range in the project area from 4.6 to 15.2 meters (15–50 feet) above sea level. Land use is a mix of agriculture, silviculture, and small residential areas. Several large agricultural areas producing grain and peanuts, coupled with vast areas of contiguous forests, provide ideal habitat conditions to support high wildlife densities, including American black bears (*Ursus americanus*), white-tailed deer (*Odocoileus virginianus*), red wolves (*Canis rufus*), and bobcats (*Lynx rufus*).

Black bear densities in the area represent some of the highest reported in the literature (up to 1.5 bears per square kilometer [3.9 bears per square mile]; Nicholson 2009). White-tailed deer densities are among the highest one-third reported in the state (NCWRC, unpublished data).

At the time of construction, the human population was 13,723 for Washington County and 4,149 for Tyrrell County (U.S. Census Bureau 2002). In Washington County, 21.8% of its residents lived below the poverty level, whereas 23.3% of Tyrrell county residents lived below the poverty level. Washington County was expected to lose 8.2% of its residents by 2010 (North Carolina Department of Transportation 1999). Because Washington and Tyrrell counties are economically depressed and have little industry other than agriculture and silviculture, the new roadway was viewed by local and state officials as an important economic stimulus because it would link coastal tourist areas with population centers and Interstates 40 and 95 further inland. The roadway was also viewed as a way to develop ecotourism in these counties.

Recreational hunting is extremely popular in this region (NCWRC, unpublished data), and the area is popular with hunters from other areas of North Carolina and other states because of special hunting opportunities for prized species such as black bear and waterfowl. Most private lands in the area are leased to individuals or hunt clubs. There are several local guide services. Overall, hunting contributes a substantial amount of money to the economies of these counties.

Rationale for the Project

In the 1990s, North Carolina's governor made improving the state's highway infrastructure a top priority for economic development and driver safety. By January 2003, North Carolina's 126,535-kilometer (78,642 mile), state-maintained highway system was the nation's second largest behind Texas (Federal Highway Administration 2004). NCWRC and NCDOT have long been concerned about the impacts of highway modification and expansion on wildlife populations. The new highway diverged from the existing two-lane and spanned a new route through important wildlife habitat. This, coupled with four travel lanes, a 14-meter (46 feet) grass median, cable guardrails, and right-of-way fences was expected to reduce permeability for wildlife. Although widespread and relatively abundant in the coastal plain, North Carolina's black bear population is discontinuous in distribution (Jones et al. 1998, Jones and Pelton 2003). Therefore, based on existing literature (e.g., Beringer et al. 1990, Wooding and Maddrey 1994, Proctor et al. 2002), the potential impacts of the new section of U.S. 64 on the viability of black bear populations was an important cause for concern. Additionally, although rare in the project study area, red wolves were a species of concern because of the federally endangered status of this species. No such concerns existed for white-tailed deer (McCullough et al. 1997, Peck and Stahl 1997) because they occur statewide and population density in the project area was high (NCWRC, unpublished data). However, white-tailed deer are frequently involved in vehicle collisions and represent a major driver safety hazard (Conover et al. 1995, Romin and Bissonette 1996). Because the black bear is the largest terrestrial species of wildlife in North Carolina, they also pose an increased risk to driver safety.

Responding to Environmental Assessment comments, primarily from NCWRC, and in an attempt to reduce wildlife–vehicle collisions, NCDOT agreed to build three fenced wildlife underpasses along a 24.1 kilometer section (15 miles) of U.S. 64 where the new highway would be diverted from the original highway route. The underpass and fencing designs that NCDOT and NCWRC agreed on (fig. 11.2) were patterned after structures built on Interstate Highway 75 in south Florida (Evink 1997, see

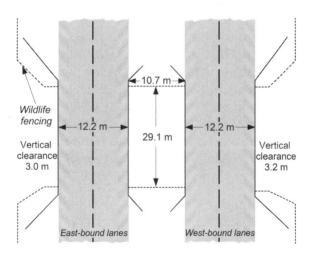

FIGURE 11.2. Design and dimensions of the eastern underpass (aerial view) for a new section of U.S. Highway 64, Washington County, North Carolina. Dimensions for the central and western underpasses were slightly different from this diagram. Measurements provided by A. Burroughs and D. Herman, North Carolina Department of Transportation.

chapter 10 this volume) because of their documented use by black bears, deer, and other wildlife species (Foster and Humphrey 1995, Land and Lotz 1996) and because the geography and habitats of the two areas were similar. Underpasses and wildlife-proof fences are part of a system, and neither work without the other. Therefore, 3-meter-high (9.8 feet) chain-link fencing extended at least 800 meters (2,625 feet) in both directions from all underpass structures.

At the request of NCDOT, NCWRC initiated a study to determine optimal locations for the wildlife underpasses along the new section of U.S. 64 (Scheick and Jones 2000). Construction deadlines limited data collection to 10 months and ruled out the use of radiotelemetry, so the study was primarily based on sign surveys and focused on black bears but also included white-tailed deer, red wolves, coyotes (*Canis latrans*), and several medium-sized mammals.

Following establishment of the underpass locations and focal species, NCWRC desired to gain a better understanding regarding the effectiveness of the underpasses and the impacts of the four-lane, divided highway on black bears. Researchers from the University of Tennessee/U.S. Geological Survey were contracted to determine if spatial use and demographic patterns of the black bear population changed after the new section of U.S. 64 was completed (van Manen et al. 2001). That study was also designed to determine if the three underpasses mitigated the potential barrier effect of the highway to bear movements and if the underpasses reduced wildlife–vehicle collisions, particularly with white-tailed deer. Early planning by NCWRC personnel allowed a rigorous design for this study (before–after control-impact experiment [Green 1979]), including two field sampling phases, one before construction of the highway (2000–2001) and one after the highway was opened to vehicles (2006–2007). The study design included a control area to ensure that observed treatment effects were not a result of variability in food availability, habitat changes, or other factors that may influence black bear ecology (van Manen et al. 2001).

Roadway and Environmental Issues

Several factors influence the ability to construct effective wildlife underpasses in eastern North Carolina. The low elevation and flat terrain can be challenging for engineering wildlife underpasses large enough to accommodate large mammals. Little to no topographic relief, coupled with the concerns of transportation planners regarding driver expectations, necessitated the use of long, wide "fill approaches" as the highway reaches the underpass. Thus the driver experiences a gradual rise of the roadway in an otherwise flat terrain, rather than an abrupt increase in elevation on a flat highway. Providing a minimum of 2.4 meters (8 feet) of vertical clearance for the underpass structures required considerable fill material to build these approaches. Costs were increased because of limited supplies of quality fill material in the area and wetland impacts associated with the raised fill height. Granular fill (i.e., sand) allowed 0.9 meters (3 feet) of horizontal distance to 0.3 meters (1 foot) of vertical rise without requiring the use of permanent soil erosion protection such as riprap. With large wetlands occurring throughout the U.S. 64 project area, using fill heights with a 1:3 side slope increased impacts to wetlands regulated under the Federal Clean Water Act. That impact was somewhat mitigated by placing the underpass structures in areas that spanned those jurisdictional wetlands. However, the structures could not be placed in wetland types that are subject to seasonal inundation because standing water can deter use by certain animals.

Critical Factors

During the transportation planning process, three important factors influenced the decision to consider wildlife mitigation measures: driver safety, underpass construction costs, and permitting by the U.S. Fish and Wildlife Service and U.S. Army Corps of Engineers.

Driver Safety

One of the most important factors influencing the decision of NCDOT to construct the wildlife underpasses was the concern for driver safety. Vehicle collisions with deer and bears along the existing two-lane portion of U.S. 64 had been documented. However, the speed limit for the new highway would be higher and was expected to increase the risk of wildlife–vehicle collisions (Gunther et al. 1998). When planning for the U.S. 64 project began, the U.S. Centers for Disease Control and Prevention had just published new data on driver injuries and fatalities sustained in animal collisions in Kentucky (U.S. Centers for Disease Control and Prevention 1991). Like other states in the southeastern United States, North Carolina has abundant white-tailed deer populations. With deer densities in Washington County ranking in the top third of all counties in North Carolina, NCWRC biologists were concerned that the new section of highway would pose a substantial risk for deer–vehicle collisions. Although black bears are less numerous than deer, NCWRC personnel pointed out that vehicle collisions could pose a greater risk to driver safety because of the large body mass of black bears in the project area (> 350 kg [772 pounds] for some bears). Thus both species were perceived as major threats to driver safety.

Cost

During the initial planning stages of the project, the cost of the wildlife underpasses were considered to be an impediment (estimated cost of $1.2 million per underpass, three total underpasses). However, when considered within the context of the overall project costs ($195 million), construction of the underpasses represented approximately 1.85% of the entire project.

Regulatory Commitment

The inclusion of wildlife underpasses on the new section of U.S. 64 would not have been possible without the support of the U.S. Fish and Wildlife Service and the U.S. Army Corps of Engineers. These two agencies had permit authority related to NCDOT's highway construction and evaluated the project in the context of a broad range of natural resource conservation issues in the region and not solely on the area of wetlands to be impacted. The commitment of the individual staff members of the agencies involved with project planning was crucial to the success of this project.

Outcomes of the Project

The project provided important information regarding the effectiveness of the underpasses to maintain habitat connectivity and to reduce wildlife–vehicle collisions.

Maintaining Habitat Connectivity

Locations of the three underpasses were determined based on a study conducted two years prior to construction and involved comparisons of track surveys, ditch crossing surveys, remote camera surveys, and land-cover maps (Scheick and Jones 1999). Subsequent research established the effectiveness of that approach. Based on 1,771 telemetry locations from 35 black bears, analysis confirmed that the western and eastern underpasses were appropriately located within important areas of habitat connectivity. Additionally, potential function of the underpasses may extend to a larger area (Kindall and van Manen 2007). The western underpass may not only facilitate travel between areas just north and south of the new highway, but also with areas west of that underpass. Furthermore, habitat linkages between the Roanoke River (west of the project area) and Bull Bay Swamp (northeast of the project area) converged with habitat linkages near the westernmost underpass. Similarly, the eastern underpass is located within a regional habitat connection between the Big Swamp and areas of core black bear habitat south of the project area.

The frequency of use of the underpasses provides a good indication of their effectiveness for many wildlife species. Monitoring of the three underpasses using remote cameras and track-count surveys from July 2006 to July 2007 produced 3,614 track counts of animal crossings and 2,053 photographs, representing at least 20 different animal species (see fig. 11.1, McCollister 2008). A minimum of 15 mammal species were identified from photographs and tracks, representing almost all medium and large mammals known to exist in the area. Additionally, at least three bird species (eight records of a great blue heron [*Ardea herodias*], two records of wild turkey [*Meleagris gallopavo*], and several records of unidentified species) and numerous reptiles were recorded (52 records of snakes, four records of turtles). The majority of track counts (88 percent) and photographs (92 percent) represented white-tailed deer. These data indicate that the underpasses provide habitat connectivity for many wildlife species, particularly for white-tailed deer.

The track and camera surveys indicated that the underpasses were used by black bears on 17 occasions, representing approximately ten individuals. Despite this relatively low frequency, use by 10 bears suggests sufficient demographic and genetic exchange, thus reducing the barrier effect of the highway (Nicholson 2009). Additionally, collection of these data started less than one year after the highway was opened for traffic and some species may require more time to learn to use underpasses (Foster and Humphrey 1995, Land and Lotz 1996, McCollister 2008). As female bears using these structures raise cubs and other bears learn their locations, bear use of the underpasses may increase.

Reducing Wildlife–Vehicle Collisions

Wildlife mortality surveys from July 2006 through July 2007 recorded 196 mortalities representing 41 species, of which 15 were mammals (n = 127), 12 were birds (n = 16), and 14 were reptiles and amphibians (n = 53) (McCollister 2008). No difference in the frequency of mortalities between the three sections of the highway with the underpasses and fencing and the four adjacent, unfenced sections were observed (see fig. 11.1). However, many of the records represented animals that could pass under or through the fence, which likely influenced this finding (McCollister 2008). Current research is focusing on statistical comparisons of mortality frequencies of larger mammals and reptiles among fenced, unfenced, and underpass segments of U.S. 64.

The frequency of deer–vehicle collisions in sections with underpasses was compared with similar sections of U.S. 64 from 2006 to 2008, accounting for differences in traffic volume and the habitat matrix (McCollister 2008). Collision frequencies for the sections immediately west (Williamston to Plymouth) and east (Creswell to Columbia) were greater compared with the new section of U.S. 64. Given the frequent use of the underpasses by deer, numerous deer–vehicle collisions were probably avoided in the new section of U.S. 64, thus increasing human safety and reducing economic losses from damage to vehicles (McCollister 2008).

Although the frequency of deer–vehicle collisions may have been reduced because of the underpasses, this may not be the case for black bears. Radio-collared bears that crossed the highway did not regularly use the underpasses (McCollister 2008), which may have increased the risk of collision with vehicles. At least eight bears were reported killed due to vehicle collisions from May 2007 to November 2008 (McCollister 2008, C. Olfenbuttel, NCWRC, pers. comm.). Seven of those occurred within unfenced sections of the highway (i.e., away from underpass areas), of which three were near the edge of the wildlife fencing associated with the underpasses. Two of the eight mortalities involved bears that were monitored via radiotelemetry and were part of a group of only four adult females near the highway that had locations both north and south of the highway (McCollister 2008). One bear, a young male, was killed within 200 meters (656 feet) of the eastern underpass, after crawling under the fencing and heading south across the highway. Without the three underpasses, the number of bear mortalities may have been greater, so we do not suggest that the underpasses are ineffective to prevent black bear–vehicle collisions. Rather, the effective area of each underpass may be limited, and bears may be more likely than deer to climb over or crawl underneath the fencing. This interpretation was supported by telemetry data: of the four resident female bears that crossed the highway only one of 36 documented crossings was near an underpass (McCollister 2008). Because the mortality rate due to vehicle collisions may have contributed to decline of the population within the U.S. 64 project area (Nicholson 2009), additional mitigation measures may need to be considered to reduce the frequency of bear–vehicle collisions for future projects.

Lessons Learned

The U.S. 64 underpasses were the first in North Carolina designed specifically for wildlife and according to specifications provided by NCWRC staff. As such, they were a landmark development in the history of highway construction and wildlife mitigation in the state. The collaborative nature of this process provided important lessons for NCDOT and NCWRC professionals.

Initial public reaction to the establishment of wildlife underpasses was affected by media accounts portraying "bridges for bears" published as part of a series on "Government waste" by a major state newspaper. NCWRC was able to regain public support only after years of disseminating factual information about the relative cost ($\approx 1.85\%$ of the entire road project) and the ecological and driver safety benefits of the project. Future efforts to establish underpasses would benefit from a concerted educational and outreach effort to inform the public about both the costs and the benefits of wildlife underpasses.

Since the new section of U.S. 64 was completed in 2005, management of the vegetation within the underpasses and along the fences and maintenance of the fences and gates have become issues at various times. In the humid, temperate climate of eastern North Carolina, vegetation can quickly overtake underpass openings, potentially forming a barrier to wildlife movements. Therefore, NCWRC's written agreement with NCDOT requires regular maintenance of the fencing, the gates, and the vegetation growing in the underpass openings. Additionally, further research may be necessary to determine how vegetation type and structure influence the effectiveness of the underpasses for different wildlife species.

Initial public comments on the project referred to the passageways as "black bear underpasses" despite NCWRC attempts to clarify that the underpasses were intended and designed for all wildlife species. NCWRC staff

234 EFFECTIVE PARTNERSHIPS

predicted deer collisions would be reduced because of the underpasses, and the effectiveness was even greater than expected. Thousands of documented deer crossings and relatively few deer–vehicle collisions have demonstrated that increased driver safety may be the greatest public benefit from these underpasses.

Early NCWRC efforts to determine underpass locations (Scheick and Jones 1999, Scheick and Jones 2000) were questioned because of the extra expenses. Empirical evidence indicates that early research efforts were justified because the chosen locations of the underpasses provided important habitat and landscape connectivity (Kindall and van Manen 2007). In other words, a small early investment in research dollars paid off in the long term. Similarly, the early commitment of NCWRC and NCDOT to integrate research objectives into the overall project allowed the development of a rigorous experimental design (pre- and post-construction periods, treatment and control areas) for the impact assessment study.

Finally, early planning was critical to the success of this project. The engineering specifications of highways are often developed years in advance by state departments of transportation. Incorporating underpasses into design plans and in the most desirable locations can be difficult when wildlife agencies enter the discussion late in the planning process. Clearly, good communication and collaboration between departments of transportation and state or provincial wildlife agencies is essential for effective wildlife mitigation of highway infrastructure.

New Standards

Underpass specifications including size, control of public access, fencing, gates, and maintenance have proved critical for effective use by wildlife in many studies. Recommendations for future projects follow based on our experiences.

Size

We suggest minimum design criteria of 3 meters (9.8 feet) vertical clearance and a 36.6 meter (120 feet) opening for large mammal underpasses. Our data, and those of other studies, indicate that underpass structures with these dimensions are used by a wide array of animal species and address permeability in areas with abundant black bear and white-tailed deer populations. Structures of this size can maintain connectivity prioritized in natural resource planning efforts. One new issue that NCWRC is addressing with NCDOT is the potential incorporation of smaller passage structures into future projects. These smaller structures, such as drainage culverts, would facilitate the passage of species with small home ranges, such as small mammals, reptiles, and amphibians.

Public Disturbance Issues

Clevenger and Waltho (2000) demonstrated the importance of human activity in impacting the use of underpasses by large mammals in Alberta, Canada. Because the underpasses were installed in an area with an active outdoor recreation culture, NCWRC biological staff felt it would be imperative to control unregulated use of the passageways and the underpass access areas. Therefore, in locations not adjacent to public land, a minimum easement of 2.02 hectares (5 acres) was purchased to protect each entrance from human disturbance and development and to restrict hunting at underpass openings. Additionally, concrete bollards placed 0.9 meters (3 feet) apart prevent vehicle and ATV access into the underpass areas (fig. 11.3).

FIGURE 11.3. Central wildlife underpass on U.S. Highway 64, Washington County, showing vegetation structure and bollard placement near the northern entrance.

Fencing

The chain-link fence used to direct wildlife to the underpasses was 3 meters (9.8 feet) high. This fence extended a minimum of 800 meters (2,625 feet) from the underpass entrances in both directions. Whereas the height of this fencing was effective, four of seven deer mortalities from vehicle collisions occurred within the fenced sections, of which three were within 300 meters (984 feet) of the fence edge (McCollister 2008). These observations suggest that some animals may have been directed away from the underpasses and accessed the fenced sections from the edge. Additionally, seven of eight bear–vehicle collisions we reported previously occurred in unfenced sections. Therefore, fencing may be more effective if it were continuous throughout the project area. Additional, but costly, improvements would be to add barbed-wire outriggers at the top of fencing (Clevenger et al. 2001a) and bury the bottom of the fencing to deter animals from climbing over or crawling underneath the fence.

Gates

The only gates that have been tested to date in North Carolina are standard fence gates at multiple locations along the fencing. Those gates are used to access underpass structures and to allow authorized personnel to open the gates in case an animal is trapped between the fences and the roadway. One disadvantage is that the gates are sometimes left open, therefore leaving a gap in the fence. Locks have been placed on the gates, but this required distributing keys or lock combinations to appropriate personnel. For future projects, alternatives to the standard wildlife release gates may need to be reviewed.

Maintenance Issues

An underpass vegetation management plan was part of the agreement between NCDOT and NCWRC to ensure proper functioning of the underpass from a wildlife use and a maintenance standpoint. The objective of that plan was to provide sufficient cover for wildlife by allowing shrubby, weedy, and grassy vegetation to grow in the underpasses and at the entrances while allowing NCDOT to maintain their right-of-way by keeping large woody vegetation suppressed (see fig. 11.3). This vegetation structure may be best maintained with a combination of mowing and herbicide applications targeted at trees and larger woody species. Fence maintenance and repair are conducted on an as-needed basis by reporting damage to NCDOT division of maintenance personnel. Agencies would benefit from developing specific protocols for monitoring, reporting, and maintaining all aspects of vegetation management associated with wildlife passages.

Conclusions

The process that resulted in the establishment and research of the U.S. 64 underpasses has set a positive precedent for future road construction in North Carolina. Since the completion of the U.S. 64 underpasses, NCDOT routinely considers wildlife passageways for road projects in the state. Moreover, NCWRC and NCDOT staff now communicate regarding wildlife mitigation measures when suggestions can still be incorporated at the engineering and planning stages of road projects. NCWRC incorporated scientific components into this project from the beginning to guide the planning process and to document the potential impacts of the highway and the mitigation value of underpasses. Integrating scientific objectives as part of this project allows NCWRC staff to bring important credibility to the planning process on current and future NCDOT transportation infrastructure plans. Finally, the U.S. 64 project was different from most other highway projects in North America because of our ability to obtain biological data before highway construction began. Although not all studies are completed at this point, we anticipate that the research component of this project will make an important contribution to worldwide knowledge of wildlife responses to highways and the mitigation value of wildlife underpasses.

Acknowledgments

This project was funded by the North Carolina Department of Transportation, North Carolina Wildlife Resources Commission, and Weyerhaeuser Company. B. Scheick, J. Kindall, L. Thompson, M. McCollister, J. Nicholson, and numerous field technicians collected much of the research data. Logistical support was provided by B. Barber, A. Burroughs, D. Evans,

J. Folta, B. Hulka, D. Miller, C. Olfenbuttel, T. L. Riddick, C. Smith, W. Wescott, and North Carolina Wildlife Resources Commission's District 1 personnel. Finally, we appreciate local landowners and state and federal agencies for providing access to properties for the collection of research data.

Chapter 12

Strategic Wildlife Conservation and Transportation Planning: The Vermont Experience

JOHN M. AUSTIN, CHRIS SLESAR, AND FORREST M. HAMMOND

This chapter presents a strategic planning framework for state governments to address wildlife conservation and transportation development needs. In 2001, the State of Vermont established an interagency working group of wildlife scientists, transportation planners and engineers, and administrators to consider the public's mutual interests in effective wildlife conservation and transportation planning and development. This chapter describes the process through which this interagency collaborative effort was formed, as well as many of the outcomes from this effort that illustrate the course for long-term wildlife conservation and transportation planning. The chapter highlights several key projects that illustrate potential outcomes achieved by state agencies with different missions but common goals and recognition that both serve the public in similar ways. Many of the details related to these projects, such as scientific methods, are not included here, but are available in reports or published papers.

Vermont is a northern New England state bordered to the north by Canada (province of Quebec), to the south by the state of Massachusetts, to the west by the state of New York, and to the east by the state of New Hampshire. Vermont's landscape ranges from low-elevation valleys, agricultural lands, and wetland habitats along Lake Champlain and the Connecticut River to high elevation subalpine ridgelines along the spine of the

239

Green Mountains. The northeast highland region of Vermont largely comprises boreal forest habitats and large expanses of northern hardwood forests that have long been part of a traditional forest products economy. The southwest part of the state is more similar to southern New England with warmer temperatures, talus slope habitats, forests of hickory and oak that support wildlife, such as timber rattlesnakes, found nowhere else in Vermont.

The human population of Vermont has changed significantly over the past 30 years at a rate of approximately 8.2 percent to the present size of 620,000 (Austin et al. 2004). While the change has been dramatic for such a small and rural state, it is still a small population relative to neighboring states such as Massachusetts and New Hampshire. Nevertheless, Vermont's close proximity to the enormous population centers found in Massachusetts, New York, Connecticut, and others, combined with its outstanding scenic and recreational values, creates great challenges for wildlife conservation and accommodating demands for easy, effective transportation. The growth of infrastructure and development to support the current population has increased as well. According to a report by the Vermont Forum on Sprawl (1999), Vermont is developing its landscape at a rate of 3.2% annually. In 2001, the U.S. Environmental Protection Agency reported that Vermont was losing 2,630 hectares (6,500 acres) of open space to development each year (Austin et. al. 2004). Transportation infrastructure has expanded to suit the demands of dispersed residential and commercial development.

Fragmentation of habitats and the overall landscape are considered the most significant threat to fish and wildlife conservation in Vermont as referenced in the state's wildlife action plan. As a result of these ongoing changes associated with development, Vermont loses approximately 202 hectares (500 acres) of significant wildlife habitat each year to regulated development alone (not all development is subject to state or federal regulation in Vermont, Austin et al. 2004). The state is rich with roads, and the overall network of transportation infrastructure and dispersed development is a leading cause of habitat fragmentation. Vermont has recognized that this will be a critical issue to address with respect to climate change adaptation. Conserving important linkage habitats associated with roads represents one of the primary objectives to maintain native wildlife populations in the face of changing climatic conditions.

As in every state, wildlife–vehicle collisions in Vermont represent a serious public safety issue as well as a wildlife conservation consideration. Also, some species such as reptiles and amphibians are impacted by road mortality at a population level (e.g., North American racer snake [*Coluber* spp.]). However, the preponderance of wildlife species killed on roads in Vermont are white-tailed deer (*Odocoileus virginianus*) and moose (*Alces alces*). These are two common species in Vermont and both are involved in many vehicle accidents each year. Black bears (*Ursus americanus*) are occasionally hit by vehicles while crossing roads in Vermont. They are an important part of the Vermont fauna and generally cross roads in very specific locations based on habitat, development, and road conditions (Hammond 2002). Both agencies take the issue of public safety with respect to wildlife–vehicle collisions seriously and have worked to educate the traveling public on wildlife crossing areas where moose, deer, or bear frequently cross segments of roads.

Development of an Interagency Relationship in Vermont

The Vermont Fish and Wildlife Department (VFWD) is a component of the Vermont Agency of Natural Resources and is responsible for the conservation of all species of plants, animals, and their habitats for the people of Vermont. The mission of the Vermont Agency of Transportation (VTrans) is a safe, efficient, and fully integrated transportation system that promotes Vermont's quality of life and economic well-being. VTrans's mission is to provide for the movement of people and commerce in a safe, reliable, cost-effective, and environmentally responsible manner. These two agencies have worked together regarding the permitting and construction of transportation projects as they relate to impacts on the environment for decades. Several factors influence the views of these agencies with respect to wildlife conservation and transportation planning. Title 10 V.S.A. Chapter 151, commonly referred to as Act 250, is Vermont's renowned land-use development control law. This is the regulatory tool that has facilitated what was traditionally a very adversarial relationship between the two agencies. Today, as is explained below, it is now an effective collaborative working relationship.

This law applies to all state-level transportation projects and requires that VTrans secure a permit from Vermont's Environmental Board. The department is a statutory party to the Act 250 permit process and provides guidance and biological opinions regarding the impacts of proposed development on what the law defines as "necessary wildlife habitat." This definition, in the context of transportation development, has been applied broadly to include wetlands; deer winter habitat; habitat for rare, threatened, and endangered species; feeding habitat for black bears; and wildlife travel corridors or linkage habitat for an array of species. This permit process has been the catalyst for developing a working relationship between the two agencies. Strong public support for wildlife conservation programs in Vermont has influenced the department's approach to habitat conservation through interagency coordination (Duda 2007). Concomitantly, the Vermont public continues to express strong support for transportation development that is mindful of and sensitive to habitat fragmentation and impacts to fish and wildlife (Vermont Agency of Transportation 2002). Vermont is fortunate to have consistent public support for transportation programs that are environmentally sensitive and limit impacts to fish, wildlife, habitats, and the public's interests in them.

In 1999, VTrans's director of program development was invited to participate in the European Scan Tour that traveled to various countries in Europe to observe and learn about transportation projects that incorporated wildlife crossing infrastructure. This experience occurred during a time of growing interagency collaboration as the two agencies began to shift from traditional permitting relationships to more proactive efforts to plan for transportation growth in ways that reduce impacts to wildlife and habitat fragmentation. This experience had a profound impact on the development of environmental perspectives in both agencies. During the same timeframe, both agencies were participating in the International Conferences on Ecology and Transportation, which also had a profound impact in building both awareness and appreciation for lessons learned from other states and countries, as well as providing an opportunity for building better interagency relationships. These experiences can be credited, in large part, for facilitating the effective interagency relationships that exist today.

Early Projects Leading to a Broad-Based Plan for Wildlife and Transportation in Vermont

VTrans and VFWD have endeavored to bring the concept of habitat connectivity beyond regulatory compliance. Since the early 1990s, VTrans has included wildlife crossing structures into the design of three large-scale infrastructure projects. These projects, Chittenden County Circumferential Highway (VT Route 289), the Bennington Bypass (VT Route 279), and the Vermont Route 78 upgrade project are framed within a regulatory context and should not be viewed as proactive environmental initiatives. Nevertheless, these projects brought wildlife crossings into the vocabulary of

VTrans project managers and onto the drafting tables of VTrans engineers and designers; and were the organizational foundation for the development of Vermont's reputation as a leader in addressing wildlife and transportation issues. The subsequent development and implementation of collaborative and proactive stewardship initiatives focusing on interactions between wildlife and transportation have solidified a working relationship between VTrans and VFWD that has become an unexpected model for how two state agencies, with very different agendas, can work toward a common goal.

Overall, Vermont residents consider *safety and security* and *environmental protection* as the most important issues for the state's transportation system to address (Wilbur Smith Associates 2006). In terms of which one is most important, *safety and security* have a slight edge over *environmental protection* (Wilbur Smith Associates 2006). These philosophies are reflected in VTrans's mission, and the commitment to environmental responsibility is specifically articulated in VTrans's goal regarding planning, design, construction, and maintenance of all projects in compliance with federal and state environmental laws; adherence to the agency's environmental stewardship policy; and collaboration with other Vermont agencies and entities to develop effective and efficient ways to protect or enhance the environment. Expanding on that philosophy, VTrans's 2004 Environmental Policy Statement articulates a commitment to reach beyond regulatory compliance: "The Agency will aim to be a positive force in supporting the state's environmental quality and unique sense of place, and will strive to exceed state and federal environmental laws when practicable."

Perhaps the earliest proactive environmental stewardship effort undertaken by VTrans was the installation of kestrel nesting boxes on the back of interstate signs starting in 1995. This project was developed by VTrans staff in the environmental, construction, and operations sections in response to reports of declining American kestrel populations and the decline of available grassland habitat. In consultation with the Vermont Institute of Natural Sciences, ten boxes were built and installed along Interstate 89 in central Vermont in areas where relatively large tracts of grassland were being maintained by VTrans along the Interstate Right of Way. This low-cost initiative has become an ongoing project that has brought VTrans much positive public relations, and it has led to the successful fledging of over 100 American kestrel chicks.

In 2002, VTrans kicked off another relatively low-cost environmental initiative that, like the kestrel boxes, some constituents of the Vermont public have come to expect year after year. The US Route 2 Frog Fence is a collaborative venture between VTrans and VFWD. It developed after strong public reaction from the traveling public in response to massive amounts of northern leopard frogs (*Lithobates pipiens*) killed by vehicles on a very busy stretch of US Route 2 through the Sandbar Wildlife Management Area in the northern Champlain Valley. During the summer months, if conditions are right, swarming northern leopard frogs can virtually coat the roadway surface, eventually creating a grisly, and reputedly slippery, spectacle. The road-kill problem had the potential to become a safety issue for humans, as VTrans maintenance personnel reported that some drivers were so upset by the carnage that they would attempt to stop traffic—on a road with an average daily traffic count of 10,200. A public outcry reached VTrans and VFWD asking the two state agencies to do something to reduce the amount of road-kill at this site. While no permanent changes have been made to the road infrastructure, the interagency team that collaborated to address the problem developed a creative and inexpensive solution that has come to be known as the Frog Fence.

From June through October, VTrans and VFWD install 305 meters (1,000 feet) of construction silt fencing along the roadway edge. The silt fence creates a temporary barrier to prevent movement of northern leopard frogs onto the road. It includes funnel-like openings to allow wildlife to escape from the roadway if they get on the roadway between the fencing. Based on data collection efforts in 2003, project organizers concluded that the silt fence barrier reduced the road-kill numbers by over 80 percent compared to an untreated control area. In addition to a significant reduction in road-kills, VTrans and VFWD succeeded in addressing a public concern over wildlife–vehicle collisions. This project generated an overwhelmingly positive public response and a great deal of positive media coverage. In doing so it unexpectedly contributed to bringing the concept of habitat connectivity to mainstream Vermont (Hoffman 2003).

Proactive and voluntary stewardship opportunities have a remarkable ability to bring habitat connectivity issues to light for the public as well as transportation professionals. VTrans has successfully combined regulatory obligations with voluntary stewardship efforts to develop a reputation as a leader among state departments of transportation in terms of habitat connectivity. The Federal Highway Administration has called Vermont "an eastern state on the cutting edge" of wildlife crossings; it featured the collaborative efforts of VTrans and VFWD in a 2005 cover article of *Public Roads* (Levy 2005). University of Utah researcher Patricia Cramer, recognized for her work evaluating the integration of wildlife crossings into transportation plans and projects in North America, has often used the collaborative relationship between VTrans and VFWD as an example of how a small state can make significant contributions toward addressing wildlife crossing issues (Cramer and Bissonette 2007). This type of recognition adds a sense of legitimacy to a subject that has previously appeared somewhat "fringe" within the traditional perception of the mission of a department of transportation, and perhaps even a fish and wildlife department; which in turn fuels future collaborations.

Developing an Interagency Steering Committee

An interagency steering committee was established in 2003 in an effort to provide structure and continuity to the collaborative working relationship that VTrans and VFWD enjoy. This group, the Interagency Wildlife Crossing Steering Committee, meets quarterly to discuss the progress of ongoing wildlife crossing initiatives and to introduce new projects, ideas, and research initiatives. The group focuses on addressing proactive initiatives and avoids specific project and regulatory discussions. This simple tenet of keeping the project and permit discussions out of the committee meetings fosters the spirit of collaboration by allowing participants to focus on shared goals and objectives. The group consists primarily of staff biologists and environmental specialists who are implementing the initiatives. However, the co-chairs of the committee are the VTrans director of program development and the commissioner of VFWD, who provide the decision-making roles for the group. The VTrans VFWD geographic information system (GIS)-based Wildlife Habitat Linkage Area Assessment (a tool to help the two agencies set priorities) and the Northeast Transportation and Wildlife Conference are two examples of the type of projects initiated by the steering committee.

By early 2005, both agencies recognized that the Interagency Wildlife Crossing Steering Committee was an exceptional way to maintain communication on wildlife and transportation issues, not project or permit related. Furthermore as the steering committee continued its work, administrations within Vermont state government were changing. This motivated the group to formalize the relatively informal relationship that made up the steering committee. To this end, a memorandum of agreement (MOA) was drafted outlining a shared recognition that both VTrans and VFWD "desire to improve accommodations of wildlife and aquatic organism movement around and through transportation systems, and to minimize fragmentation resulting from the presence of transportation infrastructure; and the parties' desire to reduce the potential for wildlife collisions along transportation infrastructure through improved planning for fish and wildlife impacts from transportation." VTrans and VFWD staff presented this MOA to administration officials from the Vermont Agencies of Natural Resources, VTrans, and VFWD. The MOA was signed by the respective secretaries and commissioners and the state assistant attorney general. This MOA memorialized and formalized the formerly ad hoc steering committee, and habitat connectivity officially became part of the organizational agenda of VTrans and VFWD.

Northeast Transportation and Wildlife Conservation Consortium

At the 2003 International Conference on Ecology and Transportation conference in Lake Placid, New York, participants from Maine, New Hampshire, and Vermont gathered to discuss opportunities for interstate coordination on efforts regarding wildlife conservation and transportation planning. That discussion highlighted a strong interest to create a formal regional dialogue among the northeast states regarding wildlife and transportation. The result of that conversation was the Northeast Transportation and Wildlife Conference cosponsored by the Federal Highway Administration and the host state. The conference was inaugurated in 2004 in Vermont, and has subsequently been held in 2006 in Maine and in 2008 in New Hampshire. This biannual conference (scheduled between International Conference on Ecology and Transportation years), is an opportunity for representatives from the northeast states, and beyond, to "cross-pollinate" over regional habitat connectivity issues (Levy 2005).

Wildlife and Transportation Awareness and Education Course

Before the development of the Linkage Area Assessment, VTrans was interested in developing a method for identifying important wildlife crossing locations in order to guide appropriate infrastructure design decisions that were environmentally sensitive and reduced the costs of securing the necessary permits for projects. A training course titled Road Ecology: Habitats and Highways was developed to prepare a core group of individuals within VTrans to systematically collect field data on wildlife movement along and across Vermont roads. The course consists of eight day-long sessions from January through June. These sessions involved field excursions as well as classroom seminars, and are co-taught by Susan C. Morse of Keeping

Track, James Andrews of the Vermont Reptile and Amphibian Atlas Project, as well as biologists from VFWD. Almost immediately, the course generated a tremendous degree of enthusiasm and interest from staff within VTrans and the Vermont Agency of Natural Resources. It quickly became apparent that the greatest value of these trainings was in developing awareness of habitat connectivity among transportation professionals and the agency staff who work on transportation issues. Because of this, the American Association of State Highway and Transportation Officials Center for Environmental Excellence nominated the training program for a 2003 Environmental Stewardship Best Practices award. The format for the course has developed since its inception but maintains much of its original form. Punctuated by classroom sessions, much of the training takes place in the field, both in pristine examples of habitat as well as along the highway edge.

By the end of the first year of the Road Ecology course, a waiting list had developed in anticipation of a second year of the training. Through research and planning funds from the Federal Highway Administration, the original course was refined with a focus on reaching out to all transportation professionals; not just those involved in the environmental process of project development. Since 2003, the course has become a yearly training and has had participation from program managers, project designers, operations and maintenance staff, planning staff, project managers, construction, and roadway and safety engineers from VTrans. Also participating in the course have been VANR staff, transportation consultants, as well as transportation staff from neighboring states. At the time of writing, nearly 100 participants have been through the training, and an understanding of the value of habitat and how transportation projects can be planned, designed, constructed, and maintained with habitat in mind is beginning to permeate VTrans. In addition to better understanding road ecology and habitat, both participating agencies have developed a better sense of each other. The trainings bring together two agencies that can sometimes be at odds on certain project-related issues, outside of the context of projects and permitting. And without the baggage of projects and permits, common ground and understanding can be reached. This is not to say that discussions during the trainings are always tranquil. Many of the best moments during the trainings have spun out of spirited discussions on the challenges and conflicts of developing and managing transportation infrastructure versus protecting and regulating a natural resource. But these are the learning opportunities, and all participants are encouraged to expect to venture out of their comfort zone. Since its inception, the training has served as a model for similar trainings in New Hampshire and Maine.

The overwhelmingly positive response to Road Ecology: Highways and Habitats, demonstrates the genuine interest among transportation professionals toward environmental stewardship, many of whom never previously thought that they had opportunity or reason to consider habitat in their particular area of expertise. However, the overall philosophy of the training is founded in the notion that a transportation system is best managed by professionals who incorporate a sense of stewardship into their work. Empowered with the knowledge of the relationship between transportation and habitat, committed professionals can find ways to do their work so that transportation will fit as well as it possibly can on the landscape. One participant summed up the value of the training by asserting that, since the training, he can no longer design a project from in front of a computer without understanding the environmental context within which that project is to be constructed.

Development of New, Eastern-Based Tools and Science: Linkage Area Model

In 2000, VFWD and VTrans developed a wildlife linkage habitat model to identify potentially significant linkage habitats throughout the state. This model was designed similar to those of other states using a least cost path approach. It relied on GIS data, including data describing developed land, road data, contiguous habitat data, land cover and land type data, as well as wildlife road crossing and mortality data for black bear, moose, and bobcat. This project is a significant component of the interagency collaborative planning efforts and is now a key component for any wildlife conservation and transportation planning work within Vermont. It also provides an analysis of linkage habitats based on state-based conditions rather than relying on the models, information, and related inferences from other parts of the United States and Canada. This sort of information was lacking for the northeastern United States but is now available for Vermont and has raised the interests of neighboring states to conduct similar analysis.

This project resulted in the development of several important pieces of information and related conservation planning tools, including the following:

1. Development of a comprehensive, centralized database of all wildlife road mortality, wildlife road crossing, and related habitat data for all species for which data exist throughout the state of Vermont.

This involved updating an existing database developed for a complementary project designed to compile all existing data on black bear road mortality, road crossing, and significant habitats. It also included incorporating all data on moose collisions and deer collisions. In addition, new databases were created to record existing bobcat, amphibian, and reptile information. In order to expand and improve wildlife road mortality data, this project developed a partnership with VTrans field/district staff enabling them to record a new array of wildlife road mortality information for other species, including fisher, mink, otter, beaver, coyote, American marten, wolves, and lynx.

2. Development of a GIS-based wildlife linkage habitat analysis using landscape-scale data to identify or predict the location of potentially significant wildlife linkage habitats associated with state roads throughout Vermont. This project relied on available GIS data, including (a) land use and land cover data, (b) development density data (E911 sites), and (c) contiguous or "core" habitat data from the University of Vermont. The GIS conserved-lands data were also used for this project as a way of analyzing the feasibility of conserving or ranking potentially significant wildlife linkage habitats identified as a result of this project. These data were classified according to their relative significance with respect to creating potential wildlife linkage habitat. The elements that make up the overall GIS data layers were ranked in accordance with their relative significance to creating potential wildlife linkage habitat. The analysis, in conjunction with the newly updated wildlife road mortality data, provides a scientifically based planning tool that will assist VFWD and VTrans in understanding, addressing, and mitigating the effects of roads on wildlife movement, mortality, habitat, and public safety early in the design process for transportation projects.

Bennington Bypass Project: Vermont's First Large-Scale Wildlife Underpass Structures

An example of the current process that the two agencies are using to better understand and address the effects of highways on wildlife is the recently completed Highway 279 Bennington Bypass Project located in southwestern Vermont. It is the first of a three-phase highway project designed to move traffic around the city of Bennington. The 7 kilometer (4 mile)-long segment of highway was built through an urban area that was forested and important to a variety of wildlife, including winter habitat for white-tailed deer.

Potential impacts to wildlife resulting from the project were mitigated by VTrans purchasing adjacent habitat and transferring ownership to VFWD to become the 202 hectare (500 acre) Whipstock Wildlife Management Area. Impacts were also mitigated by the construction of three wildlife crossing structures: two expanded bridges and a long culvert. A three-year study was also conducted by a graduate student from the University of Massachusetts to monitor the effectiveness of the crossing structures and to compare rates of wildlife movement across the highway in mitigated and unmitigated sections (Bellis et al. 2008).

Conservation Planning with Vermont Communities

The VFWD plays an increasingly active leadership role in conservation planning with Vermont municipalities and regional planning commissions. At the request of the towns, the department developed a manual (see Austin et al. 2004) to offer technical guidance on conservation planning in Vermont. A key component of the publication is devoted to promoting the conservation of both unfragmented and linkage habitats by avoiding and minimizing impacts from roads and highways. In 2006, the department created a conservation planner position to work specifically with Vermont communities. Some of the major initiatives of this position have included a new Community Wildlife Program and a Web site as well as the establishment of collaborative partnerships throughout the state.

SPECIES CONSERVATION PROJECTS

North American racer snakes (*Coluber constrictor*), a state-listed threatened species, were discovered living around an old abandoned welcome center along Interstate 91 in southeastern Vermont in 2003. Excitement regarding the discovery of the population of this species was cut short upon learning that VTrans had plans to pave the site with asphalt and develop it into a truck weighing station and salt sheds. Rather than cause conflict between the two agencies, in an unprecedented spirit of cooperation engineers and biologists strategized to allow the planned development of the site while providing replacement and alternate habitat on an adjacent, state-owned wildlife management area. VTrans funded a monitoring program to learn more about the species' movements, habitat requirements, and population characteristics and then used that data in a program of forest patch cuts and annual mowing designed to create and maintain optimal travel and foraging habitat for the snakes. As a result of the monitoring program and the habitat improvements the VFWD was able to draft a long-term management plan.

Vermont Route 78

Vermont Route 78 travels through one of the largest and most significant wetland complexes in the state of Vermont. This fact is exemplified by the presence of the Missisquoi National Wildlife Refuge and the Carmens Marsh State Wildlife Management Area as the primary landowners of this large wetland system. This mosaic of wetlands offers outstanding wildlife habitat for a myriad of resident and migratory species ranging from waterfowl (e.g., black ducks [*Anas superciliosa*], wood ducks [*Aix sponsa*], goldeneye [*Bucephala* spp.]) and wading birds (e.g., great blue herons [*Ardea herodias*]—the state's largest colony of nesting great blue herons occurs in this wetland system, American bitterns [*Botaurus lentiginosus*], Virginia rail [*Rallus limicola*]), to rare, threatened and endangered species such as the black tern (*Chlidonias niger*) and spiny softshell turtle (*Apalone spinifera*). Although black bear and moose are not common species in this part of the state, vehicle collisions with those species have occurred in the project area. Each year, many white-tailed deer are killed by vehicle collisions in one area of this roadway. Numerous other species of mammals, birds, reptiles, and amphibians are killed by traffic in this area each year. In addition, the potential displacement effect of traffic on sensitive wetland-dependant wildlife may be significant.

The Missisquoi River is one of several major river systems that flow through Vermont into Lake Champlain. The Missisquoi River creates the lakeshore wetland system comprising thousands of acres. This is the second largest wetland system in the state of Vermont and possibly the most biologically diverse. Vermont Route 78 parallels approximately 4.8 kilometers (3 miles) of this river without any buffer. The lack of separation between the road and river precludes movement along the river by wildlife, creates water quality problems, and presents a serious public safety hazard.

Route 78 is a narrow, rural arterial state highway with an increasing volume of traffic, most notably commercial truck traffic coming from and going to Canada. In 1996, Route 78 was designated a national highway. Public safety concerns regarding the high traffic volume and poor road conditions have caused VTrans to pursue upgrade of the road along the

Missisquoi River and through the Missisquoi wetland system, including the Missisquoi National Wildlife Refuge. In order to address safety issues related to the road conditions and wildlife habitat and associated environmental concerns, a collaborative process was developed to identify issues and solutions. The VFWD in coordination with the Missisquoi National Wildlife Refuge, VTrans, and other government organizations (e.g., Environmental Protection Agency) identified impacts to wetland habitat, effects of traffic on sensitive wetland-dependent wildlife, and the barrier effect of the existing road conditions as primary concerns related to this project. In order to address those concerns, VTrans and VFWD evaluated landscape and habitat conditions along the road project corridor, distribution of road-related wildlife mortality, animal movement information based on evidence of animal presence and activity in habitats near the road (e.g., tracks, observations of animals), and local knowledge of animal movements and areas where there has been a high frequency of animal–vehicle collisions from Missisquoi National Wildlife Refuge biologists and VFWD wardens (Trombulak and Frissell 2000).

The assessment of wildlife movement and linkage habitat associated with the Route 78 project area consisted of (1) wildlife species inventories, (2) significant habitat inventories, (3) landscape and vegetative cover data, (4) evidence of animal movement along Route 78 (track data), and (5) road mortality data for wildlife. Aerial and orthophotography was used to identify habitat features within the road corridor (Fahrig and Merriam 1995). Areas reviewed for evidence of wildlife movement were within 91.4 meters (300 feet) on either side of the road edge. Due to the fact that much of the project area is in public ownership for conservation purposes by the U.S. Fish and Wildlife Service and the VFWD, a great deal of institutional knowledge was available to identify important wildlife species, habitats, and road crossing locations without extensive field inventory work (Clevenger et al. 2002a,b).

Another example of a species-specific project is related to water snake (*Nerodia sipedon*) conservation. Polymer-based (plastic) mesh netting, typically found in most erosion control matting, is an entanglement hazard to snakes and other wildlife (Stuart et al. 2001, Barton and Kinkead 2005). Periodic incidents of snake entanglement in plastic erosion control matting have been brought to the attention of VTrans and VFWD. In the fall of 2007, approximately 50 northern water snakes, considered a Species of Greatest Conservation Need in Vermont, were entangled in erosion control matting used to stabilize an eroding roadside bank. As a result, the Vermont Interagency Wildlife Crossing Steering Committee put this topic on their meeting agenda, and an interagency working group was assembled to address the problem. The group concluded that using natural fiber erosion control matting had several advantages. The loosely woven mesh netting is flexible enough so that wildlife does not suffer from entanglement—as opposed to plastic netting, which is "welded" to create an immovable mesh. Additionally, the natural fiber netting biodegrades and does not remain in the ground nearly as long as the polymer-based netting. Furthermore, using natural fiber, instead of a plastic product, leaves less of a footprint on the landscape.

Reviewing and selecting erosion control matting types on a project-by-project basis would be an unwieldy and daunting task for VTrans. Therefore, in the long run, it would be easier and more effective for VTrans to use only natural fiber temporary erosion control matting. To that end, VTrans has eliminated polymer-based matting for use as temporary erosion control matting on construction and maintenance projects. In the winter of 2008 VTrans rewrote its standard specifications so that contractors are required to use loosely woven natural-fiber erosion control matting for temporary erosion control on construction projects. Shortly after the rewrite of the standard specification, the VTrans Operations Division mandated the use of natural fiber products for temporary erosion control on routine maintenance projects. Thus a sweeping change in the type of product used constitutes a long-term stewardship improvement. As Vermont is home to several large-bodied snakes that are listed as threatened or endangered, the change also helps ensure that VTrans's projects do not result in an accidental "taking" of a listed species.

Vermont Wildlife Action Plan

The Vermont Wildlife Action Plan was developed in 2004 with the participation of many different organizations and interest groups, including VTrans. The plan provides a comprehensive blueprint for the conservation of all species of fish, wildlife, plants, habitats, and natural communities. Identification of threats to wildlife conservation is a key component of the plan and drives the prioritization and implementation of actions and strategies for achieving the conservation goals established in the plan. Top among those threats to wildlife conservation in Vermont and the Northeast generally is fragmentation of habitat and the attendant effects on fish and wildlife associated with roads, traffic, and related development. This threat was listed in the top five in Vermont's Wildlife Action Plan.

VTrans was an important and active participant in the development of Vermont's Wildlife Action Plan, and the goals and strategies related to addressing the effects of transportation are the outcome of this unique collaborative relationship. The plan specifies working to develop and implement landowner incentive, technical support, and education as related to management of the species of greatest conservation need. It also lays out the intention to "[p]roactively collaborate with transportation planners and engineers regarding the location and design of new and expanded roadways." A quick read of the Wildlife Action Plan reveals the following priority strategies related to transportation and transportation planning:

- "Maintain and restore habitat connectivity and minimize fragmentation of forest blocks; Identify and prioritize wildlife road crossing locations; Work with the Agency of Transportation and adjacent landowners to reduce wildlife mortality and increase the potential for movement from one side of the road to the other." (from Action Plan, Conserving Vermont's Mammals, page 4:28)
- "Encourage reports of road-killed specimens, road crossings, and road basking areas to VFWD, VTrans, and the Vermont Reptile and Amphibian Atlas Project. Develop safer crossings at significant sites when roads are being upgraded." (from Action Plan, Conserving Vermont's Herps, page 4:32)
- "Identify and prioritize, for conservation, existing contiguous forest blocks and associated linkages that allow for upward and northward movement in response to climate change" (from Action Plan, Landscape Level Forest Summary, page 4:45)
- "Work with VTrans to identify and maintain wildlife highway/road crossings" (from Action Plan, Northern Hardwood Forest Summary, page 4:52)
- "Provide technical assistance to state and federal land management agencies on riparian habitat management goals/strategies" (from Action Plan, Landscape Level Riparian Summary, page 4:70)
- "Provide technical assistance to VTrans, towns, and private landowners to identify and maintain (or restore) aquatic habitat connectivity" (from Action Plan, Fluvial [Stream] Summary, page 4:79)

Vermont's wildlife action plan combined with the long-range transportation plan establish a commitment for long-term collaboration to intelligently address the serious challenges posed to wildlife conservation by transportation and other forms of development that destroy and fragment habitat. These plans will continue to guide the state's efforts on these matters.

The VFWD is currently working to create updated maps of the major wildlife-highway crossing areas and attempting to develop a prioritization process for the crossing areas throughout the state. Following up the state's existing data layer of crossing and linkage area model and maps, VFWD has developed collaborative partnerships with conservation organizations to map the most important crossing areas while developing specific linkage habitat conservation plans. Two projects that are partially completed are (1) Critical Paths for Wildlife, in which VFWD personnel are working with The Wildlife Federation and the Vermont Natural Resources Council to develop a conservation plan for the crossing areas of the highways bisecting the Green Mountains extending north and south through the state; and (2) the Adirondacks to Green Mountains Project involving developing a conservation plan to maintain connectivity between New York and Vermont. Information gained from these two conservation projects will serve as a prototype to map and prioritize crossing areas throughout the remainder of the state.

Lessons Learned

In Vermont, strong public support for wildlife conservation programs has influenced both VFWD and VTrans approaches to habitat conservation through interagency coordination. This consistent public support for transportation programs that are environmentally sensitive and limit impacts to fish, wildlife, and habitats, and the public's interests in them have been a driving force in moving conservation forward through safe passage projects in the state.

The formation of an interagency steering committee and the subsequent drafting of an MOA, in an effort to provide structure and continuity to the collaborative working relationship that VTrans and VFWD enjoy was key in keeping the collaborations moving forward. Importantly, the steering committee focuses on addressing proactive initiatives and avoids specific project and regulatory discussions. This simple tenet of keeping the project and permit discussions out of the committee meetings fosters the spirit of collaboration by allowing participants to focus on shared goals and objectives.

It is also important to note that Vermont's statewide linkage analysis approach emphasizes how a systems approach may increase conservation effectiveness and enhance connectivity for multiple species. This statewide approach was also instrumental in maintaining predictability and transparency of transportation projects from beginning to end for both state agencies, which further fostered collaborations between VFWD and VTrans. Because of the collaborations between the two agencies, Vermont has become a leader in crafting solutions to wildlife and transportation issues.

Conclusions

The greatest asset that Vermont enjoys for the long-term conservation of wildlife relative to transportation planning and development is the collaborative interagency working relations between VTrans and VFWD. Maintaining the interagency steering committee remains a priority as a mechanism to withstand inevitable changes in administrations. This interagency effort to work collaboratively to address the common goal of the public's interests in a transportation system that is sensitive to wildlife conservation continues to stand the test of politics.

Chapter 13

Arizona State Route 260: Promoting Wildlife Permeability, Highway Safety, and Agency Cultural Change

NORRIS L. DODD AND JEFFREY W. GAGNON

The Arizona State Route 260 reconstruction project has garnered acclaim as a comprehensive approach to minimizing the incidence of wildlife–vehicle collisions and promoting wildlife permeability. With its 11 large wildlife underpasses and six bridges along 27 kilometers (17 miles) averaging 0.63 passage structure/kilometer (1.0/mile), this project compares to landmark efforts in Banff National Park, Alberta, Canada, with 24 passage structures in 45 kilometers (28 miles) or 0.53 structures/kilometer (0.86/mile; Clevenger and Waltho 2003, see chapter 7), and those ongoing on U.S. Highway 93 in Montana with 42 passage structures in 91 kilometers (56.5 miles) or 0.46 structures/kilometer (0.74/mile). The State Route 260 project received one of the first federal Exemplary Ecosystem Initiative Awards, recognizing its integration of highway reconstruction with long-term research monitoring in an adaptive management context. This research proved to be critical to the success of the reconstruction in achieving environmental objectives, generating tremendous public awareness and ultimately serving to foster Arizona Department of Transportation (ADOT) integration of road ecology (Forman et al. 2003) into its highway construction practices. Further, our research shed light on a multitude of road ecology issues, including demonstrating not only that wildlife underpasses work but *why* they work. We elucidate heretofore theoretical models of the impact of traffic volume on permeability (Iuell et al. 2003), and employ Global Positioning System (GPS)-based metrics to quantify highway permeability under a before- and after-control impact experimental design (Hardy et al. 2003, Roedenbeck et al. 2007). Given the benefits of eight years of continuous research and adaptive management, it is difficult to comprehend that this research almost did not occur. Nearly as difficult to comprehend is the subsequent phenomenal change in institutional culture that has occurred within ADOT, with the State Route 260 project as a catalyst for change.

Geographical, Historical, Political and Social Setting

Our focus was on a 27 kilometer (17 mile) stretch of State Route 260 in central Arizona, east of Payson (fig. 13.1). This highway cuts though ponderosa pine (*Pinus ponderosa*) forest at elevations from 1,590 (5,216 feet) to 2,000 meters (6,562 feet) to the base of the Mogollon Rim, the area's dominant landform that rises precipitously to 2,400 meters (7,874 feet). Numerous canyons emanate from this escarpment southward, punctuating the rugged terrain. Of particular importance in an otherwise arid landscape (annual precipitation averages 52.6 centimeters [20.7 inches], with only two-thirds of that since 2002) are numerous riparian and wet meadow habitats along the highway corridor (see fig. 13.1); some exceeding 150 hectares (370 acres) in size. Several perennial streams flow adjacent to the highway. Nearly all the corridor right-of-way is within U.S. Forest Service boundaries.

Planning the reconstruction of State Route 260 began in the early 1990s, culminating with the issuance of an environmental impact statement (FHWA and ADOT 2000). Addressing the high incidence of wildlife–vehicle collisions, especially those involving elk (*Cervus elaphus*) by widening a narrow two-lane to a four-lane divided highway was a major issue. During the design concept process, historical wildlife–vehicle collision patterns and biologist and engineer expert opinion were used to arrive at 11 passage structures to be constructed on five phased reconstruction sections (table 13.1).

Three highway sections with seven of the underpasses and six bridges over streams that accommodate wildlife passage have been completed (Dodd et al. 2009a). With the well-documented efficacy of ungulate-proof fencing in reducing wildlife–vehicle collisions, especially in conjunction with passage structures (Bissonette 2006, Clevenger et al. 2001a), the design concept incorporated fencing adjacent to 75 percent of the corridor.

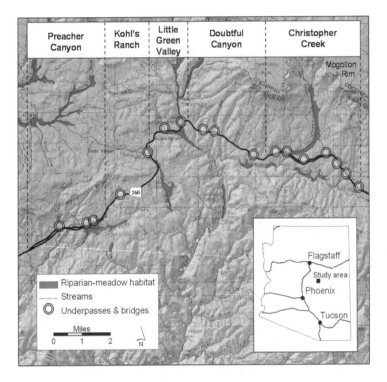

Preacher Canyon	Kohl's Ranch	Little Green Valley	Doubtful Canyon	Christopher Creek

Riparian-meadow habitat
Streams
Underpasses & bridges

Miles
0   1   2   N

Flagstaff
Study area
Phoenix
Tucson

FIGURE 13.1. Location of the State Route 260 study area and the five highway sections along which phased highway reconstruction has been ongoing since 2000, and the location of planned and completed wildlife underpasses and bridges. The shaded areas correspond to riparian and wet meadow habitats located adjacent to the highway. Topographic relief reveals the study area's proximity to the Mogollon Rim escarpment, the dominant physiographic feature within the study area.

However, in early 2000, as plans were being drafted for the first two reconstruction sections, ADOT maintenance personnel raised concerns about vehicle access adjacent to the fence and funding to accomplish fence maintenance. The Arizona Game and Fish Department research branch was summoned to play a monitoring role, as ADOT's partners agreed that the extent of fencing could be reduced *provided* monitoring was conducted. To expedite monitoring, research funding was programmed into construction for the first phased section but was eliminated when project managers changed. Also, when the bid plans for the first section were issued, 90 percent of the fencing in the previous plans disappeared. A year later, though rating well in ADOT's competitive research grant process, our proposed monitoring project went unfunded. It wasn't until the U.S. Forest Service

TABLE 13.1. State Route 260 highway reconstruction sections, reconstruction status, section length, number of average annual elk–vehicle collisions from 2001 to 2007 (Dodd et al. 2007a), and the number of wildlife passage structures implemented or planned as part of the reconstruction

Highway Section	Reconstruction Status	Length (km)	Elk–vehicle Collisions/Year	Wildlife Passages	
				Underpass	Bridges
Preacher Canyon	Completed 2001	5.0	11.7	2	1.0
Little Green Valley	Control	4.0	1.2	1	0.5
Kohl's Ranch	Completed 2006	5.5	5.5	1	1.5
Doubtful Canyon	Control	4.5	5.0	3	0
Christopher Creek	Completed 2004	8.0	19.2	4	3.0
All		27.0	42.6	11	6.0

refused to permit the second phase of reconstruction that ADOT funding was finally secured to initiate monitoring in late 2001.

Given this initial resistance to research, we were particularly diligent in sharing information in a timely manner with project engineers. We even deployed elk GPS telemetry collars with daily Argos satellite data uplink capabilities to provide immediate access to highway crossing data. As these and other data accumulated and pointed to needed modifications in the highway reconstruction, our adaptive management efforts with ADOT project managers ensued. These efforts initially met with a degree of resistance (though our data often proved compelling) but eventually led to a closer working relationship based on a mutual desire to build the most environmentally sensitive and safe highway possible, thereby merging our agencies' missions. Our project ended in 2008 after ADOT ultimately funded four phases of continuous research. Their enthusiasm increased with each subsequent phase as the growing pool of data and insights, as well as public support, underscored the benefits of the research and adaptive management.

With each phase of State Route 260 research, demonstrating a highway safety benefit through reduced wildlife–vehicle collisions was critical to obtaining ADOT funding. In a watershed event in 2006, with wildlife permeability insights from State Route 260, ADOT funded a pronghorn (*Antilocapra americana*) movements research project along U.S. Highway 89 where no safety issue exists (U.S. 89 is a near total barrier and no collisions occur; Dodd et al. 2009b). ADOT had recognized the importance of promoting permeability for this sensitive species; this project will ultimately identify passage structure locations during highway reconstruction scheduled within seven years. ADOT also has since funded wildlife telemetry research to determine passage structure locations and installed permanent traffic counters to further Arizona Game and Fish Department traffic–wildlife relationships research on three other highways; Interstate-17, State Route 64, and U.S. Highway 93. In the latter case, McKinney and Smith's (2007) study recommended that ADOT consider three overpasses to promote desert bighorn sheep (*Ovis canadensis*) permeability; construction on all three overpasses was begun in 2009 as well as additional monitoring. ADOT also provided enhancement funding for ungulate-proof fencing along the first reconstruction section of State Route 260. This fencing was integrated with an animal detection and motorist alert system at the fence terminus and has resulted in a 96 percent reduction in wildlife–vehicle collisions (Gagnon et al. 2010). Another fencing enhancement project was funded to address the worst elk collision "hotspot" along Interstate 17, and will also promote permeability. Thus, from their initial resistance to stalwartly funding research and embracing adaptive management, ADOT's culture has changed dramatically, to the point that they have become a recognized national leader among state transportation agencies in embracing road ecology principles.

There is a misconception that the 2003 *Booth v. State of Arizona* court decision (in which the state was found partially negligent in an elk–vehicle collision on Interstate 40 and ordered to pay a significant settlement) was the impetus for the wildlife measures along State Route 260. In fact, this case had no bearing on the reconstruction project, and the design concept report and plans for wildlife underpasses had been finalized well in advance of the accident that led to this case. However, the outcome of the case certainly has affirmed ADOT's overall commitment to the resolution of wildlife–highway conflicts.

Rationale for the Project

The overarching rationale for the reconstruction of State Route 260 was its substandard condition and insufficiency to accommodate current and anticipated traffic volume. The narrow, two-lane roadway twisted and climbed steep grades as it wove through mountainous terrain. The substandard conditions were exacerbated by the increasing traffic from motorists traveling to popular recreational destinations along the Mogollon Rim and beyond. Average annual daily traffic (AADT) volume on SR 260 increased nearly threefold in the 10 years from 1994 to 2003, from 3,100 to 8,700 AADT (Dodd et al. 2007a).

Highway safety was a significant impetus for the reconstruction; its accident rate was 2.4 times the rate for typical two-lane highways in Arizona (FHWA and ADOT 2000). The incidence of wildlife–vehicle collisions contributed to highway safety concerns, especially collisions involving elk, which accounted for nearly all human injuries and property damage. From 1994 to 2005, an average of 37 wildlife–vehicle collisions occurred each year, 84% involving elk, and during this period 51 percent of all single-vehicle accidents involved wildlife, underscoring the severity of the highway safety problem (Dodd et al. 2007a).

Environmental Issues

A multitude of environmental issues arose during the planning process and have been addressed with the State Route 260 reconstruction (FHWA and ADOT 2000). Issues ranged from impacts of construction on threatened spotted owl (*Strix occidentalis*) nesting to preservation of water quality and avoidance of impact to adjacent aquatic ecosystems and perennial streams.

From the perspective of wildlife–vehicle collisions and highway safety, as well as direct impact to wildlife populations (Schwabe and Schuhmann 2002), proximity of the highway immediately adjacent to numerous riparian and wet meadow habitats (fig. 13.1) was particularly problematic. Whereas previous studies reported that elk selected areas away from roadways with far less traffic than State Route 260 (Rost and Bailey 1979, Witmer and deCalesta 1985, Rowland et al. 2000), we found the frequency of elk ($n = 33$) GPS relocations within a kilometer (0.6 mile) of State Route 260 were twice that of a random distribution (Dodd et al. 2007b). Elk attraction to riparian and meadow habitats accounted for their concentration near the highway. Typical of most roads, State Route 260 was built adjacent to riparian areas, and this habitat was seven times more concentrated within a kilometer (0.6 mile) of the highway compared to the proportion within elk home ranges. Use of these lush habitats, particularly during drought conditions, were key to where elk crossed the highway and elk–vehicle collision peaks occurred (Manzo 2006, Dodd et al. 2007a). As such, the State Route 260 alignment adjacent to streams and meadows has contributed to long-term wildlife–vehicle conflict and presented a significant challenge to reducing the incidence of wildlife–vehicle collisions.

With a reconstruction footprint often exceeding 0.5 kilometers (0.3 miles) in width, reduced wildlife permeability and highway barrier effect were of paramount concern (Noss and Cooperrider 1994, Forman and Alexander 1998, Forman et al. 2003). Though numerous studies have alluded to these barrier effects, few have yielded quantitative data relative to animal passage rates, particularly in an experimental context such as State Route 260.

Key Players and Critical Factors

Though engineering and safety issues drove the reconstruction of State Route 260, many factors and key players were instrumental to its success in also addressing ecological needs, research, and adaptive management. Foremost was the growing awareness of road ecology (Forman et al. 2003), successes occurring elsewhere in North America (e.g., Clevenger and Waltho 2000, 2003), and advocacy by organizations like the International Conference on Ecology and Transportation. Stakeholder partnerships during planning *and* implementation were fundamental to the project's success. Aside from the U.S. Forest Service being instrumental in bringing about our research, as the principal corridor land manager, they also used their leverage to promote environmental excellence, including the incorporation of sufficient number and quality of wildlife passage structures. Leadership from ADOT's Natural Resources Management Section, FHWA, and Arizona Game and Fish Department management personnel fostered synergy in this pursuit. The receptiveness of key ADOT project managers to pursue new initiatives when presented with sound monitoring data was the foundation for successful adaptive management that should be integral to every major highway project. The phased reconstruction of State Route 260 facilitated both our research experimental design and adaptive management. Our success in accomplishing long-term research was predicated upon ADOT's commitment to allocating funds under a dedicated research program administered by its Arizona Transportation Research Center.

Outcomes and Lessons Learned

The SR 260 reconstruction project has served as a living road ecology laboratory, and the many insights gained from our research contribute significantly to the understanding of the relationships between wildlife and highways.

Role of Ungulate-Proof Fencing

It became clear early in our project that ADOT's "limited fencing" approach to erecting 2.5 meter (8 foot) ungulate-proof fencing in conjunction with underpasses did not yield desired results. On the first completed highway section with three passage structures and only 13 percent of the corridor fenced, the average incidence of wildlife–vehicle collisions with elk increased 58 percent after reconstruction. Though Dodd et al. (2007a) found that the annual incidence of elk–vehicle collisions was related to the combined influence of elk population numbers and average annual traffic, the 67% increase in traffic from before- to after-construction levels did not solely account for the collision increase here. In the two years (2007–2008) since the entire section was fenced under an enhancement grant, only one elk–vehicle collision occurred, a 96 percent reduction (Gagnon et al. 2010). These results underscored the role of fencing in complementing passage structures in reducing wildlife–vehicle collisions, as reported elsewhere (Ward 1982, Woods 1990, Lavsund and Sandegren 1991, and Clevenger et al. 2001a).

On the second reconstructed section with seven passage structures, ADOT planned to fence only 22 percent of the corridor, along with boulder "elk rock" adjacent to another 10 percent; elk rock was an effective alternative to fencing to deter at-grade wildlife crossings (Dodd et al. 2007a, c). Dodd et al. (2007b) estimated that the fencing would intercept only 13 percent of GPS-determined elk crossings (and riprap 27 percent). Under adaptive management, we recommended that ADOT extend fencing to encompass 49 percent of the highway corridor, bringing the total projected interception of elk crossings to 85 percent (Dodd et al. 2007b, c). This reconstructed section was opened to traffic nearly a year before it was fenced, providing an opportunity to compare before- and after-fencing wildlife–vehicle collision patterns, wildlife use of passage structures, and elk permeability. Compared to the two years before the section was opened to traffic (19 elk–vehicle collisions/year accounting for 52 percent of accidents), elk collisions increased 2.7-fold (to 52 elk collisions/year and 76 percent of accidents) once opened to traffic but before fence was erected. In four years since fencing was erected, collisions dropped to an average of 5.0/year or a 90 percent reduction from the before-fencing level, validating the approach to strategically locating fence based on GPS elk crossing data (Dodd et al. 2009a). Fencing this section yielded another benefit, improving underpass use. Before fencing, we video recorded 500 elk and deer (*Odocoileus* spp.) at two underpasses, of which only 12 percent successfully crossed via underpasses and 81 percent crossed the highway at grade. However, in the first year after fencing, we recorded 595 elk and deer, of which 56 percent crossed successfully; no animals crossed the highway at grade. The probability of an approaching animal crossing through an underpass increased from 0.09 to 0.56 with fencing (Dodd et al. 2007a, c). Without fencing, underpasses proved ineffective in promoting wildlife passage even after being in place for over 14 months before our assessment, providing time for animals to habituate to the structures (Clevenger and Waltho 2003).

Wildlife Use of Underpasses

We monitored wildlife use of six of the seven underpasses constructed on State Route 260 (table 13.2).

We used triggered four-camera video camera systems to record wildlife and calculated passage rates as our comparable metric of wildlife use;

TABLE 13.2. Physical characteristics associated with wildlife underpasses monitored by video camera surveillance along State Route 260, Arizona, including the highway section, bridge span, height above the floor, atrium width between bridges, and years that each underpass has been monitored

Wildlife Underpass	Highway Section	Span (m)	Height (m)	Length (m)[a]	Atrium (m)[b]	Monitor Years
East Little Green Valley	Preacher Canyon	41.1	6.8	52.7	11.0	6.0
West Little Green Valley	Preacher Canyon	41.1	11.5	110.6	11.0	6.0
Pedestrian–wildlife	Christopher Creek	34.2	6.8	128.0	47.9	5.0
Wildlife 2	Christopher Creek	39.9	10.0	118.8	31.9	5.0
Wildlife 3	Christopher Creek	37.7	5.1	63.9	None	4.5
Indian Gardens	Kohl's Ranch	41.1	12.5	65.6	36.5	3.0

[a]Length = distance for animals to fully negotiate passage structure, from mouth to mouth, including fill material.
[b]Atrium = width of opening between eastbound and westbound bridge spans.

passage rates were calculated from the proportion of animals that crossed through underpasses relative to those that approached (Dodd et al. 2007d). We also simultaneously recorded traffic passing above animals to assess the impact of traffic volume on underpass use (Gagnon et al. 2007a). We recorded 11 wildlife species and 15,134 animals (Dodd et al. 2009a). Of these animals, 68 percent passed through an underpass. Elk accounted for 77 percent of the animals at underpasses while white-tailed deer (*Odocoileus virginianus*) and mule deer (*O. hemionus*) accounted for 13 and 6 percent, respectively.

Our overall underpass passage rates for all species ranged from 0.46 to 0.78 crossings/approach. Elk passage rates ranged from 0.20 to 0.83 crossing/approach and averaged 0.61, and mule deer passage rates ranged from 0.42 to 0.61 crossings/approach (average = 0.55). White-tailed deer passage rates exhibited the greatest variability among underpasses, ranging from 0.06 to 0.96 crossings/approach (average = 0.39), tied largely to habitat characteristics on each side of the underpasses (Dodd et al. 2009a).

We used wildlife and traffic video data to examine the effect of traffic volume on elk underpass passage rates. Gagnon et al. (2007a) found that elk passage rates during low (one to four vehicles/minute) and higher (more than four vehicles/minute) traffic levels did not significantly differ from the passage rates when no vehicles were present (fig. 13.2). Thus increasing traffic volume had no effect on elk underpass passage rates, a significant finding toward understanding why underpasses promote wildlife passage.

Factors related to successful elk crossing through underpasses included underpass design and placement, months monitored, season, and time of day in order of importance (Dodd et al. 2007a). Day of the week (our surrogate measure for traffic level, as SR 260 traffic levels were higher on weekends than weekdays, Gagnon 2006) did not have a significant influence on crossing probabilities, consistent with Gagnon et al. (2007a).

Underpass structural design and placement were the most important factors in predicating successful elk crossings. However, our inference was tempered by limited replications of underpass type and placement when compared to other assessments (Clevenger and Waltho 2000, 2005, Ng et al. 2004), as all State Route 260 underpasses were large single-span bridges. In only one instance was placement controlled for two underpasses situated literally side by side (within 250 meters [820 feet]), allowing structure alone to be evaluated (Dodd et al. 2007a, d). Here, we concluded that concrete retaining walls at one underpass resulted in lower probability of elk crossing. Though we did not document predator–prey interactions (Lit-

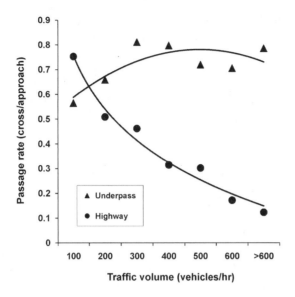

FIGURE 13.2. Comparison of at-grade and below-grade (through six wildlife under-passes) elk passage rates at varying traffic volume levels along State Route 260, Arizona (figure from Dodd et al. 2007c). At-grade passage rates were determined from Global Positioning System telemetry tracking of 44 elk from 2003 to 2006 (Gagnon et al. 2007a) and below-grade underpass passage rates were determined from video surveillance of wildlife use at underpasses from 2002 to 2006 (Gagnon et al. 2007b).

tle et al. 2002), elk appeared hypervigilant of predators potentially lurking atop the concrete walls; the ledge effect likely accounted for the differences in elk use. Original construction plans for the Indian Gardens underpass on the third State Route 260 reconstruction section entailed more than 4,000 square meters (43,000 square feet) of concrete walls. Based on our research, ADOT eliminated the concrete walls and increased floor width from 16 meters (50 feet) to 32 meters (100 feet; Dodd et al. 2007d). The subsequent elk passage rate here is the highest of any monitored underpass (0.83 crossing/approach), attained in only six months, and is testimony to the benefit of adaptive management (Dodd et al. 2007a).

Our comparison of two underpasses on the second reconstruction section provided insights into the importance of bridge placement (Dodd et al. 2007a). Though both structures were similar open span bridges with wide atria, they differed in their bridge alignment, reflected in their respective elk passage rates (0.70 versus 0.34 crossings/approach). The underpass bridges with higher passage rate were constructed in line, allowing approaching animals to see completely through the underpass. Conversely, the other underpass bridges were offset along the existing drainage alignment such that views by approaching elk were obstructed by fill slopes.

As our second most important factor influencing the probability of successful elk crossings, the length of time underpasses were monitored related to the "learning curve" associated with elk habituation. In the first year of monitoring, the mean probability of a successful elk crossing was 0.52, which increased 38 percent to a peak of 0.71 by the second year, then leveled off over the next two years at 0.69 (Dodd et al. 2007a). Clevenger and Waltho (2003) also found that elk achieved peak underpass use within two years.

Season was also important, with the highest probability of elk crossing through during summer (0.71) and the lowest during fall (0.59). Monthly passage rates over the first five years reflected a recurring pattern of declines during the fall–winter months, less than 0.40 crossings per approach, followed by recovery in spring–summer to greater than 0.80. The fall–winter passage rates coincided with migratory elk dropping off the Mogollon Rim and wintering near State Route 260 (Dodd et al. 2007d). However, by our sixth year of monitoring, the wide seasonal fluctuations had diminished, and it appears that even migratory elk were adapting to the underpasses (Dodd et al. 2009a); expanded application of ungulate-proof fencing likely facilitated this habituation (Ng et al. 2004).

Time of day influenced the probability of successful underpass crossings, highest (0.73) in the evening hours (1:600–2:200 hours). Dodd et al. (2007d) reported a bimodal crossing pattern by elk at underpasses that coincided with elk–vehicle collisions along the highway, of which 67 percent occurred within a three-hour departure from sunrise/sunset, similar to Haikonen and Summala (2001) and Gunson and Clevenger (2003).

Influence of Traffic Volume on Elk At-Grade Crossings and Distribution

Traffic can be a "moving fence" that renders highways impermeable to wildlife (Bellis and Graves 1978). One theoretical model predicts that highways become impermeable barriers to wildlife at 10,000 vehicles/day (Iuell et al. 2003). We estimated traffic volume using a permanent traffic counter installed in cooperation with ADOT in 2003, which recorded mean hourly traffic volumes. We linked traffic volume to GPS-collared elk ($n = 44$) relocations and modeled the importance of factors determining elk crossings of the highway, including traffic, presence of riparian and meadow habitats, season, sex, and time of day (Gagnon et al. 2007b) The probability of elk occurring near the highway decreased with increasing traffic volume, indicating that habitat near the highway was used by elk primarily when traffic volumes were low (< 100 vehicles/hour ≅ 2,400 average annual daily traffic, Gagnon et al. 2007b). Elk shifted distribution away from the highway as traffic volume increased, with a 0.40 mean probability of an elk occurring within 200 meters (760 feet) of the highway with traffic ≤ 100 vehicles/hour, but dropped below 0.20 when traffic was ≥ 600 vehicles/hour (≅ 14,400 average annual daily traffic), consistent with Iuell et al. (2003). Even with much higher traffic volume, we did not find a permanent shift away from the highway; rather, elk shifted away at high traffic and returned when traffic was relatively low. This was consistent with 48 percent of our > 430,000 elk (n = 100) GPS relocations occurring within a kilometer (0.6 mile) of the highway (Dodd et al. 2009a). When we calculated passage rates (Gagnon et al. 2007b) as a function of traffic volume, increasing traffic resulted in dramatically diminished passage rates when elk crossed the highway at grade, especially compared to underpass crossings (see fig. 13.2).

We found that traffic volume, presence of riparian and meadow habitat, and season jointly influenced highway crossings by elk (Gagnon et al. 2007b). As traffic volume increased from zero to 1,500 vehicles/hour, the probability of highway crossings declined by 50 percent. However, traffic volume effect was strongly influenced by both season and proximity to riparian and meadow habitat. This influence related to the motivation for animals to cross the highway and hence their tendency to tolerate higher traffic volumes while crossing (Gagnon et al. 2007b). Elk also crossed State Route 260 at higher traffic volumes during spring and fall migrations and when accessing preferred food, particularly in the spring when forage growth is most vigorous (Dodd et al. 2007a, b, Manzo 2006).

Wildlife Highway Permeability

We conducted several phases of telemetry to assess wildlife permeability under a before- and after-control impact experimental design (Roedenbeck et al. 2007). In our first phase of telemetry (2002–2004), we developed passage rate as a metric of highway permeability, calculated as the proportion of highway crossings to approaches for each animal (Dodd et al. 2007b). An approach was considered to have occurred when an animal moved to within 0.25 kilometers (0.16 mile) of State Route 260. To infer highway crossings, we drew lines connecting all consecutive GPS fixes; crossings were identified where lines between fixes crossed the highway.

Across all telemetry phases from 2002 to 2008, we instrumented 100 elk with GPS collars. Elk crossed State Route 260 11,052 times and averaged 0.50 crossings/approach (Dodd et al. 2009a). Mean elk passage rates among highway reconstruction classes ranged from 0.67 crossings/approach on sections before reconstruction (including controls) to 0.41 after-reconstruction (see table 13.3). Passage rate differences among classes were consistent with the road avoidance model developed by Jaeger et al. (2005). The 39 percent significant difference in permeability between the two-lane controls and four-lane divided highway after reconstruction was not as dramatic as that reported in other studies, such as the 89 percent decline in moose crossing rates after reconstruction of a highway in Sweden with three passage structures (Olsson 2007), and reflects the benefit associated with passage structures and fencing on reconstructed State Route 260 sections.

In our second telemetry phase (2004–2006) along the second reconstructed section, we conducted a before- and after-fencing assessment of elk

TABLE 13.3. Mean annual number of elk–vehicle collisions/km and elk passage rates by highway reconstruction class along State Route 260, Arizona (from Dodd et al. 2009a)

Highway Reconstruction Class	Mean Elk–Vehicle Collisions collisions/km (collisions/mi)[a]	Mean Elk Passage Rate (crossings/approach)[b]
Before/controls	0.68 (1.1) A	0.67 A
During	2.17 (3.5) A, B	0.64 A, B
After—no fencing	2.86 (4.6) B	0.40 C
After—fenced	0.68 (1.1) A	0.42 B, C

[a]ANCOVA differences among reconstruction classes $F_{3, 45} = 14.73, P < 0.001$.
[b]ANOVA differences among reconstruction classes $F_3, 131 = 7.20, P < 0.001$.
Elk–vehicle collision rates were determined from monitoring conducted 2001–2008. Elk passage rates were determined from GPS telemetry ($n = 100$ elk) conducted 2002–2008. Letters denote significant differences among construction classes by analysis of variance (ANOVA).

permeability. Compared to the mean elk passage rate of 0.79 when under reconstruction (Dodd et al. 2007b), permeability dropped significantly by 32 percent to 0.54 crossings/approach following reconstruction but before ungulate-proof fencing was erected (Dodd et al. 2007c). Yet once fencing was erected, the passage rate rebounded significantly by 52 percent to 0.82 (Dodd et al. 2007c). Others have alluded to the benefit of passage structures in maintaining or enhancing wildlife permeability (Romin and Bissonette 1996, Clevenger and Waltho 2000). Our study provided empirical evidence that passage structures do promote wildlife permeability. We attribute the recovery in elk passage rate with fencing to the funneling of animals toward underpasses where they were presented below-grade opportunities for crossing that ameliorated road avoidance resistance to crossing at grade (Jaeger et al. 2005). Gagnon's (2007b) finding that traffic volume affected elk crossings at grade but had minimal influence on below-grade underpass crossings (Gagnon 2007a, see fig. 13.2) thus provided insight into *why* passage structures promote permeability. We suspect that elk permeability here was also partly attributable to close spacing of passage structures along this section. Recommended spacing between structures was 3.5 kilometers (2.2 miles) to promote elk passage (Bissonette and Adair 2008), and the average spacing of passages on this section was only 1.1 kilometers (0.7 miles).

From 2004 to 2007, we tracked 13 white-tailed deer fitted with GPS collars on three sections where reconstruction was complete and two control sections (Dodd et al. 2009a). On the control sections, we recorded an average of 0.02 crossings/day, while on the reconstructed sections, the deer crossing rate was nearly 15-fold higher (0.28 crossings/day). Deer passage rates on the control sections averaged 0.03 crossings/approach, indicating that the narrow two-lane highway was a significant barrier. On the reconstructed sections, the mean passage rate was five times higher than for our control sections; 0.16 crossings/approach. Thus wildlife passage structures significantly improved deer permeability along State Route 260.

The first reconstructed State Route 260 section (only 13 percent originally fenced) was completely fenced in late 2006 due to continued high incidence of elk–vehicle collisions (Gagnon et al. 2010). After fencing, the mean elk passage rate declined to 0.09 crossings/approach (Gagnon et al. 2010), an 86 percent reduction from before-fencing levels (Dodd et al. 2009a). On this section, the average distance between passage structures was 2.4 kilometers (1.5 miles), more than twice the distance on our second reconstructed section. On the third reconstructed section with 2.1 kilometer (1.3 mile) spacing between passage structures, elk permeability averaged

0.27 crossings/approach. We found a strong inverse relationship between passage rate and passage structure spacing, pointing to the importance of passage structure spacing on elk permeability; this also suggests that the recommended spacing by Bissonnette and Adair (2008) may be inadequate to maintain permeability for elk, though other factors such as motivation to seek food and water or to migrate may drive elk to cross regardless of spacing.

Economic Benefit of Underpasses and Fencing

Not only have State Route 260 underpasses and fencing yielded benefit in reduced wildlife–vehicle collisions (see table 13.3) and improved permeability, they also provide considerable economic benefit. Costs associated with wildlife–vehicle collisions including vehicle property damage, human injuries and fatalities, carcass disposal, and loss of recreational value can be substantial; the cost of each elk–vehicle collision was estimated at $18,561 (Huijser et al. 2007). The annual economic benefit from reduced elk–vehicle collisions approached $2 million in 2006 and 2008, and exceeded $2.5 million in 2007 (Dodd et al. 2009a). Over a 20-year period, the economic benefit from reduced elk–vehicle collisions with no change in traffic and elk population levels would exceed $35 million in current U.S. dollars, or an amount exceeding the cost of constructing all wildlife underpasses and fencing.

Comparison of GPS Elk Crossing and Vehicle Collision Patterns

We had the benefit of extensive GPS telemetry along State Route 260 to recommend passage structure and fencing locations. We found strong associations between GPS elk crossings locations and collisions with vehicles at multiple scales; their strength increased with increasing scale (Dodd et al. 2007a). Though our strongest association was found at the highway section level scale, the association at the 1 kilometer (0.6 mile) scale was optimal as it afforded relatively high "power" yet was refined enough to plan and locate mitigation measures to address wildlife–vehicle collisions and permeability. These associations point to the utility of using wildlife–vehicle collision data as an inexpensive and available surrogate for costly GPS telemetry data in planning passage structures and fencing where such data exist.

New Standards

Our research underscored the ability to integrate transportation and ecological objectives into highway construction, yielding benefits to highway safety, wildlife permeability, and economics. The combined application of phased construction, adaptive management, and effective monitoring was instrumental in jointly achieving transportation and ecological objectives. We recommend a phased, adaptive management approach to highway construction and monitoring, when and where possible. And though expensive, monitoring of wildlife mitigation measures can yield significant long-term benefit in improving the efficacy of the measures via adaptive management. Effective monitoring should be incorporated and funded as part of all major construction projects, further adding to the body of knowledge on effective wildlife measures.

The presence of riparian and meadow habitats constituted the "engine" that drove conflicts between State Route 260 and wildlife. Highway planning should avoid limited, valuable habitats where possible. However, underpasses constructed adjacent to these habitats received high levels of wildlife use due to their proximity to preferred foraging areas, as well as ungulate propensity to travel along drainages. Where highway alignments near valuable habitats are unavoidable, such sites are excellent locations for wildlife passage structures.

Underpass structural characteristics and placement are important in promoting wildlife passage. Elk avoided underpasses with concrete walls erected for soil stabilization, and their use should be avoided. Visibility through underpasses should be maximized, and where they are designed for divided highways with atria between bridges, the bridges should be placed in line regardless of the drainage pattern. Although both elk and white-tailed deer permeability was improved by fenced underpasses, we found a wide disparity between passage rates for these species at individual underpasses (Dodd et al. 2009a), pointing to the need to design underpasses to accommodate multiple species. Our data relative to the distance between wildlife passages to facilitate large ungulate permeability indicates that structures should be spaced less than 2 kilometers (1.2 miles) apart for elk. This is less than that recommended by Bissonette and Adair (2008), though their recommendation for white-tailed deer was adequate to promote permeability (Dodd et al. 2009a).

Fencing in conjunction with underpasses promoted permeability as animals were funneled toward underpasses where they crossed below grade with minimal impact from traffic passing above. Fencing should be considered an integral component of measures to promote permeability. Our adaptive management process was a viable means to determine strategic placement of limited fencing to intercept elk at peak crossing zones that helped address high cost, maintenance requirements, and impact to visual quality. In the absence of GPS crossing data, wildlife–vehicle collision data can serve as a surrogate for planning. Cost-effective "retrofitting" (using both barbed-wire and electric fence) of existing right-of-way fencing along State Route 260 was an effective alternative to costly new fence (Gagnon et al. 2010). Fencing is beneficial in reducing wildlife–vehicle collisions, maximizing underpass use, and promoting permeability. Fencing nonetheless requires constant maintenance and attention to maintain its integrity and limit potential for liability. As these measures are employed on new projects, adequate funding for increased maintenance demands is an often overlooked critical consideration.

Acknowledgments

Our research was funded by the ADOT's Arizona Transportation Research Center and Arizona Game and Fish Department's Federal Aid Wildlife in Restoration Act, Project W-78-R for wildlife research. The Tonto National Forest (TNF) and FHWA also provided funding for GPS telemetry. We thank Terry Brennan, Robert Ingram, and Duke Klein of the TNF, and Paul Garrett and Steve Thomas of FHWA for their commitment to making this project possible. Many individuals at ADOT provided endless project support and guidance, including Estomih Kombe, Bruce Eilerts, Siobhan Nordhaugen, Doug Brown, Mark Catchpole, Doug Eberline, Tom Foster, Myron Robison, David Gerlach, William Pearson, James Laird, Tom Goodman, Jack Tagler, and Dallas Hammit. Within our agency, Ray Schweinsburg provided valuable research project oversight and Susan Boe conducted extensive GPS telemetry analysis.

PART IV

Effective Innovations

An understanding of science and technology is vital to North American culture and politics. Transportation agencies invest in cutting-edge technology and innovation to address challenges in the transportation sector and to develop and implement forward-looking solutions. The application of advanced sensor information and communications technologies, also known as "smart" or "intelligent" systems, is an example. We end the book by describing recent innovative developments and solutions designed to mitigate highways for wildlife populations. This part of the book exemplifies how new ideas are rapidly being integrated into road ecology research projects and transportation planning. New technologies, such as Web-based citizen science for data collection on wildlife crossings, and the passage of local tax initiatives illustrate not only innovative thinking but also some of the new pathways for implementing ecologically based solutions. In the final chapter we summarize the four main sections of the book and provide recommendations on the needs, future direction, and improvements of public–private partnerships in transportation projects seeking to mitigate impacts on wildlife and fish resources. Many of these recommendations were a result of breakout sessions at a workshop held March 29–30, 2007, at the Western Transportation Institute at Montana State University. This workshop was cosponsored by American Wildlands, the Wildlife Conservation Society, and the Yellowstone to Yukon Conservation Initiative.

Chapter 14

A Local Community Monitors Wildlife along a Major Transportation Corridor

Tracy Lee, Michael Quinn, and Danah Duke

The successful development of wildlife transportation mitigation strategies requires access to timely, accurate information on the spatial and temporal movement patterns of wildlife. Unfortunately, conventional long-term monitoring programs can be expensive and time consuming. In addition, expert-based approaches often marginalize local participation and knowledge. Alternative approaches to knowledge generation and information sharing, including mechanisms to collaboratively engage citizens, academics, and decision makers, offer innovative means to overcome the challenges associated with conventional data collection.

To address this challenge in relation to wildlife and transportation issues in the Canadian Rocky Mountains, the Miistakis Institute established a citizen science framework for wildlife and transportation issues in the Crowsnest Pass. The Crowsnest corridor consists of a two-lane highway (Highway 3), a railway line, and five principle settlements. The Road Watch model has successfully engaged citizens in data collection, generated a large dataset of wildlife observations, and informed conservation planning processes in the region. The project has also experienced challenges typical of citizen science projects, such as data accuracy concerns, the opportunistic nature of data collection, and sustaining volunteer participation. Solutions to these challenges are discussed in relation to Road Watch project goals.

Rationale for the Project

Transportation networks are designed and built to facilitate the efficient, timely, and safe movement of people and goods. However, they also present challenges for wildlife by interrupting wildlife movement patterns, alienating animals from critical habitat, and causing genetic isolation, as well as by direct mortality from collisions with vehicles or trains (Forman and Alexander 1998, Trombulak and Frissell 2000, see chapter 1). The rate of wildlife–vehicle collisions in North America continues to increase in the face of expanding road networks, upgrades to existing roads, and increasing traffic (Conover et al. 1995). The effects reach beyond individual wildlife populations and pose broader conservation, economic, and social consequences, including a considerable human safety risk from vehicle–wildlife collisions (L. P. Tardif and Associates Inc. 2003, Huijser et al. 2007).

These issues are highly significant in the Crowsnest Pass, an east–west transportation corridor through the Canadian Rocky Mountains, where wildlife–vehicle collision rates are high and the barrier effect may have significant negative consequences on movement opportunities, especially for wide-ranging carnivores such as grizzly bears (*Ursus arctos horribilis*), cougars (*Puma concolor*), and wolves (*Canis lupus*) (Carroll et al. 2001, Proctor et al. 2005, Apps et al. 2007). As the Rocky Mountain region is one of Canada's most ecologically intact ecosystems, these issues have garnered the attention of both the conservation community and the popular media at the local, regional, national, and international levels (Carroll et al. 2001, Weaver 2001, Proctor et al. 2005).

Addressing wildlife transportation issues requires access to timely, accurate information on the spatial and temporal movement patterns of wildlife. Research has emphasized that success of mitigation measures to ensure movement, while reducing collisions is highly dependent on obtaining an accurate understanding of wildlife spatial patterns (Clevenger and Waltho 2000, Farrell et al. 2002, Alexander et al. 2004, Ng et al. 2004). Unfortunately, long-term monitoring information on wildlife movement and collision patterns for highways and railways is lacking in most jurisdictions. This is partly due to the cost and complexity of more conventional expert-based methodologies (Irwin 1995, Pollock et al. 2003).

Recent critiques of the ability of science to provide information in a timely, efficient manner and of sufficient quality to address increasingly complex environmental issues, emphasize the importance of exploring alternative approaches to knowledge generation and sharing (de Neufville 1985, Fischer 2000, Hage et al. 2006). Addressing the complexity of environmental problems requires the development of new approaches and frameworks where citizens, academics, and decision makers work jointly to understand and address issues of significance. Ultimately, making science-based information more accessible and fluid requires an exploration of new approaches to integrate citizens in research. (Clark and Murdoch 1997).

To this end, the Miistakis Institute, an ecosystem-based management research institute at the University of Calgary, established a citizen science framework and program to help address wildlife and transportation issues in the Crowsnest Pass through the Rocky Mountains in southwestern Alberta (fig. 14.1).

Geographic Setting

The Crowsnest corridor consists of a two-lane highway, a railway line, and five small urban settlements. The provincial transportation authority, Alberta Transportation, has slated Highway 3 for an upgrade to four lanes due to expected increases in traffic volume. Over the last 14 years the traffic

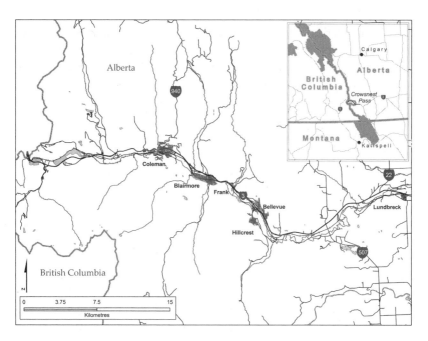

FIGURE 14.1. Map of Crowsnest Pass in southwestern, Alberta. Current study area from Alberta/British Columbia border east to Lundbreck, Alberta.

volume has increased 59 percent along Highway 3; currently the average annual daily traffic volume is 4,801 vehicle/events per day, reaching as high as 9,900 vehicles/events per day between local communities within the pass using the road as a local commuter road. Over the last 10 years the reported mean annual number of wildlife–vehicle collisions was 109 on this 44 kilometer (27.3 mile) stretch of highway, a rate of 2.47 mortalities/kilometer. From a wildlife movement perspective, Highway 3 is considered a major fragmentation zone within the Canadian Rocky Mountains. Genetic research on grizzly bears has indicated female movement is prohibited and male movement greatly reduced (Proctor et al. 2005). Apps et al. (2007) applied a modeling approach to identify important movement corridors for carnivores across Highway 3. Observations of carnivore movement across the highway are important to verify modeling results. Information on spatial and temporal movement patterns of wildlife in the region is essential for the development of effective mitigation strategies to facilitate movement and reduce collisions with vehicles on the expanded and realigned highway.

Project Description

Road Watch in the Pass (hereafter referred to as Road Watch) is an innovative means for connecting researchers, citizen volunteers, and decision makers through a citizen science project to address wildlife conservation issues in the Crowsnest Pass. Road Watch enables citizens to use an interactive, Web-based, mapping tool (see www.rockies.ca/roadwatch) to enter records of wildlife observed along Highway 3.

Road Watch was developed to accomplish three goals: (1) to create a valuable data set of large mammal observations for use by decision makers and the community, (2) to highlight the value of data collected by volunteers to the local community, decision makers, and the academic community, and (3) to create an environment where citizens can learn and share knowledge about local wildlife and conservation issues (i.e., community capacity building).

Road Watch is rooted in the concept of experiential learning and the recognized need for increased public participation in local wildlife conservation and management issues. There are many perceived benefits to integrating citizens into knowledge production: promoting awareness of local environmental issues, building community capacity to enhance public involvement in stewardship, fostering an environment for a stronger public role in decision making, incorporating local knowledge and wisdom, and generating data at a lower cost than that of conventional scientific research (Au et al. 2000, Bliss et al. 2001, Pollock et al. 2003). Road Watch provides an opportunity for scientists, citizens, and decision makers to work collaboratively and learn together about wildlife movement across Highway 3. Ultimately, this approach aims to increase community members' collective knowledge on wildlife movement as well as promote a community informed about wildlife conservation issues. We assume that such a community is more likely to take action to protect the wildlife in the region. Public engagement (observing and recording wildlife) should evolve toward the development of an ecologically literate citizenry actively pursuing community sustainability.

Contributing to Road Watch

There are currently three ways to contribute to the Road Watch project: (1) submit observations through an interactive Web-based mapping tool, (2) report through a telephone hotline; and (3) participate in systematic wildlife surveys of Highway 3 using a handheld Global Positioning System unit with a customized species key pad. This multipronged approach ensures that a diversity of users is reached because it addresses different commitment and computer literacy levels across the community. Recruitment of participants occurs through posters displayed throughout the community, local media announcements, demonstrations of the tool, attendance at local conservation workshops, and personal communication. A local project coordinator promotes Road Watch and is responsible for engaging, motivating, and addressing participants' concerns. The project Web site acts as an effective mechanism for soliciting citizen participation. In addition to participants entering their wildlife observations through the mapping tool, participants can access tutorials for using the interactive mapping tool and wildlife identification, and they can view cumulative results from the project.

The most common mechanism for participation in Road Watch is through the user-friendly Web-based mapping tool. The telephone hotline represents less than 5 percent of the observations. The phone recordings are entered through the mapping tool by the local project coordinator. A Web-based mapping tool is a system that integrates geospatial data for display on the Web. The use and role of a Web-based geographic information system (GIS) in projects involving citizens has many advantages such as efficient data collection, storage, exploration, and dissemination (Steinemann et al.

2004). Other important benefits of employing a Web-based GIS include improving communication among stakeholders, displaying data in real time, and allowing for display of multiple forms of knowledge in a single forum (Gouveia et al. 2004).

The Road Watch mapping tool runs on an Internet Information Server (IIS) using two open-source products, Map Server and Chameleon. The open-source software is freely distributed and accessible, thus allowing programmers to access and modify the code to meet their specific needs (Hall and Leahy 2006). The mapping tool consists of basic GIS functionality with the following spatial layers: roads, rivers, communities, 1 meter air photo mosaic, and local landmarks to assist participants in adding their observations accurately to the map. A customized "add observation" button enables participants to add their observation directly onto the map. Once participants enter their wildlife observation they are prompted to provide information through a pop-up form. Participants provide details on species, date, time of day, and group size. Raw data are instantly converted into a spatial layer and are displayed back to participants through the mapping tool. The online mapping tool facilitates the collection of local observations into a usable format. Once observations are entered on the Web site, the information is readily available to be displayed in maps or converted to tabular data for analysis. This allows easy access for researchers and timely feedback to participants and the community. Monthly participant updates are circulated to all participants, posted on the Web site and published in the local newspapers to increase awareness of the project and solicit new volunteers.

A systematic wildlife driving survey is the latest data collection method added to Road Watch. It was designed based on requests from dedicated users for a more efficient data entry method as well as recommendations to improve current data collection methodology (i.e., based on opportunist rather than systematic sampling). Collecting information systematically enables Road Watch to calculate the rate of movement and collisions across the transportation corridor because data represent both species presence and absence. Participants of the systematic survey are assigned a defined transect (a section of highway they regularly drive). Each user commits to driving the transect twice a week and records start and end times as well as wildlife observations that occur along their designated route. To increase the efficiency of this approach, a new device, called the Otto Wildlife Companion (http://www.myottomate.com) is used, which combines a GPS unit and species keypad to mark observations. This component of the Road Watch program has recently been implemented and results are not yet available.

Outcomes of the Project

The challenges typical of citizen science and the solutions woven into the program have established Road Watch as a successful citizen science model. Road Watch successes include (1) the generation of a large dataset, which has been used to influence land use management and planning in the region; (2) validation of citizen science data to the community, government agencies, and academic audiences; and (3) building community capacity to respond to wildlife and transportation issues in the region.

Generation and Uses of the Data

Road Watch has proven to be a successful model for engaging a local community in data collection. Over 4,044 wildlife observations have been entered into Road Watch by citizens. Of the 4,044 records, 2,246 observations (56 percent) of the observations are within 100 meters (328 feet) of the highway and 500 (12 percent of total observations) represent highway crossing observations. The data predominantly consist of ungulate observations: 70 percent deer, 13 percent bighorn sheep (*Ovis canadensis*), 7 percent elk (*Cervus elaphus*), and 3 percent moose (*Alces alces*) but also include 1 percent rare carnivore sightings (including three grizzly bear crossing observations). The remaining records consist of coyotes (*Canis latrans*) and other smaller animals, including birds. Since project inception in 2004, over 70 individuals from the region have added Road Watch observations to the project. Monitoring the number of community members who have contributed to Road Watch is complicated as individuals also contribute to the project in ways other than data collection, such as community meetings, community action events, and Road Watch contests. We are aware of at least 80 additional people who have contributed to the project in ways aside from data collection.

Road Watch observational data can be analyzed to inform wildlife conservation issues and human safety concerns in the region. For example, the location of high wildlife–vehicle collision zones, areas along the highway with important adjacent habitat or collision hotspots for species of special concern, such as bighorn sheep, can be extracted from the dataset. Figure 14.2 is a visual display of wildlife crossing observations (based on 500 crossing observations) along Highway 3. In this display the data have been summed in 250 meter (820 feet) segments along the highway; therefore higher bars represent more crossings within the 250 meter (820 feet) segment.

Road Watch data and results have already been used by community members, conservation organizations, and municipal and provincial government agencies in the following ways: (1) in combination with other datasets or models to inform conservation planning processes in the region (including provincial and municipal government agencies and land trust organizations), (2) to document the presence and absence of species, (3) to identify conservation significance of blocks of land by private landowners, and (4) as one of the datasets referred to in a regional science synthesis and mitigation report being developed by a partnership of research institutes, scientists, and conservation groups aimed at informing transportation agencies responsible for the Highway 3 transportation corridor in Alberta and British Columbia. These examples highlight the important role of Road Watch for informing land use and conservation initiatives within the region.

Validating Citizen Science

To address concerns regarding the quality of citizen science data (Heiman 1997, Engel and Voshell 2002), and specifically the opportunistic rather than systematic nature of Road Watch data, a graduate research project assessed the validity and reliability of the data (Paul 2007). The project was designed to provide a quantitative, statistical comparison between the spatial, temporal, and species distribution of data collected through Road Watch with a systematic sample. The researcher conducted driving transects to record all observed species within 100 meters (328 feet) of Highway 3 along the 44 kilometers (27.3 miles) of roadway that define the Road Watch study area. Four transects per day were conducted during the summer of 2006. Transects start times were staggered randomly to ensure adequate representation of all 24 hours within each month. Less intensive sampling continued between 15 August 2006 and 31 May 2007 with five transects conducted weekly. All species records and their associated attributes were input into ArcGIS (9.2) to enable spatial statistic comparison to Road Watch data for the same time period.

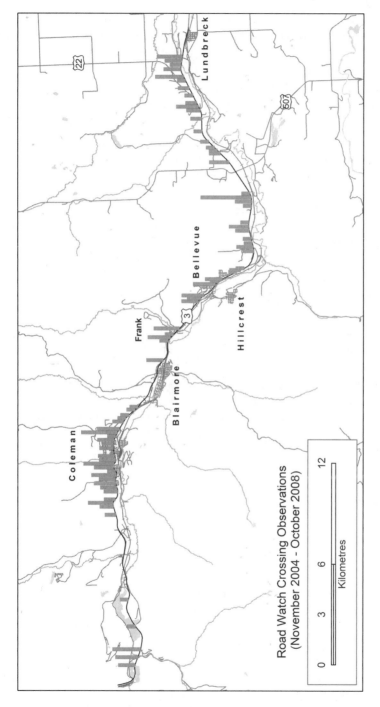

FIGURE 14.2. Road Watch crossing data per 250 meter segment along Highway 3.

The two datasets (systematic and Road Watch) were compared based on their level of spatial clustering agreement. Results indicated significant (95 percent confidence interval) spatial correlation between the two data sets. However, the systematic sample indicated some key locations that are being under-sampled by Road Watch due to regional driving patterns. Temporal comparisons were made between the data sets for time of day, period of day (i.e., dawn, dusk, day, night), and season by looking at the proportions of records in each time category. Although these were not statistical comparisons, there was a clear trend for Road Watch data to over-represent the dawn and day periods in comparison to the systematic sample. In addition, the systematic sample showed a slightly higher proportion of spring records compared to the other three seasons. There were no significant differences in species composition records between Road Watch and the systematic data when all species records were included. However, when deer (the most frequently recorded species) were removed, there was a statistically significant tendency for Road Watch data to have a high proportion of all non-deer species when compared to the systematic data. This indicates a bias for citizens to report more "novel" species.

Overall, Road Watch data showed excellent correspondence to the systematically collected data set, which indicates the value of citizen science initiatives. However, the bias toward daytime records and a potential to over-represent novel species, along with a small number of under-sampled stretches of highway, indicate areas for improvement in Road Watch. All of these concerns have now been addressed through the addition of ongoing systematic sampling methods that will occur in conjunction with the original Road Watch approach.

Building Capacity to Address Wildlife and Transportation Issues

One of the objectives of a citizen science approach is to increase individuals' knowledge on wildlife movement as well as enhancing the community's ability to address wildlife conservation and management issues in the region. In late 2007, Road Watch posted an online survey and interviewed a selection of participants to evaluate individual learning experiences, participants' descriptions of the value of Road Watch, the project's ability to create a learning environment, and tangible actions as a result of project participation.

Of the 43 responses to the online survey, 85 percent felt their knowledge of wildlife–vehicle collisions and movement zones had increased as a result of participating in the project. The other 15 percent indicated their

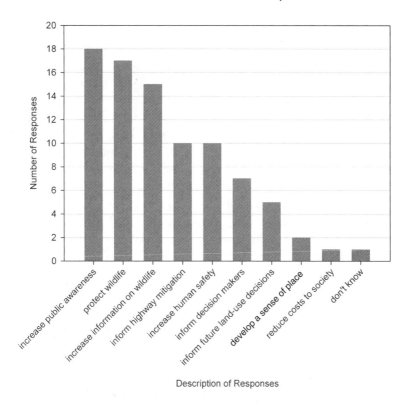

FIGURE 14.3. Public responses describing the value of the Road Watch project.

knowledge had not increased. Results from interviews suggested these individuals felt they were already knowledgeable on wildlife movement prior to project inception. Participants were asked to describe the value of the Road Watch project in an open-ended question. Results were evaluated using qualitative software. Figure 14.3 highlights the different responses and their frequency.

The responses move beyond the direct benefits of increasing the knowledge base on wildlife in the region. Participants understand the potential of the data to inform the community, decision makers, highway mitigation, and future land-use decisions to improve conditions for wildlife and human safety. Two individuals also suggested the project could foster a stronger sense of place and a conservation ethic in the region.

The project's ability to create a learning environment is difficult to measure. Individuals responding to the survey noted that 88 percent of respondents had discussed aspects of the Road Watch project with other community members, including the project objectives, data collection methods, and results. This indicates participants are spreading information through word of mouth within the community, demonstrating one-way knowledge exchange. Attempts to facilitate dialogue between participants occur through events and meetings organized by Road Watch. Attendance at the organized functions has been variable and without a consistent pattern. Regardless, the organized events provide an opportunity for researchers and participants to discuss the project and provide updates. Recommendations discussed by participants at various Road Watch meetings have resulted in changes to project direction; such as enabling participants to enter their observations of wildlife on the adjacent railway, expanding the Road Watch study area, and the implementation of the new systematic driving survey as a more efficient data collection method for dedicated participants. Feedback response and project direction recommendations illustrate an in-depth understanding of the wildlife and transportation issues for the region.

The interview process suggests that an increase in knowledge about wildlife may result in individual behavioral change, such as changes in driving behavior, including slowing down in areas of wildlife–vehicle collision hotspots. To effectively evaluate our ability to foster a learning environment leading to positive action will require a need to understand and define success. Does public engagement evolve toward the development of an ecologically literate citizenry actively addressing conservation or sustainability issues in the region? Understanding the success of Road Watch to inspire social change and result in participants taking action to address conservation issues is difficult to measure. It also depends on our definition of success. While Road Watch indicates success in many areas it is only one tool on the ground for building community capacity. Therefore understanding the place of Road Watch in relation to a community vision and other conservation initiatives provides the context for evaluating success.

Lessons Learned

Typical critiques of citizen science projects include concerns about the quality of data collected by citizens, the challenges associated with keeping citizens engaged and motivated, and the need for long-term funding (Savan et al. 2003). Road Watch faces similar challenges, including concerns in regard to citizen data collection (specifically the accuracy of species identity and location on the mapping tool), the opportunistic (i.e., nonsystematic) methodology of data collection, and the ability to engage and maintain volunteers over time.

There is a diversity of opinions regarding the accuracy and reliability of information generated by citizens (Irwin 1995). As a result, many government and regulatory agencies do not make use of citizen data (Mayfield et al. 2001). A small number of researchers have tested the quality of data collected by citizens by comparing the results with data collected by professional scientists (Au et al. 2000, Nicholson et al. 2002, Newman et al. 2003). For example, Newman et al. (2003) used volunteers to collect data on mammals; using techniques such as mark–recapture and pellet counts. The data collected by volunteers were compared to data collected by professional researchers using the same methods. Au et al. (2000) tested the ability to use high school student volunteers to collect water quality data using simplified methods. The data were compared to data collected by a microbiologist using high-quality technical equipment. Both studies concluded that volunteer data collection provided useful data for analysis when compared to data collected by professionals. In contrast, Nicholson et al. (2002) compared water quality data collected by citizens to data collected by professional scientists and found mixed results. For two variables the quality was similar while for one variable the professional samples showed less variability, possibly due to inaccuracy in volunteer data collection. Despite these successes, scientists and decision makers still question the usability of information generated by citizens.

Road Watch has implemented steps to reduce participant error in accurately identifying and marking the location of their observations, through training and by addressing error in analysis design. For example, to address concerns of wildlife identification the project Web site includes a wildlife identification tutorial to remind participants of identification features. Additionally, species commonly confused with each other are lumped together; for example, white-tailed deer (*Odocoileus virginianus*) and mule deer (*Odocoileus hemionus*) are reported as deer. Misidentification is particularly a concern for carnivore species because they are rarely observed and a single sighting is considered significant. To address this issue, carnivore observations added to the Road Watch database are automatically e-mailed to the local provincial biologist and the local project coordinator. The identity of the participant is not disclosed, but the species, location, and description are included. This enables a trained biologist to verify the observations through follow-up field investigation or through discussions with other knowledgeable individuals.

To assist participants in accurately identifying their observation on the mapping tool, a 1-meter-resolution air photo has been provided along with a spatial file of key local landmarks from the region. In addition, we address the location error in the analysis design; for example, Road Watch data are often enumerated to 250 meter (820 feet) segments along Highway 3.

Another concern is the opportunistic nature of data entry; participants are not restricted to a regular route (they are not assigned to sections of the highway), and they do not record periods when no animals are observed (i.e., absence). One of the concerns with this method of data collection is the possibility of data fragmentation. With the current method it is not possible to determine if all sections of the highway are being observed consistently throughout the year. There may be areas where the highway is over- or underrepresented by sampling, skewing our understanding of wildlife spatial and temporal patterns. The rate of wildlife movement across Highway 3 cannot be determined because time periods when no observations occurred are not recorded. Road Watch attempted to measure the impact of these challenges by designing a one-year systematic wildlife driving survey to compare to Road Watch data. A description of this analysis and results are presented in the results section of this chapter.

An additional challenge for citizen science projects is long-term engagement and consistent involvement of volunteers in the project. Engaging citizens in the project and continuing to motivate volunteers is important to ensure data collection quality and consistency (Savan et al. 2003). A lack of consistency causes data fragmentation and decreases the value of the data (Pollock et al. 2003). Road Watch uses a series of approaches to keep volunteers motivated, including monthly prize drawings, local committee meetings, local action events, wine and cheese gatherings, and regular newsletters and media releases. An online survey of participants noted that regular communication through e-mail or phone calls with the local project coordinator or via the project Web site were preferred choices of communication.

Volunteer networks require financial, technical, and inspirational support to promote long-term engagement (Savan et al. 2003, Pollock and Whitelaw 2005). Obtaining financial and technical support can be a difficult task for community groups. Road Watch is a unique partnership between local community members and a research institute affiliated with a university. Although the project is regularly guided by Road Watch participants, the support role of the Miistakis Institute has provided access to funding and expertise to ensure longevity of the project. Project fundraising, tool development, and data analysis and dissemination are the responsibility of the institute and local project coordinator. Citizens are solely responsible for collecting data and also contribute to Road Watch by attending meetings and training sessions and providing guidance and feedback. Depending on the skill set and time available for Road Watch participants, roles evolve and change over time, but Miistakis continues to play a lead role in project development since its inception.

Conclusions

Road Watch in the Pass is an innovative framework for engaging citizen volunteers, researchers, and decision makers in addressing wildlife and transportation-related issues along Highway 3, a major transportation corridor through the Canadian Rocky Mountains. Road Watch has successfully generated a large dataset of wildlife observations along Highway 3. The dataset and results have been used to inform numerous conservation planning processes, including municipal and provincial government agencies, local citizens, and nongovernmental organizations. In all occasions the Road Watch dataset was used in conjunction with other observational data or habitat/movement models for the area.

A common critique of data collected by citizens is positional and species identification accuracy. An analysis of Road Watch data in comparison with a 1-year systematic dataset highlighted the value of citizen observational data. The comparison between datasets highlighted a similar percentage of observations per species and showed a strong spatial correlation between datasets. Results highlight that Road Watch data accurately represent the spatial movement of wildlife in the region. The analysis demonstrated the need to ensure full spatial, temporal, and species coverage through the use of methods to supplement citizen records. This issue was addressed by introducing a new systematic methodology that enables Road Watch to determine rates of movement, collisions, and temporal patterns of wildlife in the region.

Road Watch is a successful model for increasing individual knowledge on wildlife movement and collision zones in the region. A qualitative study suggests participation has resulted in some behavioral change (such as self-described changes in driving behavior). Participants are also sharing information about Road Watch with the community. Results indicated a high level of understanding in regard to the value of the Road Watch project. Values were expressed beyond simply generating a large dataset, such as informing decision makers and the community, and the development of mitigation strategies to protect wildlife and increase human safety. Results indicate Road Watch success in achieving the educational goal to a certain point, certainly from the perspective of implementing a citizen science project as a framework for engaging citizens into local sustainability issues.

Acknowledgments

The authors would like to thank the community of the Crowsnest Pass and the participants of Road Watch. Thanks to Rob Schaufele, the local project coordinator, for his continued efforts in the Pass, Ken Sanderson for developing the online tool and providing technical support, and Kylie Paul for her graduate work on the Road Watch project. We would also like to thank our financial sponsors: The Woodcock Foundation, Alberta Ecotrust Foundation, Calgary Foundation, Wilburforce Foundation, Mountain Equipment Co-op, and the University of Calgary.

Chapter 15

The Sonoran Desert Conservation Plan and Regional Transportation Authority: Citizen Support for Habitat Connectivity and Highway Mitigation

CAROLYN CAMPBELL AND KATHLEEN KENNEDY

For the past 11 years, regional conservation planning has been at the fore-front of land use decisions in Pima County, Arizona. In 1998 Pima County adopted the nationally recognized Sonoran Desert Conservation Plan. The goal of the conservation plan is to "ensure the long-term survival of the full spectrum of plants and animals that are indigenous to Pima County through maintaining or improving the habitat conditions and ecosystem functions necessary for their survival" (Pima County 2008). Throughout the ensuing decade, Pima County has implemented this plan through a va-riety of measures, including revised policies in their comprehensive land use plan, open space purchases, and development of a habitat conservation plan to address endangered species issues. Another important component of the conservation plan was the identification of critical landscape linkages in Pima County. These linkages are broadly defined areas that connect pre-serve areas (such as national forest lands, national park lands, or Pima County–owned open space) but also contain existing or potential barriers to wildlife movement (Pima County 2008). These barriers include rail-roads, agricultural fields, irrigation canals, and most importantly, a network of roads.

Although Pima County contains a significant amount of public lands and open space, largely characterized by mountain ranges separated by

293

broad valley floors, there are also major roads and highways that fragment these large preserve areas. Fortunately, Pima County has benefited from the overwhelming support of its citizens for initiatives to address this issue of wildlife linkages and roadways.

In 2004, Pima County voters supported a $174.3 million habitat-based Open Space Bond that included critical wildlife linkage lands (Pima County 2004). Since then, Pima County has purchased over 78,509 hectares (194,000 acres) of important open space, including both outright acquisitions and purchase of state grazing leases (Pima County 2008). Two years later, voters adopted a 20-year transportation plan and financing tax that included $45 million to fund wildlife crossing infrastructure and related research and monitoring (Regional Transportation Authority 2009). This money is being allocated to local jurisdictions and state agencies for research and the construction of wildlife crossing structures on specific roadway projects.

This chapter explores how local jurisdictions in Pima County, Arizona, are successfully implementing a regional conservation plan, with a specific focus on wildlife linkages and transportation corridors. Through a discussion of the process by which these various implementation measures have come to fruition, we hope to provide other communities with valuable ideas and resources that could be applied in other regions.

Geographical, Historical, Political, and Social Setting

Pima County, Arizona, is partially adjacent to the border with Sonora, Mexico, and lies at the crossroads of two major ecoregions, the Apache Highlands and the Sonoran Desert (Beier et al. 2008). The Apache Highlands include the mountainous Sky Islands, which reach elevations in excess of 2,743 meters (9,000 feet). The Sonoran Desert lies in the basins between and extends west and south into Mexico. The Sonoran Desert is characterized by bajadas (a broad area of several blended alluvial fans) sloping down from the mountains and supporting forests of ancient saguaro cacti (*Carnegiea gigantean*), paloverde (*Cercidium* spp.), and ironwood (*Olneya tesota*). Creosotebush (*Larrea tridentate*) and bursage (*Ambrosia* spp.) dominate the lower desert (The Nature Conservancy 2009). More than 200 threatened species call the Sonoran Desert home and more than 500 species of birds use the region at some point in their life cycle, through migration, breeding, or permanent residence (Marshall et al. 2000).

In eastern Pima County, the Sky Islands, including the Santa Catalina, Tucson, Rincon, and Santa Rita mountains, are separated by the Santa Cruz

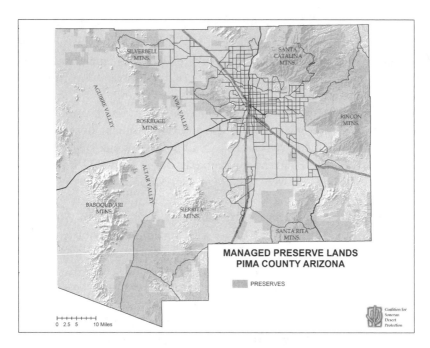

FIGURE 15.1. In eastern Pima County, the Sky Islands, including the Santa Catalina, Tucson, Rincon, and Santa Rita mountains, are separated by the Santa Cruz River valley and its extensive network of tributaries.

River valley and its extensive network of tributaries (fig. 15.1). Sky Islands are characterized by a range of biotic communities that allow both vertical and aspect migration (Warshall undated). The intervening valleys serve as bridges between the mountain ranges, allowing species to colonize, migrate, breed, and forage in both the mountains and the desert (Warshall undated).

In 2006, Pima County's population reached 1 million with growth expected to reach 2 million by 2055 (PAG 2008). The majority of this population lives in the Santa Cruz River valley. Residential and commercial development is largely concentrated within the city of Tucson and surrounding jurisdictions such as the towns of Marana and Oro Valley to the north and northwest, respectively, and the town of Sahuarita to the south. More dispersed development exists in rural areas to the west, east, and southeast of Tucson.

U.S. Interstate 10 runs through eastern Pima County from the northwest to the east, creating a barrier to wildlife moving between the Tucson and Tortolita mountains and the Rincon and Santa Rita mountains (Beier et al. 2008). Annual average daily traffic volumes in 2007 on Interstate 10 in the area of the Tucson–Tortolita wildlife linkage ranged from 41,600 to 80,500 vehicles per day and in the area of the Rincon–Santa Rita wildlife linkage from 29,300 to 42,800 vehicles per day (Arizona Department of Transportation 2008a). State Route 77 runs north through Oro Valley and bisects the linkage between the Tortolita and Santa Catalina mountains. Traffic volumes on this roadway in the area of the wildlife linkage ranged from 30,400 to 47,000 vehicles per day in 2007 (Arizona Department of Transportation 2008a). In the southern part of the county, U.S. Interstate 19 creates a barrier between the Santa Rita Mountains and the Sierrita Mountains, with 2007 traffic volumes ranging from 21,400 to 24,900 vehicles per day (Arizona Department of Transportation 2008a). Additional smaller roadways throughout the county, totaling approximately 6,437 kilometers (4,000 miles) in 2007, also act as barriers to wildlife movement (Arizona Department of Transportation 2008b).

The rich biological diversity of the Sonoran Desert and the surrounding Sky Islands, coupled with booming population growth and a growing network of roads, creates a challenging environment for the protection of important wildlife linkages. Fortunately, local governments and the citizens of Pima County have taken on this challenge over the last decade.

Endangered Species and the Pima County Multiple-Species Conservation Plan

In large part, regional conservation efforts in Pima County began in 1997. After years of habitat loss and myriad land-use practices that threatened population viability, the cactus ferruginous pygmy-owl (*Glaucidium brasilianum cactorum*) was listed as an endangered species under the federal Endangered Species Act. As a response to this listing, dozens of local environmental groups formed the Coalition for Sonoran Desert Protection, an alliance of more than 40 environmental and community organizations. The Coalition's goal is to work with local jurisdictions and the U.S. Fish and Wildlife Service to protect not only the pygmy owl but also biodiversity across the region.

The next year, after adopting the vision of the Sonoran Desert Conservation Plan and committing to the various pieces of its implementation, Pima County embarked upon a process to obtain a Section 10 incidental take permit from the U.S. Fish and Wildlife Service (referring to Section 10 of the Endangered Species Act). This permit requires the submittal of a

Multiple-Species Conservation Plan (MSCP), essentially a plan to mitigate for the sanctioned but incidental "take" of endangered species granted by the Section 10 permit. The county is currently planning to submit its permit application with a final draft MSCP in 2010 (for a summary of events related to the Sonoran Desert Conservation Plan, see box 15.1). Although this process was a direct response to endangered species liability issues countywide, in particular issues with the pygmy owl, the scientific and community responses have ultimately exceeded regulatory compliance. Two important initiatives that came out of the MSCP planning process were (1) the identification of specific critical landscape linkages in Pima County that in part served as the rationale for a $174.3 million Open Space Bond in 2004, and (2) the passage of a regional transportation plan that included $45 million for wildlife linkages in 2006.

The news of a community effort to embark upon a 1.2 million hectare (3 million acre) MSCP drew praise from then U.S secretary of interior, Arizonan Bruce Babbitt. Babbitt met with county officials over a period of months before the elected county board of supervisors kicked off a planning process, including the formation of a technical team and a stakeholder committee. The Coalition for Sonoran Desert Protection, unified in its commitment to realizing the vision of the Sonoran Desert Conservation Plan, was hailed as a key player in the success of this effort and was intimately involved in all stages of the Sonoran Desert Conservation Plan and MSCP planning processes.

In addition, two other local jurisdictions, the City of Tucson and the Town of Marana, have engaged in development of their own habitat conservation plans in conjunction with an application for a Section 10 Incidental Take Permit. These plans are smaller in scope than the MSCP, covering less land area and hence fewer species, but will be important contributions to regional conservation planning in the years ahead. Marana submitted its permit application and final habitat conservation plan draft in December 2008 and hopes to receive its Section 10 permit by late 2010. Tucson is developing two separate habitat conservation plans for different sets of land with endangered species liability issues. In 2009, one of the Tucson habitat conservation plans underwent a public scoping period prior to final submission to the U.S. Fish and Wildlife Service.

The Coalition for Sonoran Desert Protection has been intimately involved with the development of both the Marana and Tucson habitat conservation plans by serving on both technical and stakeholder advisory committees for these jurisdictions. While these habitat conservation plans do not always specifically adhere to the vision of the Sonoran Desert

BOX 15.1. ARIZONA WILDLIFE LINKAGES ASSESSMENT: FIRST-OF-ITS-KIND
REPORT IDENTIFIES HABITAT LINKAGE AREAS CRITICAL TO ARIZONA'S WILDLIFE

In 2004, a broad stakeholder group of state and federal agencies, academia, and conservation organizations formed the Arizona Wildlife Linkages Workgroup to address how Arizona's growing population and infrastructure—roads, fences, railroads, canals, and urban development—has fragmented wildlife habitat and created barriers for species' movement. A Missing Linkages workshop was organized that brought together biologists, engineers, planners, and land managers to share information and knowledge of the state's wildlife, land status, and highway plans. Based on this information, the Workgroup identified large blocks of protected habitat, the potential wildlife movement corridors through and between them, factors that could possibly disrupt the corridors, and opportunities for conservation.

Two years later in 2006, the Workgroup's efforts culminated with the release of the Arizona Wildlife Linkages Assessment report, a science-based plan that identifies 150 linkage zones in Arizona that are crucial to wildlife movement. The assessment provides a broad outline of the next steps needed to conserve and restore landscape connectivity in Arizona and is a resource for planners and engineers, providing suggestions for the incorporation of linkage zones into project planning.

The Workgroup acknowledges that the assessment is the initial step in what will be a continuing effort to identify and map potential linkage zones that are important to Arizona's wildlife and natural ecosystems.

Members of the Arizona Wildlife Linkages Workgroup are the Arizona Department of Transportation, Arizona Game and Fish Department, Bureau of Land Management, Federal Highway Administration, Northern Arizona University, Sky Island Alliance, USDA Forest Service, U.S. Fish and Wildlife Service, and Wildlands Project.

In 2006, the Arizona Wildlife Linkages Workgroup was awarded a national Exemplary Ecosystem Initiative award from the Federal Highway Administration and a state Showcase in Excellence award from the Arizona Quality Alliance. In 2007, the Workgroup was bestowed the 2007 Environmental Excellence Award by the Federal Highway Administration. The full assessment report is posted on the Arizona Department of Transportation Web site at http://www.dot.state.az.us/Highways/OES/AZ_WildLife_Linkages/assessment.asp.

Conservation Plan, the Coalition has worked tirelessly to make these habitat conservation plans as compatible with the Sonoran Desert Conservation Plan as possible.

Conservation Lands System and Critical Landscape Linkages

The broad MSCP planning process initiated in 1998 brought together scientists from state and federal agencies, scientists and advocates from nongovernmental organizations, and local county officials. Planning methodology in the general context of protecting biodiversity included development of a countywide map identifying and prioritizing biologically important lands by a science advisory team (Pima County 2005). These lands were called the Conservation Lands System. Categories developed for these lands were as follows: Important Riparian Areas, Biological Core Management Areas, Special Species Management Areas, Multiple-Use Management Areas, and Critical Landscape Linkages. Connectivity between reserves was determined to be of particular importance to a functional landscape, thus the Critical Landscape Linkages category became a focus with its own methodology for research and implementation.

Critical Landscape Linkages were defined as areas that link large reserves or preserve areas (Pima County 2005). These linkages allow wildlife to move between larger conservation areas and provide for feeding, resting, dispersal of offspring, migration, gene flow, and shifting of a species' range in response to climate change (Beier et al. 2008, see chapter 1). Linkages that provide connectivity for biological resources have either existing or potential future barriers that could isolate major preserves and open space areas. The barriers include, but are not limited to, railroads, agricultural fields, irrigation canals, and roads. The linkage definitions, maps, and land use guidelines were included in both the draft MSCP being prepared by Pima County and the county's Comprehensive Land Use Plan, with both of these plans collectively referred to as the Sonoran Desert Conservation Plan.

Pima County, in adding an environmental planning element to their Comprehensive Land Use Plan, adopted the Sonoran Desert Conservation Plan Conservation Lands System map developed by the science team for the MSCP in 2001. This was in response to the passage of the Growing Smarter Act by the Arizona Legislature in 1998. This act added four new elements to required general plans of municipalities and counties in Arizona: Open Space, Growth Areas, Environmental Planning, and Cost of Development (Arizona Department of Commerce 2008). The county also adopted natural open space requirements and appropriate land use development configuration based on the on-site biological resources. This included the Critical Landscape Linkages map and land use guidelines. Those guidelines were updated in 2005 to reflect the best available science.

Sonoran Desert Conservation Plan Stakeholder Process

In a parallel process to the development of the Conservation Lands System, a broad group of stakeholders, termed the steering committee, was convened in 1999 by Pima County to produce consensus recommendations to county and federal officials regarding Endangered Species Act compliance. During the months leading up to the decision and announcement to embark upon the conservation plan, a contentious stakeholder process to revise the county's environmental regulations polarized the community and kept interest groups locked into long-held growth versus no-growth camps.

Stakeholders included members of the development, conservation, ranching, homebuilding, property rights, and real estate communities, as well as interested public from the region. By most participants' standards, this stakeholder group was originally considered too large to be effective in reaching consensus. However, in the end, it was precisely this diverse representation that led to broad support for the Sonoran Desert Conservation Plan among Pima County citizens. All citizens who expressed interest in participating were appointed by the county, and the 84-member steering committee began its work. Pima County officials held 12 monthly educational workshops for all stakeholder participants on various elements of the Endangered Species Act; geographical, demographical, and biological resources of the region; cultural history; and reserve design. Subsequently, with the help of a meeting facilitator, three additional years of steering committee meetings were held. These meetings were regularly attended by 50 to 60 stakeholders and were highly charged and contentious. However, through the work of a smaller, ad hoc committee, the larger group began forging agreements over the course of the final six months of the four-year process.

In June 2003, the steering committee presented elected officials with a series of recommendations on plan implementation, development and mitigation standards, and proposed funding options. Key among those recommendations was acceptance of the science team's preserve design, including the Critical Landscape Linkages element, and a call for a county open space bond election to fund habitat acquisition.

Additionally, the steering committee recommended that future acquisitions (for the purposes of satisfying MSCP requirements) should be based on a subset of the biologically important set of lands, or Conservation Lands System, identified by the science team. The subset of lands, or Habitat Protection Priorities, was developed by The Nature Conservancy and Arizona Open Land Trust and based on biological goals. This criteria and set of lands was accepted by the steering committee as the best available science. Application of these goals and criteria resulted in the identification of the most important lands to protect through acquisition, provided guidance on the sequencing of land protection efforts, and has been a method for prioritizing possible goals of an adaptive conservation management program. The Habitat Protection Priorities were based on the goals and criteria outlined in box 15.2. These criteria and the associated Habitat

---

BOX 15.2. GOALS AND CRITERIA FOR THE HABITAT PROTECTION PRIORITIES[a]

Conservation Goals

1. Maximize the benefit of existing protected areas by increasing their size.
2. Emphasize protection of the rarest habitat types or "special elements" as per the Science Technical Advisory Team.
3. Maintain a network of connected protected lands where native habitat and natural corridors remain intact in perpetuity.
4. Systematically evaluate lands throughout eastern Pima County so that priorities are identified in all of the county's biologically important areas.

Selection Criteria

1. Lands identified in the most biologically important Conservation Land System categories of Biological Core, Important Riparian, and Recovery Management areas
2. Private lands equal to or greater than 4 hectares (10 acres) in size in vacant or agricultural status
3. State Trust lands within the priority Conservation Land System categories of Biological Core, Important Riparian, and Recovery Management areas emphasizing lands eligible for conservation under the Arizona Preserve Initiative

---

[a]The subset of Conservation Lands System lands that was developed by The Nature Conservancy and Arizona Open Land Trust based on biological goals.

Protection Priorities map became the basis for future Open Space Bond elections.

Pima County's 2004 Open Space Bond

The Sonoran Desert Conservation Plan steering committee recommended to Pima County in June 2003 that an Open Space Bond election be held as soon as possible. For the environmental community, public financing of conservation lands would bring conservation status of the lands in perpetuity sooner and with more certainty compared to the smaller, piecemeal development set-asides over many years. For the development community, the more mitigation lands acquired with public dollars, the less mitigation would need to be accomplished through regulatory mechanisms. In May 2004, Pima County voters supported a $174.3 million Open Space Bond by a margin of 66 percent. The Open Space Bond won by the highest margin of any of the five bonds put before voters that year.

The success of the Open Space Bond was primarily due to two factors. First, the Sonoran Desert Conservation Plan received much publicity and was widely viewed by Pima County citizens as a positive framework for regional conservation planning. The planning process also received numerous regional and local awards for its vision and approach to dealing with growth and conservation. Secondly, national resources were brought to the local bond campaign by The Nature Conservancy and the Trust for Public Land, again because of strong support for the Sonoran Desert Conservation Plan. Local scientists from The Nature Conservancy were heavily involved in the development of the scientific underpinnings of the Multiple-Species Conservation Plan and the organization was convinced that the success of this plan could have national implications.

Since the passage of the 2004 Open Space Bond, over 54,345 hectares (159,000 acres) of open space have been purchased for conservation in Pima County. These lands include those purchased outright by the county and the acquisition of grazing leases on state trust lands. Notable purchases include the 2005 purchase of the 5,665 hectare (14,000 acre) Bar V Ranch, located in the linkage between the Rincon and Santa Rita mountains, and the 2004 purchase of the 16,187 hectare (40,000 acre) A7 Ranch between the Santa Catalina and Rincon mountains. In addition, Pima County has purchased 2,752 hectares (6,800 acres) of the Marley Ranch, with plans to purchase the entire 46,538 hectare (115,000 acre) ranch over the next decade (Coalition for Sonoran Desert Protection 2009).

Regional Transportation Authority

While the county was developing the 2004 Open Space Bond, local citizens began investigating the feasibility of forming a Regional Transportation Authority (RTA) to address comprehensive transportation funding across the region. Ballot initiatives for transportation plans funded by a sales tax had been put before Pima County voters six times between 1984 and 2003; however, all failed. Each election seemed to have voters split into thirds: pro-highway, pro–alternate modes, and anti–tax increase. Following the defeat of a light-rail initiative in November 2003, local officials realized that a new plan was needed that could include a diverse assortment of projects that would have a greater chance of strong voter support. As well, the Pima County region's funding sources were severely lacking to meet transportation needs over the long term, and the options to obtain new funding sources were limited.

Recognizing this need, the Pima Association of Governments commissioned a study by the University of Arizona to assess the feasibility of an RTA. The university brought together a diverse group of stakeholders and government representatives to approach the state legislature for statutory authority for this purpose in 2003. The RTA legislation was approved at the end of the legislative session, signed into law by the governor in April 2004, and became effective in August 2004. One of the key components of the legislation for the RTA was eliminating the veto power of the two largest jurisdictions, the City of Tucson and Pima County. This single action required the nine regional jurisdictions to work cooperatively to develop the RTA plan.

The nine-member RTA board, representing the local, state, and tribal governments in the region, met for the first time in September 2004 and formed two committees to develop the plan. One was a 35-member citizens' advisory committee representing a diverse group of citizens, including representatives from the chamber of commerce, a car dealership, a light-rail organization, real estate, homebuilding, and conservation, many of whom had formally opposed each other in prior failed transportation initiatives. The other committee was a technical/management committee with transportation experts and engineers from both the public and private sectors. Their role was to advise the citizens' advisory committee on technical components of the plan being developed. Development of the plan lasted 10 months, whereby both committees met twice a month and also participated in joint, facilitated meetings, including a weekend retreat. During the final joint session, participants worked and reached consensus on a final recommendation to the RTA board.

During the 10-month process, information about how wildlife is affected by transportation activity was presented to both the citizens' advisory and technical/management committees by the executive director of the Coalition for Sonoran Desert Protection (also a member of the citizens' advisory committee). These presentations included information on projects in other communities addressing transportation infrastructure and safe passage for wildlife as well as a rationale for future wildlife linkages infrastructure in Pima County. These presentations were informed by the prior Sonoran Desert Conservation Plan process, involvement in a national road ecology conference and information sharing, and the science-based conclusions reached about the Critical Landscape Linkages present within Pima County.

The funding commitment for wildlife linkages within the $2.1 billion RTA plan was initially supported at $10 million. This grew to $45 million after stakeholder education and a presentation by the Coalition of a detailed countywide project needs list and budget totaling $90 million. Final approval of the proposed plan by the citizens' advisory and technical/management committees included $45 million for the Critical Landscape Linkages funding category in the RTA plan. This was based on a required match within each funding category. For example, development impact fees matched highway and road project funding, and farebox revenues matched transit funding. The match for the wildlife linkages category was less prescriptive, however, with proposed funding coming from federal, state, and other grants and possible in-kind contributions from local jurisdictions and nongovernmental organizations. During an 18-month public outreach process, the RTA conducted 27 open houses, presented to well over 400 neighborhood, business, and civic groups and held three major news conferences. The RTA conducted two rounds of meetings with editorial boards at all the major media outlets. All the RTA board and committee meetings were open to the public. The RTA also had informational exhibits at many public venues.

Once the RTA committees received public feedback, they incorporated 14 major changes to the final RTA plan recommended to the RTA board. The board unanimously approved the plan, which also received unanimous support from all of the governing bodies of the member jurisdictions. This was a historic first for the region and is a direct reflection of the cooperative spirit of the jurisdictions that was developed as part of the RTA process. In May 2006, Pima County voters approved the RTA plan by a 3–2 margin.

As was instrumental in the Sonoran Desert Conservation Plan steering committee process, an important part of this process was stakeholder education and commitment to consensus as well as external communications and public engagement. Citizens' advisory committee meetings had extensive participation by members of the public.

Disbursement of RTA Wildlife Linkages Funding

After the RTA plan was approved, subcommittees were formed to assist in implementation of the various funding categories. A wildlife linkages working group was charged with coordinating the disbursement of the $45 million in wildlife linkages funding. Representation on the committee includes representatives from local jurisdictions, state agencies, local tribes, and nongovernmental organizations.

The committee's first task was to discuss how funding could be used. The committee decided that appropriate uses included research on wildlife crossing structures in the Sonoran Desert region, the construction of wildlife crossing structures on new or expanding roads, and retrofits on existing roads. The next task was to devise an application process for local jurisdictions and a system for evaluation and decision making. Lastly, while wildlife linkages were identified on a statewide level in Arizona (Arizona Department of Transportation 2009), a more detailed assessment is needed in Pima County. This regional assessment, while not conducted yet, will provide a more detailed map of important wildlife linkages and allow for better prioritization of wildlife linkage projects.

Since the wildlife linkages working group was formed, numerous projects have been funded. The Arizona Game and Fish Department received funding to study the efficacy of specific crossing structures for the species of the Sonoran Desert. The Town of Marana received $20,000 in funding to conduct wildlife crossing research on a 3.2 kilometer (2 mile) stretch of road scheduled for lane expansion. The results of this project are supposed to inform the placement and design of wildlife crossing structures during the expansion project, but the project has been delayed indefinitely. The Town of Marana also received funding to construct additional wildlife crossing structures along a section of new road, including two large underpasses for mule deer (*Odocoileus hemionus*) and a series of culverts for smaller species such as snakes and javelina (*Tayassu tajacu*). Lastly, in December 2009, the Arizona Department of Transportation received approval for $8.2 million of RTA funding to construct one overpass and two underpasses specifically designed for wildlife movement across State Route 77 in the Tortolita–Santa Catalina wildlife linkage. The structures will be built as part of a larger road expansion project on State Route 77.

Jurisdictions match funds for their projects as required by the RTA, and the wildlife linkages working group will continue to evaluate project proposals as they are developed. After the wildlife linkages working group approves projects for funding, the proposals then make their way up a hierarchy of RTA committees, culminating with final approval by the RTA board, made up of elected officials from all the local jurisdictions.

Lessons Learned

Numerous lessons can be learned from regional conservation planning in Pima County, Arizona:

1. When establishing a regional conservation plan, include everyone that wants to be involved in the planning process. Pima County officials included everyone that expressed an interest in serving on the steering committee. This led to an inclusive planning process and broad support for the SDCP recommendations, including those related to wildlife linkages.

2. Educate involved stakeholders on a broad range of topics related to regional conservation planning. Debate and decisions will then be based on a common foundation of knowledge. Members of Pima County's 84-member SDCP Steering Committee were educated on various elements of the Endangered Species Act; geographical, demographic, and biological resources of the region; cultural history; and reserve design over a 12-month period at the beginning of the planning process.

3. When developing a regional transportation plan that involves both large and small jurisdictions, all jurisdictions should have an equal seat at the table. In Pima County, this occurred when the Arizona Legislature granted Pima County statutory authority to create a Regional Transportation Authority (RTA) in 2004. One of the provisions of this law revoked any veto power from the two largest jurisdictions, the City of Tucson and Pima County, thus forcing all of the jurisdictions to work cooperatively in the development of the RTA.

4. Build on the successes of other communities. When various funding categories were being debated for inclusion in the RTA, the Coalition for Sonoran Desert Protection provided information on what other communities have funded in relation to wildlife linkages and transportation infrastructure. This helped educate the citizens' advisory committee on the importance of addressing wildlife linkages and transportation planning in tandem.

5. Gather specific information with dollar amounts about wildlife linkage project needs in your region. The executive director of the Coalition for Sonoran Desert Protection interviewed staff at local jurisdictions and state and federal agencies to determine what types of projects could be funded by a wildlife linkages component in the RTA. This resulted in a list of potential projects with an estimated cost of $90 million. Since all RTA projects are required to have matching funds, $45 million was the final amount approved by voters.

Conclusions

Regional conservation efforts in Pima County over the last decade have led to successful integration of habitat conservation, transportation, and land use planning on a multijurisdictional level. Methodology to design, implement, and preserve wildlife connectivity across transportation infrastructure has been multifaceted, complex, and science based. First, nongovernmental organizations, such as the Coalition for Sonoran Desert Protection, were able to bring attention to the importance of these issues to local officials. These officials responded by adopting the linkages in local public documents such as Pima County's Comprehensive Land Use Plan. Pima County also adopted environmentally sensitive roadway design guidelines in 2005.

In 2004, voters approved an Open Space Bond of $174.3 million, which included acquisition of biologically important lands. The success of this bond can largely be attributed to robust publicity and public investment in the Sonoran Desert Conservation Plan, along with outside funding and resources from national conservation groups such as The Nature Conservancy and the Trust for Public Land. Following this success, Pima County voters approved a sales tax for a Regional Transportation Authority in 2006 that included $45 million for wildlife linkages funding. The RTA came about from an intensive stakeholder process that sought to bring together previously opposed groups and create a diverse transportation plan that satisfied multiple interests. The Coalition for Sonoran Desert Protection was instrumental in educating the RTA citizens' advisory committee about the importance of wildlife linkages in transportation planning. This led to the $45 million allocation in wildlife linkages funding that is currently being disbursed to local jurisdictions and state agencies for research and infrastructure.

Both the 2004 Open Space Bond acquisitions and the RTA funding for wildlife linkages will continue to be implemented in the years ahead. These programs are being integrated not only with each other but with multi-jurisdictional land use planning decisions, along with ongoing research and monitoring. Pima County, Arizona, comprising a diverse set of local jurisdictions, has much to be proud of as it continues to implement the visionary Sonoran Desert Conservation Plan. It will only be able to do this with the ongoing support of its citizens for the conservation of important wildlife habitat, open spaces, and the linkages that bind them together.

Acknowledgments

The development of the Sonoran Desert Conservation Plan and the Regional Transportation Authority involved the hard work and dedication of countless people. In particular, the staff and elected officials of Pima County have shown true leadership in the development of a groundbreaking and visionary regional conservation plan. The late Maeveen Behan was a tremendous force in the development of the SDCP and MSCP, and we will be eternally grateful for her contributions to these efforts. We appreciate the support and expertise of the 35-member groups of the Coalition for Sonoran Desert Protection. Sky Island Alliance and Tucson Audubon Society have both served as our fiscal sponsor during our 12-year existence, and have proved invaluable in their support of our work. Expertise provided by Coalition member group Defenders of Wildlife has informed and guided the Coalition's work on wildlife-related transportation infrastructure. Lastly, without the citizens of Pima County and their support at the ballot box, neither of these plans would have come to fruition.

Chapter 16

Current and Developing Technologies in Highway–Wildlife Mitigation

MARCEL P. HUIJSER, DOUG E. GALARUS,
AND ANGELA V. KOCIOLEK

Current and developing technologies with application in highway–wildlife mitigation are varied. This chapter describes a select few. These technologies have the potential to benefit transportation research, planning, and human safety through better data collection, better analysis tools, new technologies, influencing driver perception, and the use of innovative material and construction techniques.

Data Collection, Management and Analysis Tools for Agencies and Research Organizations

Data on the number and location of animal–vehicle collisions are essential when identifying and prioritizing road sections that may require mitigation measures to reduce collisions. In the United States and Canada these data are typically collected by transportation agencies, law enforcement agencies, and natural resource management agencies (Huijser et al. 2007). These activities result in two types of data: Data from accident reports (crash data) and data based on animal carcass counts (carcass data). However, not all transportation agencies, law enforcement agencies, and natural resource management agencies record these types of data. Furthermore, the organizations that record such data often use different methods, resulting in challenges with data integration and interpretation, and ultimately with the usefulness of the data. Specific problems with these crash and carcass data include underreporting, poor data quality (consistency, accuracy, and completeness), and delays in data entry (Huijser et al. 2007). The use of tools and procedures that are more standardized, Global Positioning System (GPS) technology, faster data entry, centralized databases, and geographic information systems (GIS) were specifically identified to help address some of these problems (Huijser et al. 2007). For example, the use of handheld computers with integrated GPS in the field, when recording crash or carcass data, can help standardize data collection. In addition, they improve the spatial accuracy of the data and allow for downloading into central databases, faster data integration, and standardized data analyses (Ament et al. 2007, Donaldson and Lafon 2008, fig. 16.1). Furthermore, they provide the opportunity to track the route of the observers, providing information about the road sections that were monitored, and thus the

FIGURE 16.1. A handheld computer with integrated Global Positioning System and software specifically developed to standardize animal–vehicle collision data collection with high spatial accuracy. (© WTI-MSU)

search and reporting effort. Underreporting is especially problematic with crash data, which usually represent only a fraction (14–32 percent) of animal carcass count data (Donaldson and Lafon 2008, Huijser et al. 2008).

The use of GIS and associated spatial analysis tools can help with data integration allowing for the identification and prioritization of road sections that may require mitigation measures to reduce animal–vehicle collisions (e.g., Kolowski and Nielsen 2008, Ng et al. 2008, Langen et al. 2009). Such tools may not only be used for crash and carcass data, but they may also be used to model animal movements across the landscape (e.g., Jaarsma et al. 2007, Clevenger et al. 2002a) and estimate the viability of animal populations in areas affected by roads before and after the implementation of mitigation measures (e.g., Haines et al. 2006, Beaudry et al. 2008, Grift et al. 2008, fig. 16.2). These types of exercises are especially helpful to identify and prioritize areas where connectivity across roads is most needed, as these are not necessarily the same locations that are identified by crash or carcass data (Clevenger et al. 2002b). If the use of such spatial databases and information systems were fully institutionalized and if crash and carcass data and connectivity data would be integrated with other infrastructure-related data, highway mitigation could be included from the earliest planning stages of an infrastructure project onward (e.g., Smith 2006).

Data Collection by the Public

Animal carcass data are typically collected by transportation agencies and natural resource management agencies. The search and reporting effort varies, but transportation agency personnel often remove animal carcasses and record carcass data as part of their regular inspection of the road sections for which they are responsible. They typically inspect a road section multiple times per week, sometimes nearly every day (Huijser et al. 2007). However, most maintenance personnel only remove carcasses that are on the actual road surface or carcasses that are highly visible and that could be a distraction to drivers. Carcasses that may be present alongside the road or further away from the road and that are not highly visible are usually not removed and remain unrecorded (Romin and Bissonette 1996, Knapp et al. 2004, Huijser et al. 2006a, 2007, Donaldson and Lafon 2008).

Reports of wildlife seen alive on or adjacent to roads can help identify and prioritize areas where connectivity across roads is most needed as these are not necessarily the same locations that are identified based on crash or

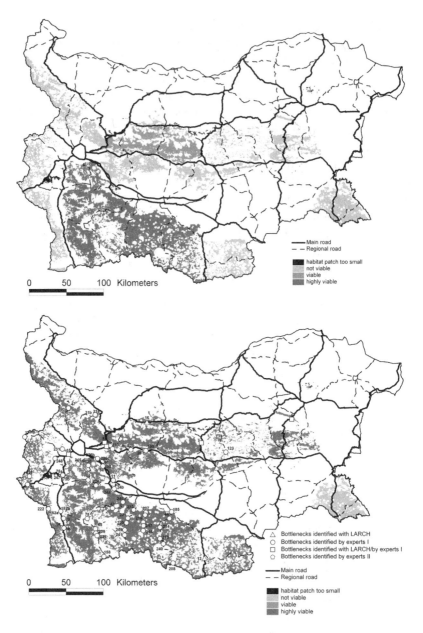

FIGURE 16.2. Population viability (dark gray area is viable, light gray is not viable) of the brown bear before (top) and after (bottom) mitigation of the barrier effect of infrastructure at all bottlenecks identified through population viability analyses modeling. Note: the bottlenecks are numbered according to their priority. (Reprinted with permission from van der Grift 2008)

carcass data (Clevenger et al. 2002b). Agency personnel may not see many live animals on or adjacent to the road since they may only check a road section once a day or several times a week. Citizen science programs are often Web based and can generate a substantial number of observations in the short term (Lee et al. 2006, Lee 2007, Paul 2007, Huijser et al. 2008, see chapter 14). In addition, such programs provide an opportunity for better communication and exchange of information between scientists, policy makers, and the public with regard to understanding the problem of wildlife–vehicle collisions and the need for habitat connectivity for wildlife, including safe opportunities to cross the road (Lee 2007). In this context, it is advisable to allow for data summary tools on the Web site that allow the public to see what their efforts have amounted to at any given point in time (see also Lee et al. 2006, Lee 2007, see chapter 14).

A specific example of a project where the public participated in data collection comes from a 41.8 kilometer (26 mile) section of State Highway 75 in Blaine County, Idaho. The public was asked to participate in collecting information about animals seen dead or alive on or immediately adjacent to the road (Huijser et al. 2008). The public was asked to participate through the local and regional media and variable message signs along the road section concerned (fig. 16.3).

The public could submit wildlife observations through a Web site that depicted the road section concerned, with names of cross streets, and satellite imagery as background (fig. 16.4).

Every 160 meters (0.10 mile) was marked, and the public could click on the tenth of a mile marker that was closest to their observation to enter further information (figs. 16.5 and 16.6).

FIGURE 16.3. Two rotating screens of the variable message sign deployed by the Idaho Transportation Department. (© Angela Kociolek, WTI-MSU)

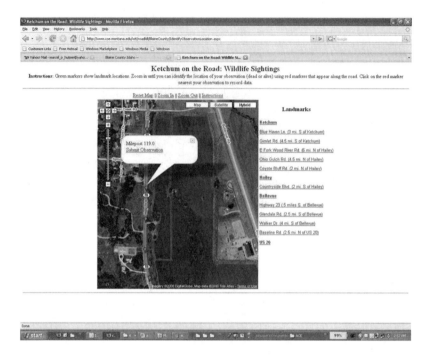

FIGURE 16.4. Screenshot of Google Earth map depicting landmarks in study area and surrounding topography. This screenshot shows 0.1 mile posts (red markers) around the 119.0 mile post.

In previous years, agency crash and carcass databases showed that 30 to 50 deer (*Odocoileus* spp.) and elk (*Cervus elaphus*) were killed annually on this road section (Huijser et al. 2008). With increased search and reporting effort, and accounting for replicate reports from different sources, including from the public and researchers, a minimum of 134 deer and elk were recorded on the same road section in 2007 (Huijser et al. 2008). Only 51 percent of these were reported in agency databases, and the public reported 38 unique deer and elk carcasses (28 percent of total) not accounted for by any other method. These data suggest that crash and carcass reports from agencies alone can seriously underestimate the number of ungulate–vehicle collisions. Data collected by the public ("citizen science") can add substantially to crash and carcass data collected by agencies, resulting in a more accurate minimum number of ungulate–vehicle collisions and better decisions with regard to the need for mitigation measures. The search and reporting effort of the public may not be constant for the road section of in-

FIGURE 16.5. Screenshot of the data form for submitting a live or dead animal observation. Note: the tenth of a mile location obtained when clicking on a red marker on the previous page was automatically stored in the data form.

terest, so care must be taken when using these data to identify potential hotspots that have a concentration of animal–vehicle collisions.

As an alternative to Web-based entry of observations, the public may also use devices specifically designed to record the location of animal–vehicle collisions. For example, trucks in the Prince George area, British Columbia, Canada, were equipped with GPS, and each time a truck driver saw a moose (*Alces alces*) or deer (dead or alive), the location and time were recorded (Moneo 2006). Nowadays it is also possible to use cell phones with integrated GPS and equip them with software to record and transmit animal–vehicle collision data in real time. The latter option has much potential because a large percentage of the human population in North America have cell phones and regularly update them with newer models. In addition, by combining different functions in the same device most people carry with them anyway, investments in specific equipment are not necessary, and the burden of carrying multiple devices is eliminated. These advantages do

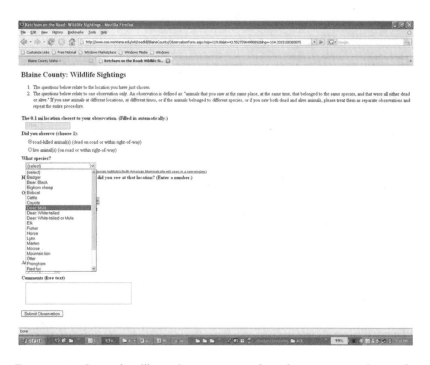

FIGURE 16.6. Screenshot illustrating easy-to-use drop-down menus and space for free-form comments.

not only benefit data collection programs that involve the public, but also data collection programs by agencies. The use of cell phones or other devices while driving is not encouraged though.

Animal Detection Systems

Animal detection systems are an example of a developing technology aimed at reducing animal–vehicle collisions (see also chapter 3). Animal detection systems use sensors to detect large mammals that approach the road. Once a large mammal has been detected, warning signs are activated urging drivers to slow down and be more alert (fig. 16.7). Since driver response depends on reliable warning signs, an effective animal detection system needs to detect large animals reliably. While many roadside animal detection systems experience reliability problems after installation (review in Huijser et al. 2006b), some systems are extremely reliable in detecting large ungulates

FIGURE 16.7. An animal detection system along Highway 160 near Durango, Colorado. (© Marcel Huijser)

(Dodd and Gagnon 2008, Huijser et al. 2009a). Minimum standards for reliability have been suggested (Huijser et al. 2009a) and are aimed at providing guidance to manufacturers of animal detection systems and agencies that may consider implementing a roadside animal detection system.

Data on the effectiveness of animal detection systems are few. Previous studies have shown variable results with regard to the effect of activated warning signs on vehicle speed:

- Substantial decreases in vehicle speed ($\geq$ to 5 kilometers per hour [$\geq$ 3.1 miles per hour]) (Kistler 1998, Muurinen and Ristola 1999, Kinley et al. 2003, Dodd and Gagnon 2008)
- Minor decreases in vehicle speed (< 5 kilometers per hour [< 3.1 mile per hour]) (Kistler 1998, Muurinen and Ristola 1999, Gordon and Anderson 2002, Kinley et al. 2003, Gordon et al. 2004, Hammond and Wade 2004)
- No decrease or even an increase in vehicle speed (Muurinen and Ristola 1999, Hammond and Wade 2004).

This variability in results is likely related to various conditions, including the type of warning signs, whether the warning signs are accompanied with advisory or mandatory speed limit reductions, road and weather conditions, whether the driver actually sees an animal, whether the driver is a local resident that may be familiar with the purpose and reliability of the system, possibly the length of road over which the animal detection system is installed and the zone to which the warning signs apply (the more location-specific the better), and possibly cultural differences that may cause drivers to respond differently to warning signs in different regions.

Minor reductions in vehicle speed may not seem meaningful, but the relationship between vehicle speed and the risk of a fatal accident (for humans) is exponential (Kloeden et al. 1997). This means that at a high vehicle speed a small decrease in speed results in a disproportionately large decrease in the risk of the severity of a potential accident. Thus a relatively small reduction in vehicle speed can be very important. However, the relationship between vehicle speed and the risk of fatal accidents has not specifically been tested with respect to large animals in rural areas.

The ultimate measure of the effectiveness of animal detection systems is whether they result in fewer collisions with large mammals. The number of vehicle collisions with roe deer (*Capreolus capreolus*) and red deer (*C. elaphus*) decreased by an average of 82 percent after seven detection systems were installed in Switzerland (Kistler 1998, Romer and Mosler-Berger 2003, Mosler-Berger and Romer 2003). All seven sites showed a reduction in collisions after an animal detection system was installed, and three of the seven sites had no collisions six to seven years after system installation. Collisions that occurred during the day when the systems were not active were excluded from the analysis. While sites with an animal detection system showed a strong reduction in the number of animal–vehicle collisions, the total number of animal–vehicle collisions in the wider region remained constant (Kistler 1998). This is further evidence that the reduction in collisions was indeed related to the presence of the animal detection systems and not the result of potential reductions of the ungulate populations or major changes in traffic volume and time of travel. Furthermore, detection data stored by the systems and tracking data confirmed that ungulates still frequented the sites (Mosler-Berger and Romer 2003).

Monitoring in other regions also suggests that these systems can decrease wildlife–vehicle collisions. In Arizona elk–vehicle collisions were reduced from 11.7 per year on average to 1.0 per year after an animal detection system was installed in a gap in an electric fence (one year of data post-installation). This was a 91 percent reduction in collisions with large animals (Dodd and Gagnon 2008). Similarly, a site in Montana showed that ungulate–vehicle collisions were 66.7 percent lower than before the system became operational (Huijser et al. in 2009c). The number of ungulate–vehicle collisions from the treatment section after the system became operational was 57.6 percent lower than in comparable control sections. However, study duration and sample size were low.

Anecdotal data from projects near Sequim, Washington, Clam Lake, Wisconsin, and Marshall, Minnesota, and preliminary results from Pinedale, Wyoming, also suggest success in using wildlife detection systems to decrease wildlife–vehicle collisions (Clam Lake Elk News 2007, CBS 2008, Dai et al. 2009, Shelly Ament, Washington Department of Fish and Wildlife, pers. comm., David Rubin, Sequim Elk Habitat Committee, pers. comm. Robert Weinholzer, Minnesota Department of Transportation, pers. comm.).

Animal detection systems may not only activate warning signs installed along a road, but could also send a warning signal to devices in cars, perhaps even cell phones equipped with GPS, when they are within a certain distance of the location of where detection has occurred. Furthermore, animal detection systems may also be mounted in cars (e.g., Bendix 2002, General Motors 2003, Hirota et al. 2004, Honda 2004, Omar and Zhou 2007, Bauer et al. 2008). However, the range of the detectors and warning time at high speeds in rural areas may cause technological challenges.

Influencing Driver Perception

While the reconstruction of rural two-lane roads typically leads to safer roads through wider lanes, shoulders, and clear zones, wildlife–vehicle collisions appear to increase rather than decrease after road reconstruction, probably as a result of an increase in design speed (Vokurka and Young 2008). One of the challenges is to not make drivers increase vehicle speed when presented with wider lanes, wider shoulders and clear zone, increased sight distance, and fewer curves. Making the lane width appear narrower through wider striping may result in lower vehicle speeds as it is more difficult to stay in between the lines at higher speeds. At the same time, the actual lane width has not changed and still makes for roads that are safer than narrower roads. Actual data on the effectiveness of wider striping and reduced vehicle speed and reduced wildlife–vehicle collisions are lacking at this time.

New Materials and Construction Techniques

Wildlife fencing in combination with wildlife underpasses and overpasses is considered a robust and highly effective measure to reduce wildlife–vehicle collisions while still allowing for connectivity for wildlife between areas separated by roads (see chapters 2 and 3). The most expensive component of this mitigation measure is wildlife overpasses, and this is one of the main reasons why relatively few have been built to date in North America. Dramatically reducing the construction and maintenance costs of wildlife overpasses could lead to a much wider implementation of this mitigation measure. If wildlife overpasses are also lightweight and modular they would also allow for greater mobility. This means that wildlife overpasses may be implemented at locations where they would not have been considered before, for example, because of the potential for changes in the landscape and associated changes in animal movements.

In recent years, bridges made of plastic (fiber-reinforced polymer composites) have been constructed and installed for vehicles, bicyclists, and pedestrians (BBC 2007, Lankhorst 2008, 2009, Toledo Blade 2008, Fiberline Composites 2009a, b). Sometimes the supporting beams are made of steel (Lankhorst 2008, Fiberline Composites 2009b). Some of these bridges consist of recycled plastics and require much less energy to produce with much less carbon output than concrete or steel bridges (The Economist 2005, Lankhorst 2008). Plastic bridges can be lighter and stronger than concrete bridges (BBC 2007), and they are expected to have greater longevity and require less maintenance than concrete or metal bridges, primarily because plastics do not suffer from corrosion caused by water and salt. Additionally, plastic bridges are estimated to require no maintenance in the first 50 to 60 years (The Economist 2005, Lankhorst 2008, Euro Graduate 2008), whereas concrete or steel bridges are estimated to require maintenance perhaps every 15 to 20 years (The Economist 2005, Euro Graduate 2008). Plastic bridges can also be assembled elsewhere, and can be delivered and installed more quickly than concrete or steel bridges, reducing costs further, including for traffic control (The Economist 2005).

In addition to cost reductions, plastic wildlife overpasses can be much more transportable than concrete or steel bridges. Plastic bridges are much lighter and can be modular, allowing for relatively quick disassembling and reassembling and transport should a location no longer have a certain minimum amount of wildlife use. This is important because currently wildlife overpasses are typically not considered unless the lands adjacent to the highway have been secured as wildlife habitat for at least the life span of the structure. Transportable wildlife overpasses could soften these require-ments, potentially increasing the number of locations where wildlife over-passes can be considered.

The weight that a wildlife overpass has to be able to support can poten-tially be much reduced if a wildlife overpass has no soil or vegetation, fur-ther reducing costs. If no soil or vegetation is present the animals may walk directly on the plastic surface. However, plastics are not necessarily slip-pery; some products may be substantially less slippery than wood (Lank-horst 2008). Cover may still be provided for by litter, tree stumps, or hol-low, imitation rocks made of fiberglass (e.g., for invertebrates and small mammals). Live trees may be grown in pots with an irrigation system. The absence of soil and vegetation on wildlife overpasses may affect the use by some species. Therefore the performance of such overpasses should be care-fully evaluated to ensure an attractive cost:benefit ratio.

Acknowledgments

The authors would like to thank Amanda Hardy for her efforts and ideas re-garding the application of handheld computers with integrated GPS in recording animal–vehicle collision data, Edgar van der Grift for the use of his illustrations, Tracy Lee for discussing citizen science and using the Inter-net to document wildlife seen on and adjacent to roads, and Sharon Mader for sharing her thoughts on the potential use of plastics for the construction of wildlife overpasses.

Chapter 17

The Way Forward: Twenty-first Century Roads and Wildlife Connectivity

JODI A. HILTY, JON P. BECKMANN,
ANTHONY P. CLEVENGER, AND
MARCEL P. HUIJSER

Global change and human expansion are occurring at a pace more rapid than at any other time in the history of humankind, and roads are a major component of human impacts across the world. Wildlife–vehicle collisions cause property damage, human injury, and possible human fatalities in addition to loss of wildlife. Roads also fragment once continuous wildlife populations. Our understanding of the impacts of roads on wildlife and approaches to mitigate these impacts is in its nascence.

While still an emerging field, the science of road ecology is expanding rapidly. Research on both terrestrial and aquatic species is providing insights into the likelihood of their use of different crossing structures and fish passages, and this is informing transportation and land management as they develop new approaches for helping wildlife cross roads safely. At the same time, mitigating wildlife–vehicle collisions is increasingly being prioritized by communities, nongovernmental organizations (NGOs), state/provincial, and federal agencies. As a result of these trends, the integration of concerns for wildlife into road construction and expansion is becoming more common, although not necessarily the norm that we argue it should be. The case studies presented in this book are set in unique and varied environments with different stakeholders and circumstances. Yet they share a number of common factors that likely provide a clear path to each project's successful road–wildlife mitigation effort. This chapter offers a reflection on the components that are likely to contribute to any project's successes.

Project versus System Level

A clear message that emanates from several of the case studies is the need to take a project- and systems-level approach. As was demonstrated in chapters 5 and 6, the success of a systems-level approach depends on ensuring that the right stakeholders are engaged throughout the process and that the process is transparent. Vermont's statewide linkage analysis approach emphasized how a systems approach may increase conservation effectiveness, connectivity, predictability, and transparency of projects from beginning to end (chapter 12). Such a prioritization also ensures that limited conservation dollars are spent where most needed. The Sonoran Desert Conservation Plan—developed with Pima County citizens, NGOs, and agency engagement—offers a site-specific example on how to move toward systems planning through integrating habitat conservation, transportation, and land-use planning across multijurisdictional levels (chapter 15). Such systematic approaches demonstrate the benefit of enabling states, NGOs, and collaborators to weigh project priorities across a state and direct investment toward the highest-need projects.

Though there are clearly benefits to systems planning, connectivity conservation actions occur one location at a time, meaning that engagement at specific priority sites is of utmost importance to achieve connectivity goals. For this reason—and that conservation sometimes stalls in creation and refinement at the planning stages rather than implementation—we believe that most resources should still ultimately be concentrated at the project level. That is not to say that these large context plans should not be developed, but rather they should be developed relatively quickly such that they can be easily updated and not be the major focus of ongoing human and financial investment.

The Role of Good Information

Science is viewed as an influential component that should underlay decision making. Interestingly, in the case studies science provided the inspiration for developing projects as well as guiding priorities. In the example of Arizona (chapter 13), monitoring elk movements provided the information that transformed an original partial fencing design to a more extensive fencing design despite initial resistance. Also, the wildlife monitoring in Tijeras Canyon offered a benchmark of success demonstrating that project goals are being achieved and reinforcing community, NGOs, and agencies' perspectives. Such positive feedback provides incentive to engage in additional, related projects as several case studies indicate.

Science also offers several types of input for designing road-mitigation projects. First, preconstruction planning uses data on wildlife distributions, health, population dynamics, movements, and other behaviors within the area affected by road construction. Second, the latest engineering and ecological data regarding highway mitigation performance for both terrestrial and aquatic species can inform project design and reduce impacts on the ecosystem. Finally, follow-up monitoring and research examine if the mitigation has met stated goals and is functioning as intended so that project improvements can be made to ultimately achieve project goals (e.g., chapter 13).

High-profile science-based projects, such as research in Banff National Park, Alberta (chapter 7), have informed and inspired new efforts throughout North America and even worldwide. With many projects in this book underscoring a common struggle for funding, particularly for monitoring, the Banff research is valuable because it exemplifies the importance of long-term monitoring. This level of monitoring revealed that conclusions made on wildlife use of crossing structures a few years after construction would have been different from the now long-term monitoring results. A concerted monitoring effort also potentially enables examination of longer-term questions such as the link between highway mitigation, gene flow, and population viability. Ideally, all projects should monitor several levels of biological organization from genetic, species-population to community-ecosystem functions and processes. However, in reality, most projects will have to target their monitoring. The ecology and road transportation community should prioritize investment in a number of longer-term monitoring efforts to use as case studies that can inform future projects and test collective working assumptions.

Clear Goals

The goals of each of the constituents in any transportation mitigation project may differ. In some cases, transportation agencies may focus primarily on motorist safety issues, while state/provincial wildlife agencies, private citizens, and/or environmental NGOs may prioritize wildlife connectivity or other environmental issues. Prior to the start of any proposed mitigation project, a common understanding of the issues and effective communication between transportation agencies and other stakeholder groups, including other governmental agencies and NGOs, is necessary (e.g., chapter 15). Dialogues among partners need to articulate the main concerns, how to integrate them into the project, and what the expected mitigation will look like. Such an effort early on in the planning phase yields better collaboration and a mutually agreed upon direction in planning transportation projects that incorporate wildlife conservation measures. This approach is also likely to streamline projects and permitting processes, and may ultimately reduce costs. In our opinion, the failure to define clear goals is one of the most common causes of stalled projects and friction between partners.

The case studies in this book demonstrate that human safety often serves as the initial incentive and even the funding lever that moves the project forward. That said, the opportunity exists to increase motorist safety at the same time as creating well-designed wildlife crossing systems that improve wildlife connectivity. Most of the projects discussed throughout this book demonstrate such a win–win scenario because partners identified clear goals that they together could support.

Successful Partnerships

Perhaps one of the most consistent messages across case studies is that functioning partnerships drive projects to success. Some partnerships, such as in Vermont (chapter 12), are primarily between governmental agencies, but many other entities supported their planning efforts. In Vermont, strong citizen support of wildlife conservation contributed to agencies' ability to carry out road–wildlife mitigation projects. Also, nongovernmental agencies played a key role in inspiring and educating agency personnel through their wildlife training course.

Engagement with all potential stakeholders early and throughout the project is instrumental. In North Carolina (chapter 11), late engagement of the state wildlife agency on highway 64 meant that, despite recommendations, options for underpasses were limited. Later projects in North Carolina engaged the wildlife agency earlier and throughout projects, which contributed to improved wildlife crossing structures.

The inception of some case study projects required creating new alliances and even organizations, such as in the Tijeras Canyon example (chapter 9). While other projects demonstrate that existing regional alliances, such as in Pima County (chapter 15), can provide needed structure and support. In either case, alliances are a productive mechanism to ensure ongoing input from NGOs and citizens that facilitates dialogue with agencies engaged on projects. Several case studies also indicate that once alliances are successful on one project, they often build on their experience and engage together on new projects.

Mixed partnerships offer additional potential benefits. For example, citizens and nonprofits can engage in actions not appropriate for agencies such as in lobbying the state, raising project profile or issues in the local media, and coordinating community outreach (chapter 6). While additional partners undoubtedly require more coordination, a diversity of organizations generally seems to offer strengths toward moving the project forward. Conversely, lack of engagement of partners including the local communities may threaten a project's success.

Public Support

A clear force for moving many case study projects forward is public support. In Tijeras Canyon (chapter 9), early citizen involvement helped sway initially reticent transportation officials, and ultimately Tijeras Canyon public input was accepted as peer review of the proposed project. Similarly, project ideas for Highway 93 in Montana (chapter 8) did not progress until the Confederated Salish and Kootenai agreed to a set of goals with the Montana Department of Transportation. Ultimately, community engagement on the Highway 93 project altered the initial plans for improving the highway. The changes in the plan better matched community needs, including conserving their wildlife heritage through maintaining connectivity, while still achieving safety goals. In North Carolina, lack of initial public support—because of negative media that portrayed the project as wasteful governmental spending—stalled the project (chapter 11). Those committed to the project had to communicate real costs, ecological impacts, and human safety issues clearly to the public to move the project forward again. Government agencies often require public comment periods, but the organizations arguably best suited to garner public support are nonprofits and community-based groups.

Creation of coalitions is one way to garner organized public support as discussed above, and direct citizen engagement is another. The Road Watch program in Crowsnest Pass (chapter 14) provided an opportunity for community volunteers to engage directly on road matters as a means of increasing support of potential mitigation projects in the region. As the authors point out, such efforts are only one tool to increase awareness and involvement. One challenge is measuring the short- and long-term impact of such programs. That said it is hard to believe that such efforts would not increase awareness at some level.

Other Incentives

Necessary partners for a proposed project are not always initially eager. Case studies discussed incentives ranging from training, policy approaches, and regulatory incentives (North Carolina, chapter 11). This speaks to the need to ensure that all involved agencies recognize the importance of road ecology to their respective missions and cultivate the expertise and interest.

Transportation agencies that have not addressed wildlife connectivity in past projects engage in such "new" activities either because of a concerted effort within their entity or because of outside pressure. In Vermont, the transportation agency training course helped staff understand how to gauge the effects of roads on wildlife and habitat and was in high demand. Anecdotal reports of individuals who attended suggest that the class altered the way staff thought about the environment more generally (chapter 12).

Initial engagement on wildlife connectivity issues in road projects by agencies is sometimes resisted. Case studies described external pressure from one agency to another agency as being effective. It wasn't until the U.S. Forest Service refused to permit the second phase of reconstruction in Tijeras Canyon (chapter 9) that funding was found to support the research on the project, for example. At the project level, the importance of individuals cannot be underscored enough. One individual can make it very hard for a project to move forward, and one individual can also be the driving force that pulls the pieces together to make the project move. At the same time, making sure that many different groups are partnering and have access to good information is one of the apparent shared factors in the studies that moved projects toward completion. An individual or several individuals, often external to partnering agencies, are essential to guiding participants through the process to keep the project on course.

Long-Term Maintenance

It is clear that a persistent challenge across several projects is the lack of long-term monitoring and maintenance of structures, especially fences. The

Florida I-75 project (chapter 10) illustrates how broken fences, including human cutting of fencing, may pose an ongoing problem, compromising the long-term project success. In such a case, it may be that an intensified commitment to maintenance should be paired with continuing and even expanded community outreach to limit gaps, especially human-created holes, in the fencing. Addressing the long-term monitoring and maintenance plan at the project inception would alleviate some of these issues.

The Landscape Context

Over the longer term, highway wildlife mitigation projects need not only to address the site-based project but to place these projects in the context of the larger landscape and ecological processes. This means moving away from thinking linearly (i.e., the linear road corridor). Planning should occur with awareness of roads in the context of surrounding habitats, ecosystems, and ecosystem processes as well as the infrastructure network and land use (e.g., agriculture, urbanization, sprawl). This contextual approach will also assist in making decisions about what type of mitigation will best ensure safe wildlife crossing opportunities.

The broader Florida panther monitoring demonstrates the need for addressing the larger landscape context. The I-75 project nearly eliminated panther road-kill on that highway, but road-kill of panthers is still a source of mortality for these animals on other nearby roads. Failure to address this systemic issue could ultimately threaten the long-term success of I-75 project goals for panthers, although the crossing structures will still be important for other species. It may be appropriate to assess mitigation needs for panthers on other roads in the larger landscape now or in the future to conserve this species over the long term.

A useful way to understand the context of a proposed project within a larger landscape is systems planning, as discussed earlier. If systems planning does not yet exist, other resources exist that can be tapped into relatively easily and can assist in developing a rapid assessment of landscape-level issues. For example, state wildlife action plans as well as state efforts to develop statewide connectivity plans may provide at least a view into that larger landscape context (chapter 6).

Finally, road projects should incorporate an ecosystem approach. Currently, decisions are mostly based on human safety concerns and are heavily biased toward mitigating impacts to large ungulates. In the future, transportation engineers, ecologists, wildlife managers, planners, and decision makers should make decisions based on population viability for a wide range of species and consider indicators of ecosystem health and integrity. Such a broader view will ensure that project scopes are not too narrow and will address short-term species-specific needs and longer-term ecosystem integrity.

Next Steps

This book provides ample evidence for the need to continue to integrate wildlife concerns in the planning, design, and construction of future transportation projects. These examples offer lessons on how this is currently being achieved as well as discussing the many challenges. These case studies represent attempts to utilize the best available science and project experience, and are novel efforts that should be followed over time to judge how effective they are at achieving meaningful conservation goals.

As awareness about the need to maintain connectivity or reconnect wildlife populations across highways grows, so too does the need to support such projects and programs. We make five general recommendations that would move such efforts forward:

1. *Policies*: Clear policies and transparent guidelines that consider and incorporate wildlife concerns into transportation planning at the local, state/provincial, and federal levels are needed.

2. *Ecological research*: Expanded wildlife and road ecology research that tests existing hypotheses and refines our understanding of how wildlife respond to roads, and the mitigation measures designed for them should guide future projects. Long lists of research needs could be compiled by gathering recommendations from existing papers or by convening transportation agencies, but we believe that some of the top priorities include (a) more rigorous and longer-term monitoring ($\geq$ 3 years) of wildlife use of crossing structures and study designs that will provide results with high inferential strength (i.e., using a before- and after-control-impact or control-impact assessment; (b) expanding our knowledge of not only how individuals within populations use linkage habitats, but how their populations benefit in terms of demographics (movements) and genetics (gene flow); (c) testing and developing more cost-effective monitoring approaches and measures of performance; (d) developing models that identify habitats that will best maintain the ability

of wildlife to move between core habitats and across roads through crossing structures, and validate the models with independent data; and (e) developing tools that allow land managers and stakeholders to compare habitat linkage design to alternative designs that minimize costs or achieve other ecosystem goals.

3. *Engineering*: Incentives are needed to (a) explore creative new approaches, materials, and designs that address the fundamentals of transportation engineering and road ecology; (b) increase the number of potential solutions for cost-effective, innovative crossing designs that can be replicated or modified for widespread use in other locations; (c) engage engineering and design professionals and students in the interdisciplinary nature of road ecology through informal and formal training and education.

4. Agencies should dedicate and increase funding for restoring wildlife connectivity across highways at local, state/provincial, and federal levels. Wildlife crossing projects are expensive and without proper appropriations, it will be difficult to move forward many projects in a timely way.

5. *Public support*: The public needs to understand the societal and environmental benefits of road mitigation strategies. A well-informed public can influence political decisions in local communities affected by transportation or suburban development projects. Greater awareness and transfer of technology need to be conveyed to local decision makers in areas where critical decisions are being made as to whether expanding road projects are mitigated for wildlife connectivity. Far too often, the public as well as local transportation practitioners do not fully understand the current science and benefits accrued from properly designed mitigation.

Conclusions

During the past 30 years, the environmental impacts of transportation have been addressed primarily through policy initiatives, planning and analysis, new programs, and new technologies. Wildlife conservation and habitat connectivity concerns have typically received little attention by transportation agencies until the last decade because the primary concern of most transportation agencies has been regulatory compliance with federal and state laws (e.g. in the United States, the National Environmental Policy Act and Endangered Species Act). Traffic and roads are strongly implicated in many of the major environmental problems we face today: air and water pollution, fragmented natural habitats, wildlife and biodiversity losses, and urban sprawl. During the next 25 years, significant growth and changes in North America's population and economy are expected to occur. The U.S. population alone is anticipated to increase to 438 million by 2050, more than a 40 percent increase from the 2008 population of 304 million. Both will undoubtedly require increasing infrastructure support, including roads. The impacts of roads on natural environments and the means of mitigating their effects are undoubtedly one of the most important land and wildlife conservation challenges of this new century.

Healthy and well-functioning ecosystems are vital to the protection of our diverse biological resources, and to sustaining the economies and communities that rely on their products and benefits. Sustainable transportation systems must provide effectively for healthy ecosystems and safe and efficient human mobility. The environmental planning process requires multidisciplinary involvement and input from review/permitting agencies. Federal and state/provincial transportation agencies have recognized that ecosystem approaches and early stakeholder involvement in identifying issues and areas of concern are essential if their projects are to be environmentally sustainable and streamlined. Therefore, partnering and collaborative approaches, like we have seen in this book, are essential when developing ecosystem and habitat conservation initiatives.

The future holds great promise for new and improved transportation planning. This planning can look at the past, present, and beyond, by envisioning future conditions under which ecological, economic, and social factors are integrated. However, this cannot be realized without political and agency support. The emerging principles of road ecology are providing useful guidelines and best practices for mitigating road impacts on ecological connectivity, but they ultimately need to become embedded in federal and state/provincial administrative policies and legal frameworks. The need for more science-based knowledge for decision making is urgent and unprecedented, as a vigorous transportation program is being carried out across the land. This will provide a sound scientific basis for effective planning, policy, and implementation.

There is progress being made to connect agencies, stakeholders, and other concerned citizens with a keen interest in working toward a new way of transportation planning and systems design. There is a growing trend among agencies, communities, and academia in accepting and implementing crossing structures to mitigate road impacts to wildlife populations. Only 10 years ago, there was little interest by transportation agencies in mitigating roads for wildlife (other than from a motorist safety objective), and the public was not well informed about societal and environmental benefits of road mitigation. As little as 15 years ago, environmental sciences and transportation engineering were completely isolated disciplines that never connected or communicated. The eighth International Conference on Ecology and Transportation (September 13–17, 2009 in Duluth, Minnesota) brought together more than 400 participants from transportation agencies, fish and wildlife agencies, NGOs, and engineering firms for five days to learn about many case studies in addition to the most recent advances in operating transportation systems in a landscape context. The conference was initiated in 1995 by a state transportation agency and primary sponsor in the U.S. Federal Highway Administration.

Where will road ecology and safe passages for wildlife be in the next 10 years? We hope that examples such as those shown in this book become commonplace throughout North America and that there will be new case studies to report on and newly conceived mechanisms in place to implement landscape-based conservation measures in and around roads. These projects should help populations of turtles, grizzly bear, elk, and many other species continue to thrive. We hope this book will serve as an inspiration and motivator for others to now move forward and make the need for another "Safe Passages" book possible in 2020.

JOHN M. AUSTIN is a senior wildlife biologist with the Vermont Fish and Wildlife Department. He is the coordinator for the department's habitat conservation programs and has worked for many years with the Vermont Agency of Transportation and Federal Highway Administration on various wildlife crossing projects in Vermont and New England. John has a BS degree in wildlife biology from the University of Vermont and an MS degree in environmental law and policy from the Vermont Law School.

PATRICK B. BASTING received his BS in forestry from the University of Montana in 1987. Over the past 15 years Pat has been a district biologist with the Montana Department of Transportation. Since the mid- to late 1990s he has written and submitted the first wetland mitigation banking prospectus in Montana to the Corps of Engineers, worked on several stream and river restoration projects, and been involved in various aspects of wetland mitigation. When he transferred into the Missoula District (summer 2000) wildlife connectivity issues were gaining momentum and recognition. Since that time he has been heavily involved in the entire spectrum of transportation/wildlife issues, working on placement, design, construction, and monitoring of more than eighty wildlife crossings in western Montana.

DALE M. BECKER completed his undergraduate degree in wildlife biology at the University of Montana in 1980 and his graduate work at the University of Montana in 1984. His early background in wildlife work focused on raptor biology and management. His work as the Tribal Wildlife Program Manager for the Confederated Salish and Kootenai Tribes of the Flathead Indian Reservation in western Montana since 1989 has provided him with a diverse variety of experience, including general wildlife and habitat management, rare species management and reintroduction, human dimensions in wildlife management, interagency cooperative management, and wildlife impact assessment and mitigation planning. He received his title as a Certified Wildlife Biologist from The Wildlife Society in 2000, and he has served as president of the Montana Chapter of The Wildlife Society. He received the Wildlife Biologist of the Year Award in 2004 from the Montana Chapter of The Wildlife Society.

JON P. BECKMANN is an Associate Conservation Scientist for the Wildlife Conservation Society and an affiliated professor at Idaho State University. Jon earned a BS degree in wildlife and fisheries biology from Kansas State University and a PhD in ecology, evolution, and conservation biology from the University of Nevada–Reno. He has more than 15 years of experience in wildlife research and conservation working on species ranging from black bears, mountain lions, and pronghorn, to small mammals and shorebirds, and has published more than twenty articles in peer-reviewed scientific journals. In 2004, he was nominated by his peers for the Alan T. Waterman Award, the most prestigious award from the National Science Foundation for scientists under the age of 35. His primary interests are population ecology and the impacts of anthropogenic factors on behavior of mammals, particularly carnivores. Most of his research focuses on conservation issues of large mammals in the Greater Yellowstone Ecosystem and other areas of western North America.

MATTHEW D. BLANK is a research assistant professor at the Western Transportation Institute and the Department of Civil Engineering at Montana State University. Matt earned his MS and PhD degrees in civil engineering at Montana State University. He earned his BS in geological engineering at the University of Wisconsin–Madison. His research focuses on the interactions of roads and riparian corridors with an emphasis on aquatic connectivity. He also does water resource consulting work with OASIS Environmental, Inc.

CAROLYN CAMPBELL is the executive director of the Coalition for Sonoran Desert Protection, an organization she helped found in 1997. She has worked on environmental policy issues for over 20 years, as a legislative aide to both Congressman Morris K. Udall (D-AZ) and Tucson City Council member Molly McKasson. Carolyn was a member of Pima County's Sonoran Desert Conservation Plan Steering Committee, and currently serves on citizen stakeholder committees for both the City of Tucson's and Town of Marana's habitat conservation plans. She serves as vice-chair of Pima County's Bond Advisory Committee and has been involved in transportation planning as a member of the Pima Association of Governments' Regional Task Force for Long Range Transportation Planning and the 2006 Regional Transportation Authority's Citizen Advisory Committee.

ANTHONY P. CLEVENGER is a senior wildlife research scientist at the Western Transportation Institute at Montana State University. He has been with the institute since 2002. Beginning in 1996, Tony directed research assessing the performance of mitigation measures designed to reduce habitat fragmentation on the Trans-Canada Highway in Banff National Park, Alberta. Tony served as a member of the U.S. National Academy of Sciences Committee on Effects of Highways on Natural Communities and Ecosystems. He has published more than forty articles in peer-reviewed scientific journals and has coauthored three books, including *Road Ecology: Science and Solutions*. Tony has a master's degree from the University of Tennessee, Knoxville and a doctoral degree in zoology from the University of León, Spain.

DAVID R. COX supervises the Technical Guidance Section within the Division of Inland Fisheries for the North Carolina Wildlife Resources Commission. He supervises eight biologists across North Carolina that review and comment on an average of 3,100 development projects yearly. The comments provided by the Technical Guidance Section focus on the conservation of terrestrial and aquatic habitats. At the time of the research for the contributed chapter, he was the Eastern Region Highway Project Coordinator and worked closely with regulatory agencies and the North Carolina Department of Transportation to get the wildlife crossings included in the U.S. 64 project. He earned BS degrees in chemistry and biology from the University of North Carolina at Greensboro. David has a background in surveying, heavy construction, and monitoring of airborne and waterborne pollutants.

NORRIS L. DODD recently retired from the Arizona Game and Fish Department after 29 years in various capacities. As a wildlife research biologist for the past 14 years, he focused on wildlife relationships to forest restoration and highways. His extensive highway research helped raise recognition of the ability to meet both transportation and ecological objectives in highway reconstruction. Norris is now employed by AZTEC Engineering as a senior natural resource specialist/project manager. He received BS and MS degrees from Arizona State University. He is past president of the Arizona chapter of The Wildlife Society, sits on several nonprofit organization boards, the Pinetop-Lakeside Town Council, and the Governor's Forest Health Council.

DANAH DUKE is the executive director of the Miistakis Institute. She oversees multiple transboundary research projects focused on addressing sustainable land use throughout the Crown of the Continent. Danah completed her MSc in environmental biology and ecology at the University of Alberta.

ELIZABETH FLEMING is the Florida representative for Defenders of Wildlife. She develops conservation objectives and strategies and works with partners to protect and restore Florida's wildlife and their habitats and establish a state ecological network. Her areas of expertise include Florida imperiled species policies, conservation of wide-ranging species, such as the Florida panther and Florida black bear, and conservation planning. Elizabeth received a bachelor's degree in political science and a minor in biology from Tufts University.

ADAM T. FORD is a wildlife research associate working with the Western Transportation Institute from Montana State University and is based in Banff National Park, Alberta, Canada. He is conducting research on the effects of roads on wildlife populations within Banff, as well as monitoring wildlife movement through crossing structures along the Trans-Canada Highway. Adam has worked for government and nonprofit agencies on wildlife and ecology research issues for 10 years. He graduated from the geography program at the University of Victoria, British Columbia, and earned a master's in biology from Carleton University in Ottawa.

JEFFREY W. GAGNON is a research biologist for the Arizona Game and Fish Department and has worked for the department for over 10 years, primarily on wildlife–highway interaction research projects, including Arizona State Route 260 and U.S. Highway 93. He completed his undergraduate and master's degrees at Northern Arizona University studying the effects of traffic volumes on elk movement and has expanded his research to the evaluation of traffic on various wildlife species throughout Arizona.

DOUG E. GALARUS has 20 years of experience in information technology development, testing, implementation, and management. He has extensive experience as the project leader for mobile data communications systems, database-driven Web sites, Web site design, desktop applications, kiosk development, PDA and Tablet PC–based development, and interactive CD-ROMs. At the Western Transportation Institute, he has applied his technical expertise to the development of specific applications for transportation safety, including improved tools for road weather management and road ecology. Mr. Galarus holds master's degrees in computer science and mathematics education, and is program manager for the Systems Engineering,

Development and Integration Program at the Western Transportation Institute.

FORREST M. HAMMOND is currently a wildlife biologist with the Vermont Fish and Wildlife Department. He works primarily doing environmental reviews of development projects and is chair of the department's wildlife habitat team. He also serves on their deer, moose, and bear management teams. He received both a BS in zoology and an MS in wildlife management from the University of Wyoming, where he studied the food habits of black bears in the Salt River Range of Wyoming. He worked as a district wildlife biologist and then as the grizzly bear biologist for the Wyoming Game and Fish Department before returning to his home state of Vermont.

AMANDA HARDY obtained her BS and MS in fish and wildlife management from Montana State University. As the first ecologist hired at the university's Western Transportation Institute in 2001, she oversaw several road ecology projects including facilitating the development of the Integrated Transportation and Ecosystem Enhancements for Montana process. She has served as a member of the National Research Council Transportation Research Board's Ecology and Transportation Committee. Amanda is currently pursuing her PhD at Colorado State University in the graduate degree program in ecology. Her dissertation research is focused on understanding interactions between wildlife, transportation systems, and visitors in national parks.

JODI A. HILTY is the director of the Wildlife Conservation Society's North America Program. She oversees a staff of approximately fifty individuals working to conserve the last wild places in North America and the wildlife inhabiting these landscapes. Trained as a conservation biologist at the University of California, Berkeley, her work addresses thresholds of human impact on biodiversity, and ensuring that science guides policy and practice. She has authored or coauthored more than a dozen scientific and popular articles and is lead author of a 2006 book titled *Corridor Ecology: The Science and Practice of Linking Landscapes for Biodiversity Conservation*.

MARCEL P. HUIJSER is a research road ecologist with over 15 years of experience in both Europe and the United States working as an applied ecologist, specializing in road–wildlife interactions with an emphasis on evaluating and documenting methods and technologies that reduce wildlife–vehicle collisions. Dr. Huijser is the co-chair for the Subcommittee on Animal–Vehicle Collisions and served as a committee member of the

National Academies' Transportation Research Board's Committee on Ecology and Transportation through 2006.

DEBORAH JANSEN is a wildlife biologist for the National Park Service who has worked in Big Cypress National Preserve for the past 28 years. She is the NPS Panther Capture Team leader and has represented the preserve in all aspects of panther recovery. In addition, she has conducted research and monitoring of red-cockaded woodpeckers, Big Cypress fox squirrels, bald eagles, wood storks, and white-tailed deer. She received her bachelor's degree in biology from the University of Wisconsin–Eau Claire and her master's degree in wildlife from the University of Wisconsin–Stevens Point.

MARK D. JONES is a Supervising Wildlife Biologist with the Private Lands Program for the North Carolina Wildlife Resources Commission. He supervises habitat management programs for private lands throughout North Carolina. At the time of the research covered in the contributed chapter, Mark was the Black Bear Project Leader for the Surveys and Research Section of the North Carolina Wildlife Resources Commission. He earned a BS in wildlife management from Virginia Tech and an MS in wildlife and fisheries science from the University of Tennessee–Knoxville. Mark has a background in research and management issues throughout North Carolina with a heavy focus on habitat-related issues for diverse species from bobwhite quail and songbirds to black bears and other game species.

KATHLEEN KENNEDY is the program associate for the Coalition for Sonoran Desert Protection. She holds a BS in geology from Northern Arizona University and an MS in environmental studies from the University of Montana. At the Coalition, Kathleen works on a variety of local conservation projects in Pima County, Arizona. These projects address issues such as water resources conservation, endangered species protection, wildlife linkage preservation, and riparian habitat restoration. Kathleen serves on environmental advisory committees for local jurisdictions and provides written comments and recommendations on conservation policy initiatives.

ANGELA V. KOCIOLEK is a research scientist in the Road Ecology Program Area of the Western Transportation Institute–Montana State University. She holds a master's degree in biological sciences from Montana State University–Bozeman and has varied field biology experience. As a Peace Corps volunteer in northeast Thailand, she focused on environmental education and community outreach. Angela's multidisciplinary background lends itself to her current position in the field of road ecology.

TRACY LEE is a senior project manager with the Miistakis Institute, a research institute affiliated with the University of Calgary. Tracy completed her master's degree at the University of Calgary on the Road Watch program and continues as the project manger.

PAT T. MCGOWEN obtained his BS and MS in civil engineering from Montana State University, and his PhD in transportation systems engineering from the University of California Irvine. He has been a licensed professional civil engineer in Montana since April 2000. He is an assistant professor with the Western Transportation Institute and the Civil Engineering Department at Montana State University, where he has worked on projects relating to rural ITS, transportation impacts to wildlife, safety, and travel and tourism. He has been involved in projects including the Roadside Animal Detection System Testbed, the National Wildlife Vehicle Collisions Study, and Habitat Connectivity and Rural Context Sensitive Design. He is the founder and co-chair of the Transportation Research Board subcommittee on animal vehicle collisions (ANB20-2). Along with other colleagues at the Western Transportation Institute, Pat was awarded the 2008 Best of ITS Award for Best New Innovative Practices for Partnerships for Deploying Animal Vehicle Crash Mitigation Strategies.

KURT A. MENKE is a geographic information system specialist who has been working in the field for over a decade. He founded Bird's Eye View to apply his expertise with GIS technology toward solving the world's mounting ecological and social problems. Kurt has worked with numerous community and environmental organizations. He also volunteers his time for many organizations. Kurt has served as president of the New Mexico Geographic Information Council and is on the board of trustees for the Grand Canyon Wildlands Council. He is one of the cofounders of the Tijeras Canyon Safe Passage Coalition and currently serves as chairman. Kurt also teaches GIS and cartography at the University of New Mexico's Division of Continuing Education, and is a member of the Society for Conservation GIS. Kurt is certified as a GIS professional through the GIS Certification Institute.

MICHAEL QUINN is an associate professor in the faculty of environmental design at the University of Calgary. His research and teaching are focused on maintaining resilient landscapes in the Crown of the Continent Ecosystem.

KATHY RETTIE is a social scientist for the Parks Canada Agency and an adjunct professor in geography at the University of Calgary. Her work focuses on the human aspects of protected areas, in particular Canada's mountain national parks. From 2005 to 2008 she was a project manager for the highways research and monitoring in Banff National Park. As such her role included raising public awareness of the effectiveness of the wildlife crossings structures on the Trans-Canada Highway.

KRISTA SHERWOOD is a professional transportation planner assisting Big Cypress National Preserve as a National Park Foundation Transportation Scholar. In joining the effort to help resolve transportation-related challenges in national parks, her transportation projects include protection of the endangered Florida panther and other wildlife from highway mortality. She received her Master of Urban Planning degree from the University of Utah and a BS degree in sociology from Texas Woman's University.

CHRIS SLESAR is an Environmental Specialist with the Vermont Agency of Transportation. He has a BA from Rutgers College and an MA in environmental studies from Antioch University Seattle. Chris is chair of the Monkton Conservation Commission, on the board of the Lewis Creek Association, and a member of the Vermont Reptile and Amphibian Scientific Advisory Committee.

FRANK T. VAN MANEN is a research ecologist with the U.S. Geological Survey's Southern Appalachian Research Branch in Knoxville, Tennessee. He holds an adjunct professor appointment with the Department of Forestry, Wildlife and Fisheries at the University of Tennessee. Frank earned an MS degree in biology from Wageningen University in the Netherlands and a PhD in ecology from the University of Tennessee. Frank blends his research interest in mammals, particularly carnivores, with landscape ecology. Much of his research is devoted to predicting species distributions and habitat use, determining wildlife responses to landscape changes, landscape genetics, and population estimation.

DEB WAMBACH moved to Helena, Montana in 1997 after graduating from the University of Wisconsin–Madison with a BA in conservation biology and a BS in wildlife management and ecology. She has since been employed as a district biologist with the Montana Department of Transportation. Her duties include biological resource analysis, threatened and endangered species coordination, wetland and aquatic resource delineations and assessments, wildlife and aquatic species connectivity, and impact analysis and mitigation. She serves as the Integrated Transportation and Ecosystem Enhancements for Montana Pilot Study project manager and is an active member of several interagency collaborations, advisory committees, and working groups.

MARK L. WATSON has been a habitat specialist with the Conservation Services Division of the New Mexico Department of Game and Fish for 12 years. Mark's primary duties include environmental review of terrestrial projects on public lands, assisting state and federal agencies such as the New Mexico Department of Transportation, Department of Defense, and U.S. Forest Service to implement mitigation to minimize adverse effects on wildlife and habitats. Mark is a department representative on the New Mexico Endemic Salamander Team, is the counter for the annual statewide aerial waterfowl surveys, and assists with state-listed species programs such as aerial Mexican wolf monitoring and conducting rare bird surveys. Mark received a BS degree in biology, with emphases in ecology and zoology, from the University of New Mexico in Albuquerque.

PATRICIA A. WHITE began Defenders' Habitat and Highways Campaign in 2000 to address impacts of highways on our nation's wildlife and to encourage transportation and community planning that incorporates wildlife conservation. Her first report, *Second Nature: Improving Transportation Without Putting Nature Second* was awarded the 2004 NRCA Award of Achievement for best publication. Trisha is a member of the International Conference on Ecology and Transportation steering committee, a founding member of the Transportation Research Board Committee on Ecology and Transportation, and proud founder of the TransWild Alliance. Prior to Defenders, Trisha spent three years with World Resources Institute's Biological Resources program and one year as environment policy consultant to USAID's Global Environment Center. In 2000, she received her master's degree in environment and resource policy from the George Washington University.

TRAVIS W. WILSON is a biologist within the Technical Guidance section of the Division of Inland Fisheries for the North Carolina Wildlife Resources Commission. As the Eastern Region Highway Project coordinator, he is responsible for assessing impacts from transportation projects and their effect on the fish and wildlife resources of North Carolina. Tasks include working with the North Carolina Department of Transportation, regulatory agencies, and other stakeholders in the planning, design, construction, and maintenance of wildlife passage structures to address habitat fragmentation and highway permeability issues for aquatic and terrestrial wildlife passage. He earned a BS in aquaculture, fisheries, and wildlife biology with a minor in forestry from Clemson University. To date, his coordination efforts have included projects for large and small mammals, herpetofauna, and diadromous fish species.

LITERATURE CITED

Adams, L. W., and A. D. Geis. 1983. Effects of roads on small mammals. *Journal of Applied Ecology* 20:403–415.

Alexander, S. M., N. M. Waters, and P. C. Paquet. 2004. A probability-based GIS model for identifying focal species linkage zones across highways in the Canadian Rocky Mountains. Pages 223–255 in G. Clarke and J. Stillwell, eds. *Applied GIS and Spatial Modeling*. University of Leeds, United Kingdom.

Allen, R. E., and D. R. McCullough. 1976. Deer–car accidents in southern Michigan. *Journal of Wildlife Management* 40:317–325.

Ament, R., D. Galarus, H. Richardson, and A. Hardy. 2007. *Roadkill Observation Collection System (ROCS). Development of an integrated Personal Digital Assistant (PDA) with a Global Positioning System (GPS) to gather standardized digital information*. Western Transportation Institute, Montana State University, Bozeman, Montana.

Andreassen, H. P., H. Gundersen, and T. Storaas. 2005. The effect of scent-marking, forest clearing, and supplemental feeding on moose–train collisions. *Journal of Wildlife Management* 69(3): 1125–1132.

Apps, C., J. L. Weaver, P. C. Paquet, B. Bateman, and B. N. McLellan. 2007. Carnivores in the southern Canadian Rockies: Core areas and connectivity across the Crowsnest Highway. Wildlife Conservation Society, Canada. Conservation Report No. 3.

Aresco, M. 2005. Mitigation measures to reduce highway mortality of turtles and other herptofauna at a north Florida lake. *Journal of Wildlife Management* 69:549–560.

Arizona Department of Commerce. 2008. Growing Smarter Legislation. Accessed October 28, 2008, at http://www.azcommerce.com/CommAsst/GrowSmart/Growing+Smarter+Legislation.htm.

Arizona Department of Transportation. 2006. Arizona wildlife linkages assessment. Final report of the Arizona Wildlife Linkages Workgroup. Arizona Department of Transportation, Office of Environmental Services, Phoenix, Arizona.

———. 2008a. Average Annual Daily Traffic (AADT). Accessed March 13, 2009, at http://tpd.azdot.gov/data/aadt.php.

———. 2008b. 2007 Highway Performance Management System Area Wide Tables 1 and 2. Accessed April 6, 2009, at http://mpd.azdot.gov/data/documents/Areawide2007_1_2.pdf.

———. 2009. Arizona Wildlife Linkages Assessment. Accessed March 11, 2009, at http://www.azdot.gov/Highways/OES/AZ_WildLife_Linkages/assessment .asp.

Au, J., P. Bagchi, B. Chen, R. Martinez, S. a. Dudley, and G. J. Sorger. 2000. Methodology for public monitoring of total coliforms, *Escherichia coli* and toxicity in waterways by Canadian high school students. *Journal of Environmental Management* 58:213–230.

Austin, J., C. Alexander, E. Marshall, F. Hammond, J. Shippee, and E.Thompson. 2004. *Conserving Vermont's natural heritage*. VT Fish and Wildlife Department and Agency of Natural Resources.

Baker, C. O., and F. E. Votapka. 1990. Fish passage through culverts. FHWA-FL-09-006. USDA Forest Service—Technology and Development Center. San Dimas, California.

Banff–Bow Valley Study. 1996. Banff–Bow Valley: At the crossroads. Summary report for the Banff–Bow Valley Task Force. Canadian Heritage, Ottawa, Ontario.

Barlow, C. 1997. Performance evaluation of wildlife reflectors in British Columbia. Pages 60–64 in A. P. Clevenger and K. Wells, eds. *Proceedings of the second Roads, Rails, and the Environment workshop*. Parks Canada, Banff National Park, Alberta and Columbia Mountains Institute of Applied Ecology, Revelstoke, Canada.

Barnum, S. A. 2003. Identifying the best locations along highways to provide safe crossing opportunities for wildlife: A handbook for highway planners and designers. Report to Colorado Department of Transportation Research Report CDOT-DTD-UCD-2003-9.

Barton, C., and K. Kinkead. 2005. Do erosion control and snakes mesh? *Journal of Soil and Water Conservation* 60(2): 33–35

Bauer, G., F. Homm, L. Walchshäusl, and D. Burschka. 2008. Multi spectral pedestrian detection and localization. Pages 21–35 in J. Valldorf and W. Gessner, eds. *Advanced microsystems for automotive applications 2008*. Springer, Berlin, Germany.

Baxter, C. V. and F. R. Hauer. 2000. Geomorphology, hyporheic exchange, and selection of spawning habitat by bull trout (*Salvelinus confluentus*). *Canadian Journal of Fisheries and Aquatic Sciences* 57:1470–1481.

BBC. 2007. Plastic bridge first wins award. June 13, 2007. Available at http:// news.bbc.co.uk/2/hi/uk_news/england/lancashire/6748401.stm.

Beaudry, F., P. G. deMaynadier, and M. L. Hunter Jr. 2008. Identifying road mortality threat at multiple spatial scales for semi-aquatic turtles. *Biological Conservation* 141(10): 2550–2563.

Becker, D. M. 1996. Wildlife and wildlife habitat impact issues and mitigation options for reconstruction of U.S. Highway 93 on the Flathead Indian Reservation. 16 pp. in G. L. Evink, P. Garrett, D. Ziegler, and J. Berry, eds. *Trends in Addressing Transportation Related Wildlife Mortality—Proceedings of the Trans-*

portation Related Wildlife Seminar. Florida Department of Transportation. Tallahassee, Florida.

Becker, D. M., D. J. Lipscomb, and A. M. Soukkala. 1993. Evaro Corridor wildlife use study and mitigation recommendations. U.S. Highway 93 Reconstruction Project. Final Report. Confederated Salish and Kootenai Tribes. Pablo, Montana.

Beckmann, J. P., and J. Berger. 2003. Using black bears (*Ursus americanus*) to test ideal-free distribution models experimentally. *Journal of Mammalogy* 84:594–606.

Beckmann, J. P., and C. W. Lackey. 2008. Carnivores, urban landscapes, and longitudinal studies: A case history of black bears. *Human–Wildlife Conflicts* 2:168–174.

———. 2004. Are desert basins effective barriers to movements of relocated black bears (*Ursus americanus*)? *Western North American Naturalist* 64:269–272.

Behlke, C. E., D. L. Kane, R. F. Mclean, and M. D. Travis. 1991. Fundamentals of culvert design for passage of weak swimming fish. FHWA-AK-RD-90-10. Fairbanks, Alaska.

Beier, P. 1995. Dispersal of juvenile cougars in fragmented habitat. *Journal of Wildlife Management* 59:228–237.

Beier, P., D. R. Majka, and W. D. Spencer. 2008. Forks in the road: Choices in procedures for designing wildland linkages. *Conservation Biology* 22:836–851.

Beier, P., K. L. Penrod, C. Luke, W. D. Spencer, and C. Cabañero. 2006. South Coast missing linkages: Restoring connectivity to wildlands in the largest metropolitan area in the USA. Pages 555–586 in K. R. Crooks and M. Sanjayan, eds. *Connectivity conservation*. Cambridge University Press, Cambridge.

Belden, R. C., W. B. Frankenburger, and J. C. Roof. 1991. Florida panther distribution. Final Report, Study Number: 7501, Federal Number E-1 II-E-1.

Belford, D. A and W. R. Gould. 1989. An evaluation of trout passage through six highway culverts in Montana. *North American Journal of Fisheries Management* 9:437–445.

Bell, M. C. 1991. Fisheries handbook of engineering requirements and biological criteria. Fish Passage Development and Evaluation Program. U.S. Army Corps of Engineers, North Pacific Division. Portland, Oregon.

Bellis, E. D., and H. B. Graves. 1978. Highway fences as deterrents to vehicle–deer collisions. *Transportation Research Record* 674: 53–58.

Bellis, M, S. Jackson, C. Griffin, and P. Warren. 2008. Evaluating the effectiveness of wildlife passage structures on the Bennington Bypass. Final Rept. 113 pp. University of Massachusetts at Amherst.

Bender, H. 2001. Deterrence of kangaroos from roadways using ultrasonic frequencies: efficacy of the Shu Roo. University of Melbourne, Department of Zoology report. Prepared for NRMA Insurance Limited, Royal Automobile Club of Victoria, Road Traffic Authority of South Wales and Transport South Australia, Australia.

Bendix Commercial Vehicle Systems. 2002. Bendix Xvision system service data. Available at http://www.bendix.com/downloads/195160.pdf.

Bennett, A. F. 1991. Roads, roadsides, and wildlife conservation: a review. Pages 99–117 in D. A. Saunders and R. J. Hobbs, eds. *Nature conservation 2: The role of corridors*. Surrey Beatty, Sydney, Australia.

Berger, J. 2004. The last mile: How to sustain long-distance migration in mammals. *Conservation Biology* 18:320–331.

———. 2007. Fear, human shields, and the re-distribution of prey and predators in protected areas. *Biology Letters* 3:620–623.

Berger, K. M., J. P. Beckmann, and J. Berger. 2007. Wildlife and energy development: Pronghorn of the Upper Green River Basin—Year 2 Summary. Wildlife Conservation Society, Bronx, New York.

Beringer, J., L. P. Hansen, J. A. Demand, J. Sartwell, M. Wallendorf, and R. Mange. 2002. Efficacy of translocation to control urban deer in Missouri: Costs, efficiency, and outcome. *Wildlife Society Bulletin* 30(3): 767–774.

Beringer, J. J., S. G. Seibert, and M. R. Pelton. 1990. Incidence of road crossing by black bears on Pisgah National Forest, North Carolina. *International Conference on Bear Research and Management* 8:85–92.

Bertwistle, J. 1997. Performance evaluation of mitigation measures in Jasper National Park, Alberta. Pages 65–71 in A. P. Clevenger and K. Wells, eds. *Proceedings of the second Roads, Rails and the Environment workshop*. Parks Canada, Banff National Park, Alberta and Columbia Mountains Institute of Applied Ecology, Revelstoke, Canada.

———. 1999. The effects of reduced speed zones on reducing bighorn sheep and elk collisions with vehicles on the Yellowhead Highway in Jasper National Park. Pages 89–97 in G. L. Evink, P. Garrett, and D. Zeigler, eds. *Proceedings of the third international conference on wildlife ecology and transportation*. FL-ER-73–99. Florida Department of Transportation, Tallahassee.

Biota Research and Consulting, Inc. 2003. Jackson Hole Roadway and Wildlife Crossing Study, Teton County, Wyoming. Final report for Jackson Hole Wildlife Foundation, Jackson, Wyoming.

Bissonette, J. A. 2006. Evaluation of the use and effectiveness of wildlife crossings. Final report for Project NCHRP 25-27 FY04, National Cooperative Research program, Transportation Research Board, Washington, DC.

———. 2007. Evaluation of the use and effectiveness of wildlife crossings. National Cooperative Highway Research Program (NCHRP) 25-27 final report. Transportation Research Board, Washington DC.

Bissonette, J. A., and W. Adair. 2008. Restoring habitat permeability to roaded landscapes with isometrically scaled wildlife crossings. *Biological Conservation* 141:482–488.

Bissonette, J. A., and P. Cramer. 2008. Evaluation of the use and effectiveness of wildlife crossings. National Cooperative Highway Research Program, Report 615. Transportation Research Board of the National Academies.

Bissonette, J. A., and M. Hammer. 2000. Effectiveness of earthen return ramps in reducing big game highway mortality in Utah. UTCFWRU Report Series 2000. No. 1:1–29. Utah State University, Logan.

Blank, M. 2008. Advanced studies of fish passage through culverts: 1-D and 3-D hydraulic modeling of velocity, fish energy expenditure, and a new barrier assessment method. PhD diss., Montana State University, Bozeman.

Bliss, J., G. Aplet, C. Hartzell, P. Harwood, P. Jahnige, D. Kittredge, S. Lewandowski, and M. Soscia. 2001. Community-based ecosystem monitoring. *Journal of Sustainable Forestry* 12:143–167.

Bollinger, E. K., and T. A. Gavin. 2004. Responses of nesting bobolinks (*Dolichonyx oryzivorus*) to habitat edges. *Auk* 121:767–776.

Boone, J. L., and R. G. Wiegert. 1994. Modeling deer herd management: Sterilization is a viable option. *Ecological Modeling* 72(3–4): 175–186.

Brandt, M. M., J. P. Halloway, C. A. Myrick, and M. C. Kondratieff. 2005. Effects of waterfall dimensions and light intensity on age-0 brook trout jumping performance. *Transactions of the American Fisheries Society* 134:496–502.

Brody, A. J., and M. R. Pelton. 1989. The effects of roads on black bear movements in western North Carolina. *Wildlife Society Bulletin* 17:5–10.

Brown, J. W. 2006. *Eco-Logical: An ecosystem approach to developing infrastructure projects*. Federal Highway Administration, Washington, DC. FHWA-HEP-06-011.

Brown, R. D., and S. M. Cooper. 2006. The nutritional, ecological, and ethical arguments against baiting and feeding white-tailed deer. *Wildlife Society Bulletin* 34(2): 519–524.

Brown, T. L., D. J. Decker, S. J. Riley, J. W. Enck, T. B. Lauber, P. D. Curtis, and G. F. Mattfeld. 2000a. The future of hunting as a mechanism to control white-tailed deer populations. *Wildlife Society Bulletin* 28(4): 797–807.

Brown, W. K., W. K. Hall, L. R. Linton, R. E. Huenefeld, and L. A. Shipley. 2000b. Repellency of three compounds to caribou. *Wildlife Society Bulletin* 28(2): 365–371.

Brownlee, L., P. Mineau, and A. Baril. 2000. Canadian Environmental Protection Act priority substances list: Supporting documents for road salts: Road salts and wildlife: An assessment of the risk. Report submitted to the Environmental Resource Group on Road Salts, Commercial Chemicals Evaluation Branch, Environment Canada: Hull, Quebec.

Bunn, A. G., D. L. Urban, and T. H. Keitt. 2000. Landscape connectivity: A conservation application of graph theory. *Journal of Environmental Management* 59:265–278.

Bunyan, R. 1990. Monitoring program of wildlife mitigation measures. Trans-Canada Highway twinning—Phase II. Final report to Parks Canada, Banff National Park.

Burford, D. D., T. E. McMahon, J. E. Cahoon, and M. Blank. 2009. Assessment of trout passage through culverts in a large Montana drainage during summer low flow. *North American Journal of Fisheries Management* 29:739–759.

Caro, T. M., L. Lombardo, A. W. Goldizen, and M. Kelly. 1995. Tail-flagging and other antipredator signals in white-tailed deer: New data and synthesis. *Behavioral Ecology* 6(4): 442–450.

Carroll, C. 2006. Linking connectivity to viability: Insights from spatially explicit population models of large carnivores. Pages 369–389 in K. Crooks and M. Sanjayan, eds. *Connectivity conservation*. Cambridge University Press, New York.

Carroll, C., R. F. Noss, and P. C. Paquet. 2001. Carnivore as focal species for conservation planning in the Rocky Mountain Region. *Ecological Applications* 11:961–980.

Cassady St. Clair, C. 2003. Comparative permeability of roads, rivers, and meadows to songbirds in Banff National Park. *Conservation Biology* 17:1151–1160.

Castro-Santos, T. 2005. Optimal swim speeds for traversing velocity barriers: an analysis of volitional high-speed swimming behavior of migratory fishes. *Journal of Experimental Biology* 208:421–432.

CBS. 2008. New ideas to stop car–deer crashes. 10 January 2008. Available at http://www.cbsnews.com/stories/2008/01/10/eveningnews/main3698120 .shtml.

Cerulean, S. 2002. Killer roads: Roadkill, habitat destruction and fragmentation count among the harms roads inflict on wildlife. *Defenders Magazine* (Winter).

Chetkiewicz, C-L. B., C. Cassady St. Clair, and M. S. Boyce. 2006. Corridors for conservation: Integrating pattern and process. *Annual Review of Ecology and Systematics* 37:317–342.

Chruszcz, B., A. P. Clevenger, K. Gunson, and M. Gibeau. 2003. Relationships among grizzly bears, highways, and habitat in the Banff–Bow Valley, Alberta, Canada. *Canadian Journal of Zoology* 81: 1378–1391.

Clam Lake Elk News. 2007. Clam Lake Elk News, July through September 2007, 7(3). Available at http://www.dnr.state.wi.us/org/land/wildlife/elk/q3.pdf.

Clark, J., and J. Murdoch. 1997. Local knowledge and the precarious extension of scientific networks: A reflection on three case studies. *Sociological Ruralis* 37(1): 38–60.

Clark, W. R., J. J. Hasbrouck, J. M. Kienzler, and T. F. Glueck. 1989. Vital statistics and harvest of an Iowa raccoon population. *Journal of Wildlife Management* 53:982–990.

Clay, C. 1995. *Design of fishways and other fish facilities*, 2nd ed. CRC Press. Boca Raton, Florida.

Clevenger, A. P. 2005. Conservation value of wildlife crossings: Measures of performance and research directions. *GAIA* 14:124–129.

Clevenger, A. P., B. Chruszcz, and K. Gunson. 2001a. Highway mitigation fencing reduces wildlife–vehicle collisions. *Wildlife Society Bulletin* 29:646–653.

———. 2001b. Drainage culverts as habitat linkages and factors affecting passage by mammals. *Journal of Applied Ecology* 38:1340–1349.

Clevenger, A. P., B. Chruszcz, K. Gunson, and J. Wierzchowski. 2002a. Roads and wildlife in the Canadian Rocky Mountain Parks—movements, mortality and mitigation. Final Report (October 2002). Report prepared for Parks Canada, Banff, Alberta.

Clevenger, A. P., A. Ford, and M. Sawaya. 2009. Banff wildlife crossings project: Integrating science and education in restoring population connectivity across transportation corridors. Final report to Parks Canada Agency, Radium Hot Springs, British Columbia, Canada. 165 pp.

Clevenger, A. P., and A. V. Kociolek. 2006. Highway median impacts on wildlife movement and mortality: State of the practice survey and gap analysis. Final report. Research report number F/CA/MI-2006/09. Western Transportation Institute, Montana State University, Bozeman.

Clevenger, A. P., R. Long, and R. Ament. 2008. I-90 Snoqualmie Pass East wildlife monitoring plan. Prepared for Washington State Department of Transportation, Yakima, Washington.

Clevenger, A. P., and M. A. Sawaya. 2010. Piloting a non-invasive genetic sampling method for evaluating population-level benefits of wildlife crossing structures. *Ecology and Society* 15(1): 7. [online] URL: http://www.ecologyandsociety.org/vol15/iss1/art7/

Clevenger, A. P., and N. Waltho. 2000. Factors influencing the effectiveness of wildlife underpasses in Banff National Park, Alberta, Canada. *Conservation Biology* 14:47–56.

——. 2003. Performance indices to identify attributes of highway crossing structures facilitating movement of large mammals. *Biological Conservation* 121:453–464.

——. 2005. Performance indices to identify attributes of highway crossing structures facilitating movement of large mammals. *Biological Conservation* 121:453–464.

Clevenger, A. P., and J. Wierzchowski. 2006. Maintaining and restoring connectivity in landscapes fragmented by roads. Pages 502–535 in K. R. Crooks and M. Sanjayan, eds. *Connectivity conservation*. Cambridge University Press, Cambridge.

Clevenger, A. P., J. Wierzchowski, B. Chruszcz, and K. Gunson. 2002b. GIS-generated expert based models for identifying wildlife habitat linkages and mitigation passage planning. *Conservation Biology* 16:503–514.

Coalition for Sonoran Desert Protection. 2009. 2004 Bond acquisition properties by name. Accessed April 6, 2009, at www.sonorandesert.org/properties.

Cochran, M. 2006. "Bugle Corps" clears the way for randy elk. *USA Today*, September 24.

Coe, D. 2004. The hydrologic impacts of roads at varying spatial and temporal scales: A review of published literature as of April 2004. Unpublished report.

Coffman, J. S. 2005. Evaluation of a predictive model for upstream fish passage through culverts. Master's thesis, James Madison University, Virginia.

Confederated Salish and Kootenai Tribes. 1994. Flathead Indian Reservation

Comprehensive Land Use Plan. Confederated Salish and Kootenai Tribes. Pablo, Montana.

———. 1996. U.S. Highway 93 land use and growth study. Confederated Salish and Kootenai Tribes. Pablo, Montana.

———. 2000. Wetlands conservation plan for the Flathead Indian Reservation. Confederated Salish and Kootenai Tribes. Pablo, Montana.

Conover, M. R. 1997. Monetary and intangible valuation of deer in the United States. *Wildlife Society Bulletin* 25:298–305.

Conover, M. R., W. C. Pitt, K. K. Kessler, T. J. DuBow, and W. A. Sanborn. 1995. Review of human injuries, illnesses, and economic losses caused by wildlife in the United States. *Wildlife Society Bulletin* 23:407–414.

Cordell, L. S., ed. 1980. *Tijeras Canyon: Analysis of the past*. University of New Mexico Press, Albuquerque.

Corridor Design. 2008. Arizona linkage design reports. Accessed March 20, 2009, at http://www.corridordesign.org/arizona/download.php.

Côté, S. D., T. P. Rooney, J. P. Tremblay, C. Dussault, and D. M. Waller. 2004. Ecological impacts of deer overabundance. *Annual Review of Ecology, Evolution, and Systematics* 35:113–147.

Cottrell, B. H. 2003. Technical assistance report: Evaluation of deer warning reflectors in Virginia. Virginia Transportation Research Council in cooperation with U.S. Department of Transportation and Federal Highway Administration. VTRC 03-TAR6. Charlottesville, Virginia.

Cramer, P. C., and J. A. Bissonette. 2005. NCHRP 25-27 update: Wildlife crossings in North America. Deer–vehicle crash reductions: Setting a strategic agenda workshop. University of Wisconsin–Madison.

———. 2007. Integrating wildlife crossings into transportation plans and projects in North America. *International Conference on Ecology and Transportation* 328–334.

Craven, S., T. Barnes, and G. Kania. 1998. Toward a professional position on the translocation of problem wildlife. *Wildlife Society Bulletin* 26(1): 171–177.

Craven, S. R., C. M. Pils, and R. E. Rolley. 2000. The impacts of deer population management and roadside vegetation management on deer vehicle collisions in Wisconsin. Pages 148–150 in T. A. Messmer and B. West., eds. *Proceedings of the 7th annual meeting of the Wildlife Society: Wildlife and highways—seeking solutions to an ecological and socio-economic dilemma*. Nashville.

Cromwell, J. A., R. J. Warren, and D. W. Henderson. 1999. Live-capture and small-scale relocation of urban deer on Hilton Head Island, South Carolina. *Wildlife Society Bulletin* 27(4): 1025–1031.

Crooks, K. R., and M. Sanjayan. 2006. *Connectivity conservation*. Cambridge University Press, Cambridge.

Crooks, K. R., and M. E. Soulé. 1999. Mesopredator release and avifaunal extinctions in a fragmented system. *Nature* 400:563–566.

Cushman, S. A., K. S. McKelvey, and J. Hayden. 2006. Gene flow in complex landscapes: Testing multiple hypotheses with causal modeling. *American Naturalist* 168:486–499.

Dai, Q., R. Young, and S. Vander Giessen. 2009. Evaluation of an active wildlife-sensing and driver warning system at Trapper's Point. FHWA-WY-09/03F. Wyoming Department of Transportation, Cheyenne.

Damarad, T., and G. J. Bekker. 2003. COST 341—Habitat fragmentation due to transportation infrastructure: Findings of the COST Action 341. Office for Official Publications of the European Communities. Luxemborg.

Damas and Smith. 1982. Wildlife mortality in transportation corridors in Canada's National Parks. Report for Parks Canada. Volume 1, Main report, Ottawa, Ontario.

Damschen, E. I., N. M. Haddad, J. L. Orrock, D. J. Levey, and J. J. Tewksbury. 2006. Corridors increase plant species richness at large scales. *Science* 313: 1284–1286.

D'Angelo, G. J., J. G. D'Angelo, G. R. Gallagher, D. A. Osborn, K. V. Miller, and R. J. Warren. 2006. Evaluation of wildlife warning reflectors for altering white-tailed deer behavior along roadways. *Wildlife Society Bulletin* 34(4): 1175–1183.

D'Angelo, G. J., A. R. De Chicchis, D. A. Osborn, G. R. Gallagher, R. J. Warren, and K. V. Miller. 2007. Hearing range of white-tailed deer as determined by auditory brainstem response. *Journal of Wildlife Management* 71(4): 1238–1242.

Darimont, C. T., P. C. Paquet, T. E. Reimchen, and V. Crichton. 2005. Range expansion by moose into coastal temperate rainforests of British Columbia, Canada. *Diversity and Distributions* 11:235–239.

Davenport, J., and J. L. Davenport, eds. 2006. *The ecology of transportation: Managing mobility for the environment*. Springer, London.

Delaware Water Gap National Recreation Area. 2003. Cover me! "Nights of Migrating Dangerously" end as the park closes River Road to protect amphibians. U.S. Dept. of the Interior National Park Service. *Spanning the Gap* 25(1): Spring 2003. 234.

Delgado, J. D., J. R. Ar'evalo, and J. M. Fern'andez-Palacios. 2001. Road and topography effects on invasion: edge effects in rat foraging patterns in two oceanic island forests (Tenerife, Canary Islands). *Ecography* 24:539–546.

de Neufville, J. I. 1985. *Knowledge and action: Making the link*. Institute of Urban and Regional Development, University of California, Berkeley.

DeNicola, A. J., K. C. VerCauteren, P. D. Curtis, and S. E. Hygnstrom. 2000. Managing white-tailed deer in suburban environments: A technical guide. A publication of the Cornell Cooperative Extension, the Wildlife Society–Wildlife Damage Management Working Group, and the Northeast Wildlife Damage Research and Outreach Cooperative. Cornell University, Ithaca, New York.

Department of Administration. 1973. Final report and recommendations for the

Big Cypress Area of Critical State Concern to the State of Florida Administration Commission. Report CA-73-2.

Ditchkoff, S. S., S. T. Saalfeld, and C. J. Gibson. 2006. Animal behavior in urban ecosystems: Modifications due to human-induced stress. *Urban Ecosystems* 9(1): 5–12.

Dodd, N., and J. Gagnon. 2008. Preacher Canyon Wildlife Fence and Crosswalk Enhancement Project State Route 260, Arizona. First year progress report. Project JPA 04-088. Arizona Game and Fish Department, Research Branch.

Dodd, N. L., J. W. Gagnon, S. Boe, A. Manzo, and R. E. Schweinsburg. 2007a. Evaluation of measures to minimize wildlife–vehicle collisions and maintain wildlife permeability across highways—State Route 260, Arizona, USA. Final report 540 (2002–2006). Arizona Transportation Research Center, Arizona Department of Transportation, Phoenix.

Dodd, N. L., J. W. Gagnon, S. Boe, and R. E. Schweinsburg. 2007b. Assessment of elk highway permeability by Global Positioning System telemetry. *Journal of Wildlife Management* 71:1107–1117.

——. 2007c. Role of fencing in promoting wildlife underpass use and highway permeability. Pages 475–487 in C. L. Irwin, P. Garrett, and K. P. McDermott, eds. *2007 Proceedings of the International Conference on Ecology and Transportation*. Center for Transportation and the Environment, North Carolina State University, Raleigh.

Dodd, N. L., J. W. Gagnon, S. Boe, R. E. Schweinsburg, and K. Ogren. 2009a. Effectiveness of wildlife underpasses in minimizing wildlife–vehicle collisions and promoting wildlife permeability across highways—State Route 260, Arizona. Final report 603 (2002–2008). Arizona Transportation Research Center, Arizona Department of Transportation, Phoenix.

Dodd, N. L., J. W. Gagnon, A. L. Manzo, and R. E. Schweinsburg. 2007d. Video surveillance to assess highway underpasses by elk in Arizona. *Journal of Wildlife Management* 71:637–645.

Dodd, N. L., J. W. Gagnon, S. Sprague, S. Boe, and R. E. Schweinsburg. 2009b. Assessment of pronghorn movements and strategies to promote highway permeability—U.S. Highway 89. Final project report 619, Arizona Transportation Research Center, Arizona Department of Transportation, Phoenix.

Dodson Coulter, E., J. Sessions, and M. G. Wing. 2006. Scheduling forest road maintenance using the analytic hierarchy process and heuristics. *Silva Fennica* 40:143–160.

Doerr, M. L., J. B. McAninch, and E. P. Wiggers. 2001. Comparison of four methods to reduce white-tailed deer abundance in an urban community. *Wildlife Society Bulletin* 29(4): 1105–1113.

Donaldson, B. M. 2005. The use of highway underpasses by large mammals in Virginia and factors influencing their effectiveness. Final Report: Virginia Transportation Research Council.

Donaldson, B. M., and N. W. Lafon. 2008. Testing an integrated PDA-GPS system to collect standardized animal carcass removal data. Report no. FHWA/VTRC 08-CR10. Virginia Transportation Research Council, Charlottesville.

Duda, M. 2007. Public Opinion on Wildlife Species Management in Vermont. Responsive Management National Office. Harrisonburg, Virginia.

Duever, M. J., J. E. Carlson, J. F. Meeder, L. C. Duever, L. H. Gunderson, L. A. Riopelle, T. R. Alexander, R. L. Myers, and D. P. Spangler. 1986. The Big Cypress National Preserve. Research Report No. 8 of the National Audubon Society. New York.

Dunning Jr., J. B., R. Borgella Jr., K. Clements, and G. K. Meffe. 1995. Patch isolation, corridor effects, and colonization by a resident sparrow in a managed pine woodland. *Conservation Biology* 9:542–550.

Dyer, S. J., J. P. O'Neill, S. M. Wasel, and S. Boutin. 2002. Quantifying barrier effects of roads and seismic lines on movements of female woodland caribou in northeastern Alberta. *Canadian Journal of Zoology* 80:839–845.

Ead, S. A., N. Rajaratnam, and C. Katopodis. 2002. Generalized study of hydraulics of culvert fishways. *Journal of Hydraulic Engineering* 128:1018–1022.

Edwards, R. T. 1998. The hyporheic zone in Robert J. Naiman and Robert E. Bilby, eds. *River ecology and management: Lessons from the Pacific Coastal Ecoregion*. Springer-Verlag, New York.

Enck, J. W., D. J. Decker, and T. L. Brown. 2000. Status of hunter recruitment and retention in the United States. *Wildlife Society Bulletin* 28(4): 817–824.

Enders, E. C., D. Boisclair, and A. G. Roy. 2003. The effect of turbulence on the cost of swimming for juvenile Atlantic salmon (*Salmo salar*). *Canadian Journal of Fisheries and Aquatic Sciences* 69:1149–1160.

Engel, S. R., and J. R. Voshell Jr. 2002. Volunteer biological monitoring: can it accurately assess the ecological condition of streams? *American Entomologist* 48:164–177.

Epps, C. W., P. J. Palsball, J. D. Wehausen, G. K. Roderick, R. R. Ramey II, and D. R. McCullough. 2005. Highways block gene flow and cause a rapid decline in genetic diversity of desert bighorn sheep. *Ecology Letters* 8:1029–1038.

Epps, C. W., J. D. Wehausen, and V. C. Bleich. 2007. Optimizing dispersal and corridor models using landscape genetics. *Journal of Applied Ecology* 44:714–724.

Euro Graduate. 2008. Engineering : Europe's first plastic bridge. August 12, 2008. Available at http://www.eurograduate.com/arch_article.asp?id=2021.

Evink, G. L. 1997. Florida Department of Transportation program to reduce highway mortality of Florida panther. Pages 510–513 in D. B. Jordan, ed. *Proceedings of the Florida Panther Conference*. Florida Panther Interagency Committee, Fort Myers, Florida.

———. 2002. *NCHRP synthesis 305: Interaction between roadways and wildlife ecology: A synthesis of highway practice*. Transportation Research Board. National Academies, Washington, DC.

Evink, G. L., P. Garrett, and D. Ziegler, eds. 1999. Proceedings of the Third International Conference on Wildlife Ecology and Transportation. Publication FL-ER-73-99. Florida Department of Transportation. Tallahassee.

Executive Order No. 13,352: 2004. Facilitation of cooperative conservation. *Federal Register* Vol. 69, No. 167.

Fahey, J. 1974. *The Flathead Indians*. University of Oklahoma Press. Norman.

Fahrig, L., and G. Merriam. 1995. Conservation of fragmented populations. Pages 16–25 in D. Ehrenfeld, ed. *Readings in Conservation Biology: The Landscape Perspective*. Society for Conservation Biology and Blackwell Science Inc., Cambridge.

Fahrig, L., and T. Rytwinski. 2009. Effects of roads on animal abundance: An empirical review and synthesis. *Ecology and Society* 14(1): 21. Available at http://www.ecologyandsociety.org/vol14/iss1/art21/.

Farrell, J. E., L. R. Irby, and P. T. McGowen. 2002. Strategies for ungulate–vehicle collision mitigation. *Intermountain Journal of Sciences* 8:1–18.

Federal Environmental Assessment Review Office (FEARO). 1979. Banff Highway Project: East Gate to km 13. Report of the Environmental Assessment Panel, Ottawa.

———. 1982. Banff Highway Project: km 13 to km 27. Report of the Environmental Assessment Panel, Ottawa.

Federal Highway Administration (FHWA). 1998. First revised record of decision for the improvement of U.S. Highway 93—Evaro to Polson-Missoula and Lake Counties, Montana. FHWA-MT-EIS-95-01-F. Federal Highway Administration. Helena, Montana.

———. 2001. Second revised record of decision for improvement of U.S. Highway 93—Evaro to Polson-Missoula and Lake Counties, Montana. FHWA-MT-EIS-95-01-F. Federal Highway Administration. Helena, Montana.

———. 2004. Highway statistics 2003. Section V: Roadway extent, characteristics and performance. Accessed June 16, 2009, at http://www.fhwa.dot.gov/policy/ohim/hs03/re.htm.

———. 2008. Roadway extent, characteristics, and performance. Highway statistics. Federal Highway Administration, Washington, DC. Available at http://www.fhwa.dot.gov/policyinformation/statistics/.

Federal Highway Administration and Arizona Department of Transportation. 2000. Final Environmental Impact Statement for State Route 260—Payson to Heber. Arizona Department of Transportation, Environmental Planning Group, Phoenix.

Federal Highway Administration and the Montana Department of Transportation. 1995. Draft Environmental Impact Statement and Section 4(f) Evaluation. F 5-1(9) 6 U.S. Highway 93, Evaro–Polson, Missoula and Lake Counties, Montana. Montana Department of Transportation. Helena.

Feldhamer, G. A., J. E. Gates, D. M. Harman, A. J. Loranger, and K. R. Dixon. 1986. Effects of interstate highway fencing on white-tailed deer (*Odocoileus virginianus*) activity. *Journal of Wildlife Management* 50(3): 497–503.

Fernald, A. G., P. J. Wigington Jr., and D. H. Landers. 2001. Transient storage and hyporheic flow along the Willamette River, Oregon: Field measurements and model estimates. *Water Resources Research* 37:1681–1694.

Fiberline Composites. 2009a. M6 plastic bridge scoops technology award. Available at http://www.fiberline.com/gb/newsroom/news6032.asp.

——. 2009b. German highways agency combats traffic queues with innovative bridge concept. Available at http://www.fiberline.com/gb/newsroom/news 6435.asp.

Fischer, F. 2000. *Citizens, experts and the environment, The politics of local knowledge.* Duke University Press, Durham, NC.

Flosi, G., S. Downie, J. Hopelain, M. Bird, R. Coey, and B. Collins. 2002. *California salmonid stream habitat restoration manual.* 3rd ed. Vol. 2, chap. 9. State of California, The Resources Agency, California Department of Fish and Game, Inland Fisheries Division. Sacramento.

Flygare, H. 1978. Ungulate mortality and mitigation measures—Trans-Canada Highway, Banff National Park. Report to Environment Canada, Canadian Parks Service, Banff, Alberta.

Ford, A. T., A. P. Clevenger, and A. Bennett. 2009. Comparison of motion-activated camera and trackpad methods of monitoring wildlife crossing structures on highways. *Journal of Wildlife Management* 73(7):1213–1222.

Ford, A. T., and L. Fahrig. 2008. Movement patterns of eastern chipmunks (*Tamias striatus*) near roads. *Journal of Mammalogy* 89:895–903.

Ford, S. G., and S. L. Villa. 1993. Reflector us and the effect they have on the number of mule deer killed on California highways. Final report. Report FHWA-CA-PD94-01. California Department of Transportation, Sacramento.

Forman, R. T. T. 1987. The ethics of isolation, the spread of disturbance, and landscape ecology. Pages 213–229 in M. G. Turner, ed. *Landscape heterogeneity and disturbance.* Springer-Verlag, New York.

——. 1999. Estimate of the area affected ecologically by the road system in the United States. *Conservation Biology* 14:31–35.

Forman, R. T. T., and L. E. Alexander. 1998. Roads and their major ecological effects. *Annual Review of Ecology and Systematics* 29:207–231.

Forman, R. T. T., and R. D. Deblinger. 2000. The ecological road-effect zone of a Massachusetts (U.S.A.) suburban highway. *Conservation Biology* 14:36–46.

Forman, R. T. T., D. Sperling, J. Bissonette, A. Clevenger, C. Cutshall, V. Dale, L. Fahrig, R. France, C. Goldman, K. Heanue, J. Jones, F. Swanson, T. Turrentine, and T. Winter. 2003. *Road ecology: Science and solutions.* Island Press, Washington, DC.

Foster, M. L., and S. R. Humphrey. 1995. Use of highway underpasses by Florida panthers and other wildlife. *Wildlife Society Bulletin* 23:95–100.

Fowle, S. C. 1990. The painted turtle in the Mission Valley of western Montana. M.S. thesis. University of Montana. Missoula.

Fraser, D., and E. R. Thomas. 1982. Moose vehicle accidents in Ontario: Relation to highway salt. *Wildlife Society Bulletin* 10:261–265.

Furniss, M., M. Love, S. Firor, K. Moynan, A. Llanos, J. Guntle, and R. Gubernick. 2008. FishXing, Version 3.0. U.S. Forest Service, San Dimas Technology and Development Center, San Dimas, California.

Gagnon, J. W. 2006. Effect of traffic on elk distribution, highway crossings, and wildlife underpass use in Arizona. Thesis, Northern Arizona University, Flagstaff.

Gagnon, J. W., N. L. Dodd, S. Sprague, K. Ogren, and R. E. Schweinsburg. 2010. Preacher Canyon wildlife fence and crosswalk enhancement project evaluation—State Route 260. Final project report submitted to Arizona Department of Transportation, Phoenix.

Gagnon, J. W., T. C. Theimer, N. L. Dodd, S. Boe, and R. E. Schweinsburg. 2007b. Traffic volume alters elk distribution and highway crossings in Arizona. *Journal of Wildlife Management* 71:2318–2323.

Gagnon, J. W., T. C. Theimer, N. L. Dodd, A. Manazo, and R. E. Schweinsburg. 2007a. Effects of traffic on elk use of wildlife highway underpasses Arizona. *Journal of Wildlife Management* 71:2324–2328.

General Motors Corporation. 2003. Cadillac DeVille night vision. Available at http://www.cadillac.com/cadillacjsp/models/feature.jsp?model=devilleand feature=nightvision.

Gerlach, G., and K. Musolf. 2000. Fragmentation of landscape as a cause for genetic subdivision in bank voles. *Conservation Biology* 14:1066–1074.

Gibeau, M. L. 2000. A conservation approach to management of grizzly bears in Banff National Park, Alberta. PhD thesis. Universitry of Calgary.

Gibeau, M. L., A. P. Clevenger, S. Herrero, and J. Wierzchowski. 2002. Grizzly bear response to human development and activities in the Bow River Watershed, Alberta, Canada. *Biological Conservation* 103:227–236.

Gibson, R. J., R. L. Haedrich, and C. M. Wernerheim. 2005. Loss of fish habitat as a consequence of inappropriately constructed stream crossings. *Fisheries* 30: 10–17.

Gill, R. M. A., A. L. Johnson, A. Francis, K. Hiscocks, and A. J. Peace. 1996. Changes in roe deer (*Capreolus capreolus* L.) population density in response to forest habitat succession. *Forest Ecology and Management* 88(1–2): 31–41.

Giller, P. S. and B. Malmqvist. 1998. *The biology of streams and rivers*. Oxford University Press, New York.

Gloyne, C. C., and A. P. Clevenger. 2001. Cougar *Puma concolor* use of wildlife crossing structures on the Trans-Canada highway in Banff National Park, Alberta. *Wildlife Biology* 7:117–124.

Goodrich, J. M. 1990. Ecology, conservation, and management of two western Great Basin black bear populations. Master's thesis, University of Nevada, Reno.

——. 1993. Nevada black bears: Ecology, management, and conservation. Nevada Department of Wildlife. Biological Bulletin No. 11.

Goodwin, C. R. 1987. Tidal-flow, circulation, and flushing changes caused by dredge and fill in Tampa Bay, Florida. U.S.G.S. Water-Supply Paper 2282, U.S. Geological Survey, Washington, DC.

Goosem, M. 2004. Linear infrastructure in the tropical rainforests of far north Queensland: Mitigating impacts on fauna of roads and powerline clearings. Pages 418–434 in D. Lunney, ed. *Conservation of Australia's forest fauna*. Royal Zoological Society of New South Wales, Mosman, NSW, Australia.

Gordon, K. M., and S. H. Anderson. 2002. Motorist response to a deer-sensing warning system in western Wyoming. Pages 549–558 in *2003 Proceedings of the International Conference on Ecology and Transportation*, 24–28 September 2001. Keystone, Colorado.

Gordon, K. M., M. C. McKinstry, and S. H. Anderson. 2004. Motorist response to a deer sensing warning system. *Wildlife Society Bulletin* 32:565–573.

Gouveia, C., A. Fonseca, A. Camara, and F. Ferreira. 2004. Promoting the use of environmental data collected by concerned citizens through information and communication technologies. *Journal of Environmental Management* 71:135–154.

Government Accountability Office. 2003. Highway infrastructure: Perceptions of stakeholders on approaches to reduce project completion time. GAO-03-398, Washington, DC.

———. 2008. Highways and environment: transportation agencies are acting to involve others in planning and environmental decisions. GAO-08-512R, Washington, DC.

Graf, R. F., S. Kramer-Schadt, and N. Fernandez. 2007. What you see is where you go? Modeling dispersal in mountainous landscapes. *Landscape Ecology* 22 :853–866.

Graves, H. B., and E. D. Bellis. 1978. *The effectiveness of deer flagging models as deterrents to deer entering highway rights-of-way*. Institute for Research on Land and Water Resources, Pennsylvania State University, University Park, Pennsylvania.

Green, R. H. 1979. *Sampling design and statistical methods for environmental biologists*. Wiley, New York.

Grift, E. A. van der, V. Biserkov, and V. Simeonova. 2008. *Restoring ecological networks across transport corridors in Bulgaria: Identification of bottleneck locations and practical solutions*. Alterra, Wageningen, the Netherlands.

Groot Bruinderink, G. W. T. A, and E. Hazebroek. 1996. Ungulate traffic collisions in Europe. *Conservation Biology* 10(4): 1059–1067.

Grosman, P. D., J. A. G. Jaeger, P. M. Biron, C. Dussault, and J.-P. Ouellet. 2009. Reducing moose–vehicle collisions through salt pool removal and displacement: An agent-based modeling approach. *Ecology and Society* 14(2): 17. Available at http://www.ecologyandsociety.org/vol14/iss2/art17/.

Gulen, S., G. McCabe, and S. E. Wolfe. 2000. Evaluation of wildlife reflectors in reducing vehicle–deer collisions on Indiana Interstate 80/90. SPR-3 (076). Indiana Department of Transportation, Divisions of Research and Toll Roads, Indiana.

Guisan, A., and N. E. Zimmermann. 2000. Predictive habitat distribution models in ecology. *Ecological Modeling* 135:147–186.

Gunson K. E., and A. P. Clevenger. 2003. Large animal–vehicle collisions in the central Canadian Rocky Mountains: Patterns and characteristics. Pages 355–366 in C. L. Irwin, P. Garrett, and K. P. McDermott, technical coordinators. *Proceedings of the International Conference on Ecology and Transportation*, North Carolina Center for Transportation and the Environment, North Carolina State University, Raleigh.

Gunson K. E., A. P. Clevenger, A. T. Ford, J. Bissonette, and A. Hardy. 2009. A comparison of data sets varying in spatial accuracy used to predict the occurrence of wildlife–vehicle collisions. *Environmental Management* 44(2):268–277.

Gunther, K. A., M. J. Biel, and H. L. Robison. 1998. Factors influencing the frequency of road-killed wildlife in Yellowstone National Park. Pages 32–40 in G. L. Evink, P. Garrett, D. Zeigler, and J. Berry, eds. *Proceedings of the International Conference on Wildlife Ecology and Transportation*. FL-ER-69-98. Florida Department of Transportation, Tallahassee.

Gunther, K. A., M. A. Haroldson, K. Frey, S. L. Cain, J. Copeland, and C. C. Schwartz. 2004. Grizzly bear–human conflicts in the Greater Yellowstone ecosystem, 1992–2000. *Ursus* 15:10–22.

Hage, M., P. Leroy, and E. Willems. 2006. *Participatory approaches in governance and in knowledge production: What makes the difference?* Page 31. Research Group Governance and Places, University of Nijmegen.

Haikonen, H., and H. Summala. 2001. Deer–vehicle crashes, extensive peak at 1 hour after sunset. *American Journal of Preventive Medicine* 21:209–213.

Haines, A. M., M. E. Tewes, L. L. Laack, J. S. Horne, and J. H. Young. 2006. A habitat-based population viability analysis for ocelots (*Leopardus pardalis*) in the United States. *Biological Conservation* 132(4): 424–436.

Hall, G. B., and M. G. Leahy. 2006. Internet-based spatial decision support using open source tools. Pages 237–262 in S. Balram, and S. Dragicevic, eds. *Collaborative Geographic Information Systems*. IDEA Group Publishing, Hershey, Pennsylvania.

Hammond, C., and M. G. Wade. 2004. Deer avoidance: The assessment of real world enhanced deer signage in a virtual environment. Final report. Minnesota Department of Transportation. St. Paul.

Hammond, F. 2002. The effects of resort and residential development on black bears in Vermont - Final Report. Vermont Agency of Natural Resources, Fish & Wildlife Department.

Hansen, M., and A. P. Clevenger. 2005. The influence of disturbances and habitat on the frequency of non-native plant species along transportation corridors. *Biological Conservation* 125:249–259.

Hanski, I. 1999. *Metapopulation ecology*. Oxford University Press, Oxford.

Hanski, I., and O. E. Gaggiotti. 2004. Metapopulation biology: Past, present, and future. Pages 3–22 in I. Hanksi and O. E. Gaggiotti, eds. *Ecology, genetics, and evolution of metapopulations*. Academic Press, San Diego.

Hardy, A. R. 2008. Developing the "Integrated Transportation and Ecological Enhancements for Montana" (ITEEM) process: Applying the Eco-Logical approach. *Transportation Research Record: Journal of the Transportation Research Board. Energy and Environment Category 2011* (2007): 148–156.

Hardy, A. R., A. P. Clevenger, M. Huijser, and G. Neale. 2003. An overview of methods and approaches for evaluating the effectiveness of wildlife crossing structures: Emphasizing the science in applied science. Pages 319–330 in C. L. Irwin, P. Garrett, and K. P. McDermott, eds. *2003 Proceedings of the International Conference on Ecology and Transportation*. Center for Transportation and the Environment, North Carolina State University, Raleigh.

Hardy, A. R., J. Fuller, M. Huijser, A. Kociolek, and M. Evans. 2007. Evaluation of wildlife crossing structures and fencing on US Highway 93 Evaro to Polson, Phase I: Preconstruction data collection and finalization of evaluation plan. FHWA/MT-06-008/1744-1, Helena, Montana.

Hardy, A. R., J. Fuller, S. Lee, L. Stanley, and A. Al-Kaisy. 2006. Bozeman Pass wildlife channelization ITS project final report. Western Transportation Institute, Montana State University, Bozeman.

Hargrove, W. W., F. M. Hoffman, and R. A. Efroymson. 2005. A practical map-analysis tool for detecting potential dispersal corridors. *Landscape Ecology* 20:361–373.

Havlick, D. 2004. *Roadkill. Conservation Magazine* Vol 5. No. 1. Society for Conservation Biology, Washington DC.

Hawkins, D. 2005. Assessing the impact of pile driving upon fish. In C. L. Irwin, P. Garrett, and K. P. McDermott, eds. *Proceedings of the International Conference on Ecology and Transportation*, Center for Transportation and the Environment, North Carolina State University, Raleigh.

Hebblewhite, M., M. Percy, and R. Serrouya. 2003. Black bear (*Ursus americanus*) survival and demography in the Bow Valley of Banff National Park, Alberta. *Biological Conservation* 112:415–425.

Hedlund, J. H., P. D. Curtis, G. Curtis, and A. F. Williams. 2004. Methods to reduce traffic crashes involving deer: What works and what does not. *Traffic Injury Prevention* 5:122–131.

Heiman, M. K. 1997. Science by the people: Grassroots environmental monitoring and the debate over scientific expertise. *Journal of Planning Education and Research* 16:291–299.

Heller, N. E., and E. S. Zavaleta. 2009. Biodiversity management in the face of climate change: A review of 22 years of recommendations. *Biological Conservation* 142:14–32.

Hilty, J. A., W. Z. Lidicker Jr., and A. M. Merenlender. 2006. *Corridor ecology: The science and practice of linking landscapes for biodiversity conservation*. Island Press, Washington, DC.

Hirota, M., Y. Nakajima, M. Saito, and M. Uchiyama. 2004. Low-cost infrared imaging sensors for automotive applications. Pages 63–84 in J. Valldorf and W.Gessner, eds. *Advanced microsystems for automotive applications*. Available at http://www.springeronline.com/sgw/cda/pageitems/document/cda_down loaddocument/0,11996,0-0-45-110604-0,00.pdf.

Hoar, W. S., and D. J. Randall, eds. 1978. *Fish physiology*. Vol. 7: *Locomotion*. Academic Press, New York.

Hoffman, N. 2003. Frog fence along Vermont Rt. 2 in Sandbar Wildlife Management Area: Collaboration between Vermont Agency of Transportation and Vermont Agency of Natural Resources. *International Conference on Ecology and Transportation* 431–432.

Holling, C. S. 1978. *Adaptive environmental assessment and management*. Wiley, New York.

Holroyd, G. L. 1979. The impact of highway and railroad mortality on the ungulate populations in the Bow Valley. Report for Banff National Park. Canadian Wildlife Service, Banff, Alberta.

Honda Motor Co., Ltd. 2004. Intelligent night vision system able to detect pedestrians and provide driver cautions. Available at http://www.all4engineers.de/preview.php?cms = andlng = enandalloc = 34andid = 560.

Horne, J. S., E. O. Garton, and S. M. Krone. 2007. Analyzing animal movements using Brownian bridges. *Ecology* 88:2354–2363.

Hotchkiss, R. H., and C. M. Frei. 2007. Design for fish passage at roadway–stream crossings: Synthesis report. FHWA-HIF-07-033. McLean, Virginia.

Huijser, M. P., and P. J. M. Bergers. 2000. The effect of roads and traffic on hedgehog (*Erinaceus europaeus*) populations. *Biological Conservation* 95:111–116.

Huijser, M. P., and A. P. Clevenger. 2006. Habitat and corridor function of rights-of-way. Pages 233–254 in J. Davenport and J. L. Davenport, eds. *The ecology of transportation: managing mobility for the environment*. Environmental Pollution, Vol. 10. Springer, the Netherlands.

Huijser, M. P., and A. P. Clevenger, and C. J. F. ter Braak. 2000. Road, traffic and landscape characteristics of hedgehog traffic victim sites. Pages 108–126 in M. P. Huijser, ed. *Life on the edge. Hedgehog traffic victims and mitigation strategies in an anthropogenic landscape*. PhD thesis, Wageningen University, the Netherlands.

Huijser, M. P., J. W. Duffield, A. P. Clevenger, R. J. Ament, and P. T. McGowen. 2009b. Cost–benefit analyses of mitigation measures aimed at reducing collisions with large ungulates in the United States and Canada: A decision support tool. *Ecology and Society* 14(2): 15. Available at http://www.ecologyandsociety.org/viewissue.php?sf=41.

Huijser, M. P., K. E. Gunson, and C. Abrams. 2006a. Animal–vehicle collisions and habitat connectivity along Montana Highway 83 in the Seeley-Swan valley, Montana: A reconnaissance. Report No. FHWA/MT-06-002/8177. Western Transportation Institute, Montana State University, Bozeman. Available at http://www.mdt.mt.gov/research/docs/research_proj/seeley/final_report.pdf.

Huijser, M. P., T. D. Holland, M. Blank, M. C. Greenwood, P. T. McGowen, B. Hubbard, S. Wang. 2009a. The comparison of animal detection systems in a test-bed: A quantitative comparison of system reliability and experiences with operation and maintenance. Final report. FHWA/MT-09-002/5048. Western Transportation Institute–Montana State University, Bozeman.

Huijser, M. P., T. D. Holland, A. V. Kociolek, A. M. Barkdoll, and J. D. Schwalm. 2009c. Animal–vehicle crash mitigation using advanced technology. Phase II: system effectiveness and system acceptance. SPR3(076) & Misc. contract & agreement no. 17,363. Western Transportation Institute–Montana State University, Bozeman.

Huijser, M. P., P. T. McGowen, W. Camel, A. Hardy, P. Wright, A. P. Clevenger, L. Salsman, and T. Wilson. 2006b. Animal vehicle crash mitigation using advanced technology. Phase I: review, design and implementation. SPR-3(076). FHWA-OR-TPF-07-01, Western Transportation Institute, Montana State University, Bozeman. Available at http://www.oregon.gov/ODOT/TD/TP_ RES/ResearchReports.shtml.

Huijser, M. P., P. T. McGowen, A. P. Clevenger, and R. Ament. 2008. Best practices manual, wildlife–vehicle collision reduction study, Report to U.S. Congress. Federal Highway Administration, McLean, Virginia.

Huijser, M. P., P. T. McGowen, J. Fuller, A. Hardy, A. Kociolek, A. P. Clevenger, D. Smith, and R. Ament. 2007. Wildlife–vehicle collision reduction study. Report to U.S. Congress. U.S. Department of Transportation, Federal Highway Administration, Washington DC.

H. W. Lochner Inc. 1972. Environmental/Section 4(f) Statement, Interstate Route 75, State Road 82 near Fort Myers in Lee County to U.S. Route 27 at Andytown in Broward County. For State Job No. 99004-1401, Federal Job No. I-75-4(1)210. Prepared for Florida Department of Transportation Division of Planning and Programming.

Iowa Department of Natural Resources. 2005. White-tailed deer harvest and population trends. Available at http://www.iowadnr.com/wildlife/pdfs/deerlog.pdf.

IPCC. 2007. Climate change 2007. Intergovernmental Panel on Climate Change Fourth Assessment Report.

Irwin, A. 1995. *Citizen science: A study of people, expertise and sustainable development*. Routledge, New York.

Iuell, B., G. J. Becker, R. Cuperus, J. Dufek, G. Fry, C. Hicks, C. Hlavac, V. B. Keller, C. Rosell, T. Sangwine, N. Torslov, and B. le Maire Wanddall. 2003. *Cost 341- wildlife and traffic: A European handbook for identifying conflicts and designing solutions*. Office for Official Publications of the European Communities, Luxembourg.

———, eds. 2005. *Wildlife and traffic: A European handbook for identifying conflicts and designing solutions*. KNNV Publishers, Utrecht, the Netherlands.

Jaarsma, C. F., van Langevelde, J. M. Baveco, M. van Eupen, and J. Arisz. 2007. Model for rural transportation planning considering simulating mobility and traffic kills in the badger *Meles meles*. *Ecological Informatics* 2(2): 73–82.

Jaarsma, C. F., and G. P. A. Willems. 2002. Reducing habitat fragmentation by minor rural roads through traffic calming. *Landscape and Urban Planning* 58:125–135.

Jacobson, S. K., M. D. McDuff, and M. C. Monroe. 2006. *Conservation education and outreach techniques*. Oxford University Press, Oxford.

Jaeger, J. A. G., J. Bowman, J. Brennan, L. Fahrig, D. Bert, J. Bouchard, N. Charbonneau, K. Frank, B. Gruber, and K. T. von Toschanowitz. 2005. Predicting when animal populations are at risk from roads: An interactive model of road avoidance behavior. *Ecological Modeling* 185:329–348.

Jaeger, J. A. G., and L. Fahrig. 2004. Effects of road fencing on population persistence. *Conservation Biology* 18(6):1651–1657.

Jaren, V., R. Andersen, M. Ulleberg, P. H. Pedersen, and B. Wiseth. 1991. Moose–train collisions: The effects of vegetation removal with a cost–benefit analysis. *Alces* 27:93–99.

Johnson, D. 1988. When antelope don't roam free. *New York Times*, November 18, A16.

Jones, E. R. 2008. The effectiveness of wildlife crossing structures for black bears in Madison County, North Carolina. Master's thesis, North Carolina State University, Raleigh, North Carolina.

Jones, J. A., F. J. Swanson, B. C. Wemple, and K. U. Snyder. 2000. Effects of roads on hydrology, geomorphology, and disturbance patches in stream networks. *Conservation Biology* 14:76–85.

Jones, M. D., and M. R. Pelton. 2003. Female American black bear use of managed forest and agricultural lands in coastal North Carolina. *Ursus* 14:188–197.

Jones, M. D., G. S. Warburton, and M. R. Pelton. 1998. Models for predicting occupied black bear habitat in coastal North Carolina. *Ursus* 10:203–207.

Jones, M. E. 2000. Road upgrade, road mortality and remedial measures: Impacts on a population of eastern quolls and Tasmanian devils. *Wildlife Research* 27:289–296.

Julyan, R. 1998. *The place names of New Mexico*. 2nd ed. University of New Mexico Press, Albuquerque.

———. 2006. The mountains of New Mexico. University of New Mexico Press, Albuquerque.

Kahler, T. H., and T. P. Quinn. 1998. *Juvenile and resident salmonid movement and passage through culverts*. Washington State Tranportation Center. Seattle.

Kane, D. L., C. E. Belke, R. E. Geick, and R. F. McLean. 2000. Juvenile fish passage through culverts in Alaska. INE/WERC 00.05. Water and Environmental Research, University of Alaska–Fairbanks.

Kane, D. L., and P. M. Wellen. 1985. A hydraulic evaluation of fish passage through roadway culverts in Alaska. FHWA-AK-RD-85-24 and 85-24A. Juneau, Alaska.

Kansas Department of Transportation (DOT). 2004. Kansas deer/vehicle collisions peak in November. News release Nov. 1, 2004 (04-146). Kansas Department of Transportation. Available at http://www.ksdot.org/offTransInfo/News04/Deer_Accidents_by_County_2003.asp.

Kendall, K. C., J. B. Stetz, D. A. Roon, L. P. Waits, J. B. Boulanger, D. Paetkau. 2008. Grizzly bear density in Glacier National Park, Montana. *Journal of Wildlife Management* 72:1693–1705.

Kerley, L. L., J. M. Goodrich, D. G. Miquelle, E. N. Smirnov, H. B. Quigley, and M. G. Hornocker. 2002. Effects of roads and human disturbance on Amur tigers. *Conservation Biology* 16:97–108.

Kindall, J. L., and F. T. van Manen. 2007. Identifying habitat linkages for black bears in eastern North Carolina. *Journal of Wildlife Management* 71:487–495.

Kinley, T. A., N. J. Newhouse, and H. N. Page. 2003. Evaluation of the wildlife protection system deployed on Highway 93 in Kootenay National Park during autumn, 2003. November 17, 2003. Sylvan Consulting Ltd. Invermere, British Columbia, Canada.

Kintsch, J. 2006. Linking Colorado's landscapes. Pages 138–142 in C. Leroy Irwin, Paul Garrett, and K. P. McDermott, eds. *Proceedings of the 2005 International Conference on Ecology and Transportation*, Center for Transportation and the Environment, North Carolina State University, Raleigh.

Kirkpatrick, J. F., and A. T. Rutberg. 2001. Fertility control in animals. Pages 183–198 in *The state of the animals 2001*, D. J. Salem and A. N. Rowan, eds. Humane Society Press, Washington, DC.

Kirkpatrick, J. F., W. Turner Jr., I. K. M. Liu, R. Fayrer-Hosken, and A. T. Rutberg. 1997. Case studies in wildlife immunocontraception: wild and feral equids and white-tailed deer. *Reproduction, Fertility and Development* 9(1):105–110.

Kistler, R. 1998. Wissenschaftliche Begleitung der Wildwarnanlagen Calstrom WWA-12-S. July 1995–November 1997. Schlussbericht. Infodienst Wildbiologie und Oekologie, Zürich, Switzerland.

Klar, N., M. Herrmann, and S. Kramer-Schadt. 2006. Effects of roads on a founder population of lynx in the biosphere reserve "Pralzerwald-Vosges du Nord": A model as planning tool. *Naturschutz und Landschaftsplanung* 38(10–11): 330–337.

Kloeden, C. N., A. J. McLean, V. M. Moore, and G. Ponte. 1997. *Traveling speed and the risk of crash involvement*. Vol. 1: *Findings*. NHMRC Road Accident Research Unit, University of Adelaide, Australia.

Kloppers, E. L., C. C St. Clair, and T. E. Hurd. 2005. Predator-resembling aversive conditioning for managing habituated wildlife. *Ecology and Society* 10(1): 31.

Knapp, K., X. Yi, T. Oakasa, W. Thimm, E. Hudson, and C. Rathmann. 2004. Deer–vehicle crash countermeasure toolbox: A decision and choice resource. Final report. Report number DVCIC–02. Midwest Regional University Transportation Center, Deer–Vehicle Crash Information Clearinghouse, University of Wisconsin–Madison.

Knapp K. K., and A. Witte. 2006. Strategic agenda for reducing deer–vehicle crashes. Report No. DVCIC–04. Midwest Regional University Transportation Center, Deer–Vehicle Crash Information Clearinghouse, University of Wisconsin–Madison.

Knight, R. L., H. A. L. Knight, and R. J. Camp. 1995. Common ravens and number and type of linear rights-of-way. *Biological Conservation* 74:65–67.

Kolowski, J. M., and C. K. Nielsen. 2008. Using Penrose distance to identify potential risk of wildlife–vehicle collisions . *Biological Conservation* 141(4): 1119–1128.

Kondratieff, M. C., and C. A. Myrick. 2006. How high can a brook trout jump? A laboratory evaluation of brook trout jumping performance. *Transactions of the American Fisheries Society* 135:361–370.

Koteen, L. E. 2002. Climate Change, whitebark pine and grizzly bears. Pages 343–414 in S. Schneider, and T. Root, eds. *Wildlife responses to climate change: North American Case Studies*. Island Press, Washington, DC.

Kynard, B., and J. O'Leary. 1993. Evaluation of a bypass system for spent American shad at Holyoke Dam, Massachussetts. *North American Journal of Fisheries Management* 13:782–789.

Land, E. D., and M. Lotz. 1996. Wildlife crossing designs and use by Florida panthers and other wildlife in southwest Florida. Page 6 in G. L. Evink, P. Garrett, D. Zeigler, and J. Berry, eds. *Trends in addressing transportation related wildlife mortality*, Proceedings of the Transportation Related Wildlife Mortality Seminar. State of Florida Department of Transportation, Tallahassee.

Langen, T. A., K. M. Ogden, and L. L. Schwarting. 2009. Predicting hot spots of herpetofauna road mortality along highway networks. *Journal of Wildlife Management* 73(1): 104–114.

Langevelde, F., and C. F. Jaarsma. 2004. Using traffic flow theory to model traffic mortality in mammals. *Landscape Ecology* 19:895–907.

Langton, T. E. S., ed. 1989. *Amphibians and roads*. ACO Polymer Products Ltd., Bedfordshire, U.K.

Lankhorst. 2008. Plastic bridge. July 10, 2008. Materia. Available at http://www.materia.nl/563.0.html?&tx_ttnews[tt_news]=195&tx_ttnews[backPid]=534&cHash=b6b4f26c41.

———. 2009. Company website. Available at http://www.lankhorst-recycling.com/lankhorst.nsf/main/EnglishRecyclingGeneralIntroduction.

Lauman, J. K. 1976. Salmonid passage at stream-road crossings: A report with department standards for passage of salmonids. Oregon Department of Fish and Wildlife, Portland.

Laurance, S. G. W. 2004. Responses of understory rain forest birds to road edges in central Amazonia. *Ecological Applications* 14:1344–1357.

Laurance, S. G. W., P. C. Stouffer, and W. F. Laurance. 2004. Effects of road clearings on movement patterns of understory rainforest birds in central Amazonia. *Conservation Biology* 18:1099–1109.

Lauritzen, D. 2002. Preferences, behaviors and biomechanics of Pacific salmon jumping up waterfalls and fishladders. PhD diss., University of California, Berkley.

Lavsund, S., and F. Sandegren. 1991. Moose–vehicle relations in Sweden: A review. *Alces* 27:118–126.

Lee, T. 2007. Evaluating the contribution of citizen participation in research to understand wildlife movement across Highway 3 in Crowsnest Pass, Alberta. University of Calgary, Alberta.

Lee, T., M. Quinn, and D. Duke. 2006. Citizen, science, highways and wildlife: Using a web-based GIS to engage citizens in collecting wildlife information. *Ecology and Society* 11:11–14.

Lehnert, M. E., and J. A. Bissonette. 1997. Effectiveness of highway crosswalk structures at reducing deer vehicle collisions. *Wildlife Society Bulletin* 25(4): 809–818.

Lesbarrères, D., T. Lodé, and J. Merilä. 2003. What type of amphibian tunnel could reduce road kills? *Oryx* 38:220–223.

Levy, A. 2005. Where the wildlife meets the road. *Public Roads* 68:2–9.

Little, S. J., R. G. Harcourt, and A. P. Clevenger. 2002. Do wildlife passages act as prey-traps? *Biological Conservation* 107:135–145.

Logan, T., and G. Evink. 1985. Safer Travel for the panther. *Naturalist* (Spring): 6–7.

Long, R. A., P. MacKay, W. J. Zielinski, and J. C. Ray, eds. 2008. Noninvasive survey methods for carnivores. Island Press, Washington, DC.

Lonsdale, W. M., and A. M. Lane. 1994. Tourist vehicles as vectors of weed seeds in Kakadu National Park, northern Australia. *Biological Conservation* 69:277–283.

Lotz, M. A., E. D. Land, and K. G. Johnson. 1996. Evaluation of S.R.29 wildlife crossings. Final Report to State of Florida Department of Transportation. Study # 7583.

Lowy, J. 2001. A better answer to "why did the critter cross the road." *Naples Daily News*, April 1.

L. P. Tardif and Associates Inc. 2003. Final report: Collisions involving motor vehicles and large animals in Canada. Prepared for Transport Canada Road Saftey Directorate.

MacArthur, R. H., and E. O. Wilson. 1967. The theory of island biogeography. Princeton University Press, Princeton, New Jersey.

MacDonald, J. I., and P. E. Davies. 2007. Improving the upstream passage of two galaxiid fish species through a pipe culvert. *Fisheries Management and Ecology* 14:221–230.

Mace, R. D., J. S. Waller, T. L. Manley, L. J. Lyon, and H. Zuuring. 1996. Relationships among grizzly bears, roads and habitat in the Swan Mountains, Montana. *Journal of Applied Ecology* 33:1395–1404.

MacPhee, C., and F. J. Watts. 1976. Swimming performance of Arctic grayling in highway culverts. U.S. Fish and Wildlife Service. Anchorage, Alaska.

MacKenzie, D. I. 2005. What are the issues with presence–absence data for wildlife managers? *Journal of Wildlife Management* 69:849–860.

Maehr, D. S., E. D. Land, and M. E. Roelke. 1991. Mortality patterns of panthers in southwest Florida. *Proceedings of Annual Conference of Southeastern Fish and Wildlife Agencies* 45:201–207.

Maine Department of Transportation (DOT). 2009. Motorists need to be wary of moose. News release for April 16, 2009. Available at http://www.maine.gov/mdot/mainedot-news-release/moose04162009.php.

Maine Department of Transportation (DOT), Department of Inland Fisheries and Wildlife, Secretary of State, Maine Turnpike Authority, and Department of Public Safety. 2004. State of Maine multi-agency animal crash task force. Summary of 2004 moose collision deterrents state of Maine. Available at http://www.maine.gov/ifw/pdf/2004mooseactivities.pdf.

Majka, D., J. Jenness, and P. Beier. 2007. CorridorDesigner: ArcGIS tools for designing and evaluating corridors. Available at http://corridordesign.org.

Malo, J. E., F. Suarez, and A. Diez. 2004. Can we mitigate animal–vehicle accidents using predictive models? *Journal of Applied Ecology* 41:701–710.

Malouf, C.I., 1998. Flathead and Pend d'Oreille. Pages 297-312 In Walker, Jr., D.E., Ed. Handbook of North American Indians, vol. 12. Smithsonian Institution Plateau, Washington, DC.

Manel, S., M. K. Schwartz, G. Luikart, and P. Taberlet. 2003. Landscape genetics: combining landscape ecology and population genetics. *Trends in Ecology and Evolution* 18:189–97.

Manly, B. F. J., L. L. McDonald, and D. L. Thomas. 2002. *Resource selection by animals: Statistical design and analysis for field studies.* Chapman and Hall, London.

Manzo, A. L. 2006. An ecological analysis of environmental parameters influencing Rocky Mountain elk crossings of an Arizona highway. Master's thesis, Northern Arizona University, Flagstaff.

Mapes, L. V. 2008. Animals need bridges, too, photos show. *Seattle Times,* December 23.

Marron and Associates, Inc. 2005. Tijeras Canyon wildlife safe passage feasibility study; Bernalillo County, New Mexico. Prepared for the New Mexico Department of Transportation. Marron and Associates, Inc., Albuquerque, New Mexico.

———. 2007. Interstate 40 wildlife safe passage feasibility study. Prepared for the New Mexico Department of Transportation. Marron and Associates, Inc., Albuquerque, New Mexico.

Marshall, R. M., S. Anderson, M. Batcher, P. Comer, S. Cornelius, R. Cox, A. Gondor, D. Gori, J. Humke, R. Paredes Aguilar, I. E. Parra, and S. Schwartz. 2000. An ecological analysis of conservation priorities in the Sonoran Desert ecoregion. Prepared by The Nature Conservancy Arizona chapter, Sonoran Institute, and Instituto del Medio Ambiente y el Desarrollo Sustentable del Estado de Sonora with support from Department of Defense Legacy Program, Agency and Institutional partners.

Marshik, J., L. Renz, J. L. Sipes, D. Becker, and D. Paulson. 2001. Preserving spirit of place: U.S. Highway 93 on the Flathead Indian Reservation. Pages 244–256 in *2001 Proceedings of the International Conference on Ecology and Transportation*.

Mata, C., I. Hervás, J. Herranz, F. Suárez, and J. E. Malo. 2005. Complementary use by vertebrates of crossing structures along a fenced Spanish motorway. *Biological Conservation* 124:397–405.

Mattson, D. J. 1992. *Conversion factors for standardized calculations of roads and trail densities: Yellowstone grizzly bears*. Interagency Grizzly Bear Study Team, Forest Sciences Labs, Montana State University, Bozeman.

Mattson, D., C. Gillin, and S. Benson. 1991. Bear feeding-activity at alpine insect aggregation sites in the Yellowstone ecosystem. *Canadian Journal of Zoology* 69:2430–2435.

Mayfield, C., M. Joliat, and D. Cowan. 2001. The role of community networks in environmental monitoring and informatics. *Advances in Environmental Research* 5:385–389.

McBride, R., R. T. McBride, R. M. McBride, and C. E. McBride. 2008. Counting pumas by categorizing physical evidence. *Southeastern Naturalist* 7(3): 381–400.

McCollister, M. F. 2008. Impacts of a 4-lane highway on the spatial ecology of American black bears and the effectiveness of wildlife underpasses in eastern North Carolina. Master's thesis, University of Tennessee, Knoxville.

McCullough, D. 1996. Metapopulation management: What patch are we in and which corridor should we take? Pages 405–410 in D. McCullough, ed. *Metapopulations and wildlife conservation*. Island Press, Covelo, California.

McCullough, D. R., K. W. Jennings, N. B. Gates, B. G. Elliot, and J. E. DiDonato. 1997. Overabundant deer populations in California. *Wildlife Society Bulletin* 25:478–483.

McDonald, M. G. 1991. Moose movement and mortality associated with the Glenn Highway expansion, Anchorage, Alaska. *Alces* 27:208–219. (as cited in Biota Research and Consulting, Inc., 2003).

McGuire, T. M., and J. F. Morrall. 2000. Strategic highway improvements to minimize environmental impacts within the Canadian Rocky Mountain national parks. *Canadian Journal of Civil Engineering* 27:523–532.

McKinney, T., and T. Smith. 2006. Distribution and trans-highway crossings of desert bighorn sheep in northwestern Arizona. Final Report to the Arizona Department of Transportation and the Federal Highway Administration, Arizona Game and Fish Department, Phoenix.

——. 2007. US93 bighorn sheep study: Distribution and trans-highway movements of desert bighorn sheep in northwestern Arizona. Final Report 576, Arizona Department of Transportation, Phoenix.

McRae, B. H., and P. Beier. 2007. Circuit theory predicts gene flow in plant and animal populations. *Proceedings of the National Academy of Sciences of the United States of America* 104:19885–19890.

McRae, B. H., and P. Beier, L. E. Dewald, L. Y. Huynh, and P. Keim. 2005. Habitat barriers limit gene flow and illuminate historical events in a wide-ranging carnivore, the American puma. *Molecular Ecology* 14:1965–1977.

McRae, B. H., B. G. Dickson, and T. H. Keitt. 2008. Using circuit theory to model connectivity in ecology, evolution and conservation. *Ecology* 89:2712–2724.

McShea, W. J., S. L. Monfort, S. Hakim, J. Kirpatrick, I. Liu, J. W. Turner Jr., L. Chassy, and L. Munson. 1997. The effect of immunocontraception on the behavior and reproduction of white-tailed deer. *Journal of Wildlife Management* 61(2): 560–569.

Merrill, J. A., E. G. Cooch, and P. D. Curtis. 2003. Time to reduction: Factors influencing management efficacy in sterilizing overabundant white-tailed deer. *Journal of Wildlife Management* 67(2): 267–279.

Messmer, T. A., and B. West, eds. 2002. Wildlife and highways: Seeking solutions to an ecological and socio-economic dilemma. Proceedings of a Symposium at the Seventh Annual Meeting of The Wildlife Society at Nashville, Tennessee.

Meyer, E. 2006. Assessing the effectiveness of deer warning signs. Final report KTRAN: KU-03-6. The University of Kansas, Lawrence.

Meyer, E., and I. Ahmed. 2004. Modeling of deer vehicle crash likelihood using roadway and roadside characteristics. In Transportation Research Board 2004 Annual Meeting Compendium of Papers. On CD-ROM. Washington, DC.

Miller, B. K., and J. A. Litvaitis. 1992. Use of roadside salt licks by moose (*Alces alces*) in northern New Hampshire. *Canadian Field-Naturalist* 106:112–117.

Miller, R. A., and J. B. Kaneene. 2006. Evaluation of historical factors influencing the occurrence and distribution of *Mycobacterium bovis* infection among wildlife in Michigan. *American Journal of Veterinary Research* 67(4): 604–615.

Minor, E. S., and D. L. Urban. 2008. A graph-theory framework for evaluating landscape connectivity and conservation planning. *Conservation Biology* 22:297–307.

Moneo, Shannon. 2006. Season begins for B.C.'s bloody road show. *Globe and Mail* posted on November 14, 2006.

Montana Department of Transportation. 2004. On-site wetland mitigation plan, U.S. 93

——. 2008. Wildlife-highway crossing mitigation measures and associated costs/benefits: a toolbox for Montana Department of Transportation.

Montana Department of Transportation, Federal Highway Administration, and Confederated Salish and Kootenai Tribes. 2000. Memorandum of Agreement. U.S. Highway 93–Evaro to Polson. Skillings-Connolly Engineering. Missoula, Montana.

Montana Department of Transportation, Confederated Salish and Kootenai Tribes, and Federal Highway Administration. 2007. Final Supplemental Environmental Impact Statement and Section 4(f) Evaluation–U. S. Highway 93 Ninepipe/Ronan Improvement Project. Montana Department of Transportation and Federal Highway Administration, Helena, Montana.

Morita, K., and S. Yamamoto. 2002. Effects of habitat fragmentation by damming on the persistence of stream-dwelling charr populations. *Conservation Biology* 16:1318–1323.

Mosler-Berger, C., and J. Romer. 2003. Wildwarnsystem CALSTROM. *Wildbiology* 3:1–12.

Mumme, R. L., S. J. Schoech, G. E. Woolfenden, and J. W. Fitzpatrick. 2000. Life and death in the fast lane: demographic consequences of road mortality in the Florida Scrub-Jay. *Conservation Biology* 14:501–512.

Muurinen, I., and T. Ristola. 1999. Elk accidents can be reduced by using transport telematics. Finncontact 7(1):7–8.

National Academy of Sciences. 2005. *Assessing and managing the ecological impacts of paved roads*. National Academies Press, Washington, DC.

National Park Service. 1990. *I-75 recreational access plan/environmental assessments*. U.S. Department of the Interior. Atlanta, Georgia.

———. 2006. Park road temporarily closed for amphibian migration. NPS Digest. Delaware Water Gap National Recreation Area, U.S. Dept. of the Interior National Park Service. Press release March 15, 2006.

Naz, R. K., S. K. Gupta, J. C. Gupta, H. K. Vyas, and G. P. Talwar. 2005. Recent advances in contraceptive vaccine development: a mini-review. *Human Reproduction (Oxford)* 20(12): 3271–3283

Nettles, V. F. 1997. Potential consequences and problems with wildlife contraceptives. *Reproduction Fertility and Development* 9(1): 137–143.

Newman, C., C. Buesching, and D. Macdonald. 2003. Validating mammal monitoring methods and assessing the performance of volunteers in wildlife conservation. *Biological Conservation* 113:189–197.

Newmark, W. D. 1995. A land-bridge island perspective on mammalian extinctions in western North American national parks. *Conservation Biology* 9:512–526.

New Mexico Department of Game and Fish. 2006. *Comprehensive wildlife conservation strategy for New Mexico*. New Mexico Department of Game and Fish. Santa Fe, New Mexico.

Ng, J. W., C. Nielsen, and C. Cassady St. Clair. 2008. Landscape and traffic factors influencing deer–vehicle collisions in an urban environment. *Human–Wildlife Conflicts* 2(1): 34–47.

Ng, S. J., J. W. Dole, R. M. Sauvajot, S. P. D. Riley, and T. J. Valone. 2004. Use of highway underpasses by wildlife in southern California. *Biological Conservation* 115:499–507.

Nicholson, E., J. Ryan, and D. Hodgkins. 2002. Community data—where does the value lie? Assessing confidence limits of community collected water quality data. *Water Science and Technology* 45:193–200.

Nicholson, J. M. 2009. Population and genetic impacts of a 4-lane highway on black bears in eastern North Carolina. Master's thesis, University of Tennessee, Knoxville.

Niehaus, A. C., S. B. Heard, S. D. Hendrix, and S. L. Hillis. 2003. Measuring edge effects on nest predation in forest fragments: Do finch and quail eggs tell different stories? *American Midland Naturalist* 149:335–343.

Niemuth, N. D., and M. S. Boyce. 1997. Edge-related nest losses in Wisconsin pine barrens. *Journal of Wildlife Management* 61:1234–1239.

North Carolina Department of Transportation. 1999. State final environmental impact statement: US64, from NC 45 east of Plymouth to approximately 1.1 km (0.7 mi) east of SR 1235 (School Maintenance Road), Washington and Tyrrell Counties. North Carolina Department of Transportation, Division of Highways, Raleigh.

Noss, R. F. 1990. Indicators of monitoring biodiversity: A hierarchical approach. *Conservation Biology* 4:355–364.

Noss, R. F., and A. Y. Cooperrider. 1994. *Saving nature's legacy*. Island Press, Washington, DC.

Noss, R. F., and K. M. Daly. 2006. Incorporating connectivity into broad scale planning. Pages 587–619 in K. R. Crooks and M. Sanjayan, eds. *Connectivity conservation*. Cambridge University Press, Cambridge.

Nowak, R. M. 1973. Status survey of the southeastern puma. Pages 112–113 in *World Wildlife Fund yearbook 1972–73*. Danbury Press, Danbury, Connecticut.

Office of Environmental Protection. 1995. Memorandum of understanding to foster the ecosystem approach. The White House, Washington, DC.

O'Hanley, J. R., and D. Tomberlin. 2005. Optimizing the removal of small fish passage barriers. *Environmental Modeling and Assessment* 10:85–98.

Olsson, M. 2007. The use of highway crossings to maintain landscape connectivity for moose and roe deer. PhD diss., Karlstad University, Karlstad, Sweden.

Omar, M., and Y. Zhou. 2007. Pedestrian tracking routine for passive automotive night vision systems. *Sensor Review* 27(4): 310–316.

O'Neill, R. V., R. H. Gardner, and M. G. Turner. 1992. A hierarchical neutral model for landscape analysis. *Landscape Ecology* 78:55–61.

Opdam, P. F. M. 1997. How to choose the right solution for the right fragmentation problem? Pages 55–60 in K. Canters, ed. *Habitat fragmentation and infrastructure*. Ministry of Transportation, Public Works and Water Management, Delft, the Netherlands.

Ortega, Y. K., and D. E. Capen. 1999. Effects of forest roads on habitat quality for ovenbirds in a forested landscape. *Auk* 116:937–946.

Pafko, F., and B. Kovach. 1996. *Experience with deer reflectors: Trends in addressing transportation-related wildlife mortality*. Minnesota Department of Transportation, Office of Environmental Services, Minneapolis.

Pannell, J. R., and B. Charlesworth. 2000. Effects of metapopulation processes on measures of genetic diversity. *Philosophical Transactions of the Royal Society* 355:1851–1864.

Parks Canada. 1995. Initial assessment of proposed improvements to the Trans-Canada Highway in Banff National Park—Phase IIIA Sunshine interchange to Castle junction interchange. Parks Canada, Canadian Heritage, Ottawa, Ontario.

Parks, S. A., and A. H. Harcourt. 2002. Reserve size, local human density, and mammalian extinctions in U.S. protected areas. *Conservation Biology* 16:800–808.

Parmesan, C., and G. Yohe. 2003. A globally coherent fingerprint of climate change impacts across natural systems. *Nature* 421:37–42.

Parsons, D. R. 2003. Natural history characteristics of focal species in the New Mexico Highlands Wildlands Network. Appendix in *The Wildlands Project 2003: New Mexico Highlands Wildlands Network Vision: Connecting the Sky Islands to the Southern Rockies.* The Wildlands Project, Richmond, Vermont.

Paul, K. 2007. Auditing a monitoring program: Can citizen science represent wildlife activity along highways? M.Sc. thesis, environmental studies, University of Montana, Missoula.

P. B. Farradyne, Inc. 2006. Intelligent transportation systems and Ft. Myers Regional Transportation Management Center concept of operations. Prepared for Florida Department of Transportation, District 1.

Pearce, J.L, and M.S. Boyce. 2006. Modeling distribution and abundance with presence-only data. *Journal of Applied Ecology* 43:405–412.

Pearson, W., M. Richmond, G. Johnson, S. Sargeant, R. Mueller, V. Cullinan, Z. Deng, B. Dibrani, G. Guensch, C. May, L. O'Rourke, K. Sobocinski, and H. Tritico. 2005. Protocols for evaluation of upstream passage of juvenile salmonids in an experimental culvert test bed. WA-RD 614.1. Washington, DC.

Peck, L. J., and J. E. Stahl. 1997. Deer management techniques employed by the Columbus and Franklin County Park District, Ohio. *Wildlife Society Bulletin* 25:440–442.

Peninsula Clarion. 2008. Moose Force One set for winter action: Big snow cat ready to steer ungulates away from busy peninsula roadways. *Peninsula Clarion*, November 10. Available at http://peninsulaclarion.com/stories/111008/new_354387072.shtml.

Peterson, D. P., B. E. Rieman, J. B. Dunham, K. D. Fausch, and M. K. Young. 2008. Analysis of trade-offs between threats of invasion by nonnative brook trout (*Salvelinus fontinalis*) and intentional isolation for native westslope cutthroat trout (*Oncorhynchus clarkii lewisi*). *Canadian Journal of Fisheries and Aquatic Sciences* 65:557–573.

Peterson, M. N. 2003. Management strategies for endangered Florida key deer. MSc. thesis, Texas A&M University, College Station.

Pettorelli, N., J.-M. Gaillard, N. G. Yoccoz, P. Duncan, D. Maillard, D. Delorme, G. Van Laere, and C. Toigo. 2005. The response of fawn survival to changes in habitat quality varies according to cohort quality and spatial scale. *Journal of Animal Ecology* 74(5): 972–981.

Pima Association of Governments (PAG). 2008. Population facts. Accessed October 24, 2008, at http://www.pagnet.org/RegionalData/Population/tabid/104/Default.aspx.

Pima County. 2004. Pima County past results: Special election May 18, 2004. Accessed March 11, 2009, at http://www.pima.gov/elections/res0504.htm.

———. 2005. Conservation lands system. Accessed March 11, 2009, at http://

www.pimaxpress.com/Planning/ComprehensivePlan/PDF/CLS/CLS_Adopted _Policy.pdf.

———. 2008. Pima County multiple species conservation plan: Version 5. Available at http://www.pima.gov/CMO/SDCP/MSCP/MSCP.html.

Platte, M. 1985. Final I-75 plans worked out. *Miami Herald*, November 6.

Pojar, T. M., R. A. Prosence, D. F. Reed, and T. N. Woodard. 1975. Effectiveness of a lighted, animated deer crossing sign. *Journal of Wildlife Management* 39:87–91.

Pollock, R., and G. Whitelaw. 2005. Community-based monitoring in support of local sustainability. *Local Environment* 10:211–228.

Pollock, R., G. Whitelaw, and D. Atkinson. 2003. Appendix 1A: Literature review in support of community based monitoring conceptual framework development, evaluation and enhancement. Canadian Nature Federation, Ecological Monitoring and Assessment Network Coordinating Office, Environment Canada.

Porter, W. F., and H. B. Underwood. 1999. Of elephants and blind men: Deer management in the U.S. National Parks. *Ecological Applications* 9(1): 3–9.

Porter, W. F., H. B. Underwood, and J. L. Woodard. 2004. Movement behavior, dispersal, and the potential for localized management of deer in a suburban environment. *Journal of Wildlife Management* 68(2): 247–256.

Powers, P. D. 1997. Culvert hydraulics related to upstream juvenile salmon passage. Report prepared for the Washington State Department of Transportation.

Primack, R. B. 2006. Habitat Destruction. Pages 189–193 in R. B. Primack, ed. *Essentials of Conservation Biology*. Sinauer Associates, Sunderland, U.K.

Proctor, M. F. 2003. Genetic analysis of movement, dispersal and population fragmentation of grizzly bears in southwestern Canada. PhD thesis, University of Calgary.

Proctor, M. F., B. N. McLellan, and C. Strobeck. 2002. Population fragmentation of grizzly bears in southeastern British Columbia, Canada. *Ursus* 13:153–160.

Proctor, M. F., B. N. McLellan, C. Strobeck, and R. M. R. Barclay. 2005. Genetic analysis reveals demographic fragmentation of grizzly bears yielding vulnerably small populations. *Proceedings of the Royal Society B* 272:2409–2416.

Putman, R. J. 1997. Deer and road traffic accidents: Options for management. *Journal of Environmental Management* 51:43–57.

Quintana, F. L., and D. Kayser. 1980. The development of the Tijeras Canyon hispanic Communities. Pages 41–59 in L. S. Cordell, ed. *Tijeras Canyon: Analysis of the past*. University of New Mexico Press, Albuquerque.

Rajput, S. 2003. The effects of low-water bridges on movement, community structure and abundance of fishes in streams of the Ouachita Mountains. Master's thesis, Arkansas Tech University, Russellville.

Ramp, D., J. Caldwell, K. Edwards, D. Warton, and D. Croft. 2005. Modeling of wildlife fatality hotspots along the Snowy Mountain Highway in New South Wales, Australia. *Biological Conservation* 126:474–490.

Ramp, D., and D. Croft. 2002. Saving wildlife: Saving people on our roads: Annual report. The University of New South Wales, School of Biological, Earth and Environmental Sciences. Prepared for the International Fund for Animal Welfare, Roads and Traffic Authority and the New South Wales Wildlife Information and Rescue Service, Australia.

Rea, R. V. 2003. Modifying roadside vegetation management practice to reduce vehicular collisions with moose (*Alces alces*). *Wildlife Biology* 9(2): 81–91.

Rea, R. V., and M. P. Gillingham. 2001. The impact of timing of brush management on the nutritional value of woody browse for moose (*Alces alces*). *Journal of Applied Ecology* 38:710–719.

Reed, D. F., T. D. I. Beck, and T. N. Woodard. 1982. Methods of reducing deer vehicle accidents: Benefit–cost analyses. *Wildlife Society Bulletin* 10:349–354.

Reed, R. A., J. Johnson-Barnhard, and W. L. Baker. 1996. Contribution of roads to forest fragmentation in the Rocky Mountains. *Conservation Biology* 10:1098–1106.

Reed, D. F., and T. N. Woodard. 1981. Effectiveness of highway lighting in reducing deer vehicle accidents. *Journal of Wildlife Management* 45(3): 721–726.

Reeve, A. F., and S. H. Anderson. 1993. Ineffectiveness of Swareflex reflectors at reducing deer vehicle collisions. *Wildlife Society Bulletin* 21:127–132.

Regional Transportation Authority. 2009. RTA Wildlife Linkages Working Group. Accessed March 13, 2009, at http://www.rtamobility.com/index.php?option=com_content&task=blogcategory&id=88&Itemid=273.

Reijnen, R., R. Foppen, C. ter Braak, and J. Thissen. 1995. The effects of car traffic on breeding bird populations in Woodland, Ill: Reduction of density in relation to the proximity of main roads. *Journal of Applied Ecology* 32:187–202.

Reijnen, R., R. Foppen, and H. Meeuwsen. 1996. The effects of car traffic on the density of breeding birds in Dutch agricultural grass lands. *Biological Conservation* 75:255–260.

Reiser, D. W., C. Huang, S. Beck, M. Gagner, and E. Jeanes. 2006. Defining flow windows for upstream passage of adult anadromous salmonids at cascades and falls. *Transactions of the American Fisheries Society* 135: 668–679.

Rettie K., A. P. Clevenger, and A. T. Ford. 2009. Innovative approaches for managing conservation and use challenges in the national parks. Pages 396–415 in T. Jamal and M. Robinson, eds. *An example from Canada*. Sage Handbook of Tourism Studies. Sage Publications, Inc., Thousand Oaks, California.

Rice, C. G. 2008 Seasonal altitudinal movements of mountain goats. *Journal of Wildlife Management* 72:1706–1716

Rieman, B. E., and J. D. McIntyre. 1993. Demographic and habitat requirements for conservation of bull trout. General Technical Report INT-302. USDA Forest Service, Intermountain Research Station. Ogden, Utah.

Riley, S. J., D. J. Decker, J. W. Enck, P. D. Curtis, T. B. Lauber, and T. L. Brown. 2003. Deer populations up, hunter populations down: Implications of interdependence of deer and hunter population dynamics on management. *Ecoscience* 10(4): 455–461.

Riley, S. J., and K. Sudharsan. 2006. Environmental factors affecting the frequency and rate of deer vehicle crashes (DVCs) in southern Michigan. Final report RC-1476. Department of Fisheries and Wildlife, Michigan State University, East Lansing.

Ritters, K. H., and J. D. Wickham. 2003. How far to the nearest road? *Frontiers in Ecology and the Environment* 1:125–129.

Robbins, J. 2008. Thinking anew about a migratory barrier: roads. *New York Times*, October 14. Available at http://www.nytimes.com/2008/10/14/science/14road.html.

Roedenbeck, I. A., L. Fahrig, C. S. Findlay, J. E. Houlahan, J. A. G. Jaeger, N. Klar, S. Kramer-Schadt, and E. A van der Grift. 2007. The Rauschholzhausen agenda for road ecology. *Ecology and Society* 12:11–31.

Roever, C. L., M. S. Boyce, and G. B. Stenhouse. 2008. Grizzly bears and forestry: Grizzly bear habitat selection and conflicts with road placement. *Forest Ecology and Management* 256:1262–1269.

Rogers, E. 2004. An ecological landscape study of deer vehicle collisions in Kent County, Michigan. Report by White Water Associates Inc. Prepared for Kent County Road Commission, Grand Rapids, Michigan.

Romer, J., and C. Mosler-Berger. 2003. Preventing wildlife vehicle accidents: the animal detection system CALSTROM. In *Proceedings of the 2003 Infra Eco Network Europe Conference: Habitat fragmentation due to transport infrastructure and presentation of the COST 341 action*. Brussels, Belgium.

Romin, L. A., and J. A. Bissonette. 1996. Deer–vehicle collisions: Status of state monitoring activities and mitigation efforts. *Wildlife Society Bulletin* 24:276–283.

Romin, L., and L. B. Dalton. 1992. Lack of response by mule deer to wildlife warning whistles. *Wildlife Society Bulletin* 20:382–384.

Rooney, T. P. 2001. Deer impacts on forest ecosystems: A North American perspective. *Forestry* 74(3): 201–208.

Rosgen, D. 1996. *Applied river morphology*. Wildland Hydrology, Pagosa Springs, Colorado.

Rost, G. R., and J. A. Bailey. 1979. Distribution of mule deer and elk in relation to roads. *Journal of Wildlife Management* 43:634–641.

Rowcliffe, J. M., J. Field, S. T. Turvey, and C. Carbone. 2008. Estimating animal density using camera traps without the need for individual recognition. *Journal of Applied Ecology* 45:1228–1236.

Rowland, M. M., M. J. Wisdom, B. K. Johnson, and J. G. Kie. 2000. Elk distribution and modeling in relation to roads. *Journal of Wildlife Management* 64:672–684.

Rudolph, B. A., W. F. Porter, and H. B. Underwood. 2000. Evaluating immunocontraception for managing suburban white-tailed deer in Irondequoit, New York. *Journal of Wildlife Management* 64(2): 463–473.

Ruediger, B. 2007. Management considerations for designing carnivore highway crossings. Pages 546–555 in C. L. Irwin, D. Nelson, and K. P. McDermott, eds. *Proceedings of the 2007 International Conference on Ecology and Transportation*. Center for Transportation and the Environment, North Carolina State University, Raleigh.

Ruediger, B., J. Claar, and J. Gore. 1999. Restoration of carnivore habitat connectivity in the Northern Rockies. Pages 5–20 in G. L. Evink, P. Garrett, and D. Zeigler, eds. *Proceedings of the Third International Conference on Wildlife Ecology and Transportation*. Publication FL-ER-72-99. Florida Department of Transportation, Tallahassee.

Ruediger, B., and J. Lloyd. 2003. A rapid assessment process for determining potential wildlife, fish and plant linkages for highways. Pages 205–225 in C. L. Irwin, P. Garrett, and K. P. McDermott, eds. *2003 Proceedings of the International Conference on Ecology and Transportation*. Center for Transportation and the Environment, North Carolina State University, Raleigh.

Saaty, T. L. 1980. *The analytical hierarchy process: Planning, priority setting, resource allocation*. McGraw-Hill, New York.

——. 1990. How to make a decision-the analytic hierarchy process. *European Journal of Operational Research* 8:9–26.

Sagar, J. P., D. H. Olson, R. A. Schmitz, and J. Guetterman. 2003. Are stream crossing culverts a barrier to the movement of the Pacific giant salamander (*Dicamptodon tenebrosus*)? *Northwestern Naturalist* 84(2): 113–114.

Saltzman, W., and O. R. Koski. 1971. Fish passage through culverts. Oregon State Game Commission Special Report. Portland, Oregon.

Sanderson, E. W., M. Jaiteh, M. A. Levy, K. H. Redford, A. V. Wannebo, and G. Woolmer. 2002. The human footprint and the last of the wild. *BioScience* 52:891–904.

Savan, B., A. Morgan, and C. Gore. 2003. Volunteer environmental monitoring and the role of universities: The case of citizens' Environment Watch. *Environmental Management* 21:561–568.

Schafer J. A., and S. T. Penland. 1985. Effectiveness of Swareflex reflectors in reducing deer vehicle accidents. *Journal of Wildlife Management* 49:774–776.

Schaefer, J. M., and D. J. Smith. 2000. Ecological characterization of identified high priority highway-ecological interface zones including the inventory and evaluation of existing Florida Department of Transportation highway facilities within these zones. Florida Department of Transportation contract no. B-B120, Tallahassee.

Scheick, B. K., and M. D. Jones. 1999. Locating wildlife underpasses prior to expansion of highway 64 in North Carolina. Pages 247–251 in G. L. Evink, P. Garrett, and D. Zeigler, eds. *Proceedings of the International Conference on Wildlife Ecology and Transportation*. FL-ER-73-99, Florida Department of Transportation, Tallahassee.

——. 2000. US64 wildlife underpass placement report, North Carolina Wildlife Resources Commission, Raleigh.

Schwabe, K. A., and P. W. Schuhmann. 2002. Deer–vehicle collisions and deer value: An analysis of competing literatures. *Wildlife Society Bulletin* 30:609–615.

Schwartz, M. K., G. Luikart, and R. S. Waples. 2006. Genetic monitoring as a promising tool for conservation and management. *Trends in Ecology and Evolution* 22:25–33.

Seagle, S. W., and J. D. Close. 1996. Modeling white-tailed deer (*Odocoileus virginianus*) population control by contraception. *Biological Conservation* 76(1): 87–91.

Seamans, T. W., and K. C. VerCauteren. 2006. Evaluation of ElectroBraide fencing as a white-tailed deer barrier. *Wildlife Society Bulletin* 34:8–15.

Seiler, A. 2003. The toll of the automobile: Wildlife and roads in Sweden. Doctoral thesis, Swedish University of Agricultural Sciences, Uppsala.

Shawnee National Forest. 2006. Snake migration LaRue–Pine Hills. Shawnee National Forest Mississippi Bluffs Ranger District. Available at http://www.fs .fed.us/r9/forests/shawnee/recreation/rogs/snake-migration.pdf.

Shipley, L. A. 2001. Evaluating Wolfin™ as a repellent to wildlife on roads in Washington and the feasibility of using deer-activated warning signs to reduce deer–automobile collisions on highways in Washington. Research Project #T9902. Department of Natural Resource Sciences, Washington State University, Pullman.

Shury, T. 1996. A summary of wildlife mortality in Banff National Park, 1981–1995. Final report submitted to the Warden Service, Banff National Park, Alberta.

Sielecki, L. E. 1999. WARS-Wildlife accident reporting system: 1998 Annual report, 1994–1998 Synopsis. British Columbia Ministry of Transportation and Highways, Environmental Services Section, Victoria, British Columbia.

———. 2000. WARS 2000: Wildlife accident reporting system, 2000 annual report. British Columbia Ministry of Transportation and Highways, Environmental Services Section, Victoria, British Columbia.

———. 2004. WARS 1983–2002: Wildlife accident reporting and mitigation in British Columbia: Special annual report. Ministry of Transportation, Engineering Branch, Environmental Management Section. Victoria, British Columbia.

Singleton, P. H., W. L. Gaines, and J. F. Lehmkuhl. 2002. Landscape permeability for large carnivores in Washington: A geographic information system weighted-distance and least-cost corridor assessment. USDA Forest Service research paper, PNW-RP-549.

Singleton, P. H., and J. Lehmukuhl. 2001. Using weighted distance and least-cost corridor analysis to evaluate regional-scale large carnivore habitat connectivity in Washington. Pages 583–594 in *Proceedings of the International Conference on Ecology and Transportation*, Keystone.

Sivic, A., and L. Sielecki. 2001. Wildlife warning reflectors spectrometric evaluation. Ministry of Transportation, British Columbia, Canada.

Smith, D. 1999. Identification and prioritization of ecological interface zones on state highways in Florida. Pages 209–230 in G.L. Evink, P. Garrett, and D. Zeigler, eds. *Proceedings of the Third International Conference on Wildlife Ecology and Transportation*. FL-ER-73-99. Florida Department of Transportation, Tallahassee.

Smith, D. J. 2003. Monitoring wildlife use and determining standards for culvert design. Final report. Florida Department of Transportation Contract no. BC354-34. Department of Wildlife Ecology and Conservation, University of Florida, Gainesville.

——. 2006. Incorporating results from the prioritized "ecological hotspots" model into the Efficient Transportation Decision Making (ETDM) process in Florida. Pages 127–137 in C. L. Irwin, P. Garrett, and K. P. McDermott, eds. Proceedings of the 2005 *International Conference on Ecology and Transportation*, Center for Transportation and the Environment, North Carolina State University, Raleigh.

Smith, D. J., R. F. Noss, and M. B. Main. 2006. East Collier County wildlife movement study: SR 29, CR 846, and CR 858 wildlife crossing project. Unpublished report. University of Central Florida, Orlando.

Smith, T. R. 1998. History of the Flathead and Pend d' Oreille. Pages 305–312 in Deward E. Walker Jr., ed. Handbook of North American Indians, vol. 12. Smithsonian Institution. Plateau, Washington, DC.

Solcz, A. 2007. Assessment of culvert passage of Yellowstone cutthroat trout in a Yellowstone River spawning tributary using a passive integrated transponder system. Master's thesis. Montana State University, Bozeman.

Spellerberg, I. F. 1998. Ecological effects of roads and traffic: A literature review. *Global Ecology and Biogeography* 7:317–333.

——. 2002. *Ecological effects of roads*. Science Publisher, Plymouth, U.K.

Stanley, L. A. Hardy, and S. Lassacher. 2006. Driver responses to enhanced wildlife advisories in a simulated environment. TRB 2006 Annual Meeting CD-ROM. Transportation Research Board: Washington, DC.

Steinemann, R., A. Krek, and T. Blaschet. 2004. Can online map-based applications improve citizen participation? TED Conference on e-Government, Bozen, Italy.

Store, R., and J. Kangas. 2001. Integrating spatial multi-criteria evaluation and expert knowledge for GIS-based habitat suitability modeling. *Landscape and Urban Planning* 55:79–93.

Stuart, J. N., M. L. Watson, T. L. Brown, and C. Eustice. 2001. Plastic netting: An entanglement hazard to snakes and other wildlife. *Herpetological Review* 32:162–163.

Stuart, T. A. 1962. The leaping behaviour of salmon and trout at falls and obstructions. Freshwater Fisheries Laboratory, Department of Agriculture and Fisheries for Scotland. Pitlochry, Scotland.

Sullivan, T. L., A. E. Williams, T. A. Messmer, L. A. Hellinga, and S. Y. Kyrychenko.

2004. Effectiveness of temporary warning signs in reducing deer vehicle collisions during mule deer migrations. *Wildlife Society Bulletin* 32(3): 907–915.

Sutherland, G. D., A. S. Harestad, and K. Price. 2000. Scaling of natal dispersal distances in terrestrial birds and mammals. *Conservation Ecology* 4(1): 16.

Swanson, K., D. Land, R. Kautz, and R. Kawula. 2005. Least cost pathways to identify key highway segments for Florida panther conservation. Pages 191–200 in R. A. Beausoleil and D. A. Martorello, eds. *Proceedings of the Eighth Mountain Lion Workshop*, Olympia, Washington.

Swihart, R. K., H. P. Weeks Jr., A. L. Easter-Pilcher, and A. J. DeNicola. 1998. Nutritional condition and fertility of white-tailed deer (*Odocoileus virginianus*) from areas with contrasting histories of hunting. *Canadian Journal of Zoology* 76(10): 1932–1941.

Tardif, L.-P. and Associates Inc. 2003. Collisions involving motor vehicles and large animals in Canada. Final report. L-P Tardif and Associates Inc., Nepean, Ontario.

Taylor, P. D., L. Fahrig, and K. Henein. 1993. Connectivity is a vital element of landscape structure. *Oikos* 68:571–573.

Tayor, R. N. 2001. Final report: Del Norte County culvert inventory and fish passage evaluation. California.

Tchir, J. P., P. J. Hvenegaard, and G. J. Scrimgeour. 2004. Stream crossing inventories in the Swan and Notikewin river basins of northwest Alberta: Resolution at the watershed scale. Pages 53–62 in G. J. Scrimgeour, G. Eisler, B. McCulloch, U. Silins, and M. Monita, eds. *Forest Land–Fish Conference II—Ecosystem Stewardship through Collaboration*. Proceedings of the Forest Land–Fish Conference II, April 26–28, 2004, Edmonton, Alberta.

Terborgh, J., A. Estes, P. Paquet, K. Ralls, D. Boyd-Heger, B. Miller, and R. Noss. 1999. The role of top carnivores in regulating terrestrial ecosystems. Pages 39–64 in M. E. Soulé and J. Terborgh, eds. *Continental Conservation: Scientific Foundations of Regional Reserve Networks: The Wildlands Project*. Island Press, Washington, DC.

Tewksbury, J. J., A. B. Black, N. Nur, V. Saab, B. L. Logan, and D. S. Dobkin. 2002b. Effects of anthropogenic fragmentation and livestock grazing on western riparian bird communities. *Studies in Avian Biology* 25:158–202.

Tewksbury, J. J., D. J. Levey, N. M. Haddad, S. Sargent, J. L. Orrock, A. Weldon, B. J. Danielson, J. Brinkerhoff, E. Damschen, and P. Townsend. 2002a. Corridors affect plants, animals, and their interactions in fragmented landscapes. *Proceedings of the National Academy of Sciences* 99:12923–12926.

Theobald, D. M., N. T. Hobbs, T. Bearly, J. A. Zack, T. Shenk, and W. E. Riebsame. 2000. Incorporating biological information in local land-use decision making: designing a system for conservation planning. *Landscape Ecology* 15(1): 35–45.

The Economist. 2005. A bridge too far? *Economist*, June 9. Available at http://www.heingartner.com/econ/bridges.htm.

The Nature Conservancy. 2009. Sonoran Desert ecoregional plan: Overview. Accessed March 14, 2009, at http://www.nature.org/aboutus/howwework/cbd/science/art14338.html.

The Wildlands Project. 2003. New Mexico Highlands Wildlands Network Vision. Connecting the Sky Islands to the Southern Rockies. The Wildlands Project, Richmond, Vermont.

Thomas, C. D., A. Cameron, R. E. Green, M. Bakkenes, L. J. Beaumont, Y. C. Collingham, B. F. N. Erasmus, M. F. De Siqueira, A. Grainger, L. Hannah, L. Hughes, B. Huntley, A. S. Van Jaarsveld, G. F. Midgley, L. Miles, M. A. Ortega-Huerta, A. T. Peterson, O. L. Phillips, and S. E. Williams. 2004. Extinction risk from climate change. *Nature* 427:145–148.

Thomas, S. E. 1995. Moose vehicle accidents on Alaska's rural highways. Alaska Department of Transportation and Public Facilities, Central Region, Design and Construction Division, Alaska.

Thompson, L. M. 2003. Abundance and genetic structure of two black bear populations prior to highway construction in eastern North Carolina. Master's thesis, University of Tennessee, Knoxville.

Tijeras Canyon Safe Passage Coalition. 2005. Comments on NMDOT I-40 reconstruction project (CN G-1243) and preliminary feasibility study.

Todaro, V. 2006. Salamander season has sprung: Lizard like amphibians begin annual migration across Beekman Road. *Sentinel*, March 16. Available at http://ebs.gmnews.com/news/2006/0316/Front_Page/006.html.

Toledo Blade. 2008. Plastic bridge in Huron County may be path to the future. *Toledo Blade*, October 6. Available at http://www.toledoblade.com/apps/pbcs.dll/article?AID=/20081006/NEWS11/810060335.

Transportation Research Board. 2002a. *Environmental research needs in transportation*. Conference proceedings 28. National Academy Press, Washington, DC.

———. 2002b. *Interaction between roadways and wildlife ecology*. National Cooperative Highway Research Program. Washington, DC.

———. 2004. *Environmental spatial information for transportation: A peer exchange on partnerships*. Conference proceedings on the Web 1. National Academies Press, Washington, DC.

Travis, M. D., and T. Tilsworth. 1986. Fish passage through Poplar Grove Creek culvert, Alaska. *Transportation Research Record* 1075:21–26.

Trombulak, S. C., and C. A. Frissell. 2000. Review of ecological effects of roads on terrestrial and aquatic communities. *Conservation Biology* 14:18–30.

Turner, J. W. Jr., J. F. Kirkpatrick, and I. K. M. Liu. 1996. Effectiveness, reversibility, and serum antibody titers associated with immunocontraception in captive white-tailed deer. *Journal of Wildlife Management* 60(1): 45–51.

Turner, J. W. Jr., I. K. M. Liu, and J. F. Kirkpatrick. 1992. Remotely delivered immunocontraception in captive white-tailed deer. *Journal of Wildlife Management* 56(1): 154–157.

Turrentine, T., K. Heanue, and D. Sperling. 2001. Road and vehicle system. *Proceedings of the International Conference on Ecology and Transportation*. September 24–28, 2001, Keystone, Colorado.

Tyser, R. W., and C. A. Worley. 1992. Alien flora in grasslands adjacent to road and trail corridors in Glacier National Park, Montana (USA). *Conservation Biology* 6:253–262.

Ujvari M., H. J. Baagoe, and A. B. Madsen. 1998. Effectiveness of wildlife warning reflectors in reducing deer vehicle collisions: A behavioral study. *Journal of Wildlife Management* 62:1094–1099.

———. 2004. Effectiveness of acoustic road markings in reducing deer vehicle collisions: A behavioral study. *Wildlife Biology* 10:155–159.

University of New Mexico Center for Wildlife Law. 1992. *Wild friends: Kids bringing people together on wildlife issues: A manual for kids, teachers, volunteers and others*. University of New Mexico Center for Wildlife Law. Albuquerque.

U.S. Census Bureau. 2002. United States Census 2000. U.S. Census Bureau, Washington, DC. Available at http://www.census.gov/main/www/cen2000.html. Accessed 2 October 2008.

U.S. Centers for Disease Control and Prevention. 1991. Effectiveness in disease and injury prevention from motor-vehicle collisions with deer, Kentucky, 1987–1989. *Morbidity and Mortality Weekly Report* 40:717–719.

U.S. Department of Agriculture (USDA). 2005. National inventory and assessment procedure for identifying barriers to aquatic organism passage at road-stream crossings. San Dimas, California.

———. 2008. Stream simulation: An ecological approach to providing passage for aquatic organisms at road-stream crossings. San Dimas, California.

U.S. Department of Agriculture (USDA), Forest Service, Northern Region. 2008. Assessment of aquatic organism passage at road/stream crossings for the northern region of the USDA Forest Service. Missoula, Montana.

U.S. Department of Transportation. 2008. *Design of fish passage at bridges and culverts*. Hydraulic Engineering Circular No. 26, 1st ed. Washington, DC.

U.S. Fish and Wildlife Service (USFWS). 2003. National fish passage program: Reconnecting aquatic species to historic habitats. Accessed December 10, 2008, at http://www.fs.gov/fisheries/fwma/fishpassage/pdfs/nfpp.pdf.

Van Deelen, T. R., B. Dheuy, K. R. McCaffery, and R. E. Rolley. 2006. Relative effects of baiting and supplemental antlerless seasons on Wisconsin's 2003 deer harvest. *Wildlife Society Bulletin* 34(2): 322–328.

van Manen, F. T., M. D. Jones, J. L. Kindall, L. M. Thompson, and B. K. Scheick. 2001. Determining the potential mitigation effects of wildlife passageways on black bears. *International Conference on Ecology and Transportation* 4:435–446.

VerCauteren, K. C., J. M. Gilsdorf, S. E. Hygnstrom, P. B. Fioranelli, J. A. Wilson, and S. Barras. 2006. Green and blue lasers are ineffective for dispersing deer at night. *Wildlife Society Bulletin* 34(2): 371–374.

Vermont Agency of Transportation. 2002. Public involvement: Vermont long range transportation plan. Vermont Agency of Transportation. Burlington.

Vermont Forum on Sprawl. 1999. Exploring Sprawl #6—Economic, Social, and Land Use Trends Related to Sprawl in Vermont. Burlington.

Vokurka, C., and R. Young. 2008. Relating vehicle–wildlife crashes to road reconstruction. Proceedings from the 87th Annual Meeting of the Transportation Research Board, Washington, DC.

Votapka, F. E. 1991. Considerations for fish passage through culverts. *Transportation Research Record* 1291:347–353.

Waddell, R. B., D. A. Osborn, R. J. Warren, J. C. Griffin, and D. J. Kesler. 2001. Prostaglandin F2infin-mediated fertility control in captive white-tailed deer. *Wildlife Society Bulletin* 29(4): 1067–1074.

Wagner, P. 2006. Improving mobility for wildlife and people: Transportation planning for habitat connectivity in Washington State. Page 79 in C. Leroy Irwin, Paul Garrett, and K. P. McDermott, eds. *Proceedings of the 2005 International Conference on Ecology and Transportation*. Center for Transportation and the Environment, North Carolina State University, Raleigh.

Waller, D. M., and W. S. Alverson. 1997. The white-tailed deer: A keystone herbivore. *Wildlife Society Bulletin* 25(2): 217–226.

Waller, J. S., and C. Servheen. 2005. Effects of transportation infrastructure on grizzly bears in northwestern Montana. *Journal of Wildlife Management* 69:985–1000.

Walters, C. J. 1986. *Adaptive management of renewable natural resources*. McGraw-Hill Publishers, New York.

Ward, A. L 1982. Mule deer behavior in relation to fencing and underpasses on Interstate 80 in Wyoming. *Transportation Research Record* 859:8–13.

Ward, R. L., and J. T. Anderson. 2008. Effects of road crossings on stream and streamside salamanders. *Journal of Wildlife Management* 72: 760–771.

Wardle, C. S. 1980. Effects of temperature on the maximum swimming speed of fishes. Pages 519–531 in M. A. Ali, ed. *Environmental Physiology of Fishes*. Plenum, New York.

Waring, G. H., J. L. Griffis, and M. E. Vaughn. 1991. White-tailed deer roadside behavior, wildlife warning reflectors, and highway mortality. *Applied Animal Behavior Science* 29:215–223.

Warren, M. L., and M. G. Pardew. 1998. Road crossings as barriers to small-stream fish movement. *Transactions of the American Fisheries Society* 127:637–644.

Warshall, P. Undated. A Madrean Sky Island archipelago: A planetary overview. Accessed October 24, 2008, at http://www.skyislandalliance.org/publications .htm.

Washington Department of Fish and Wildlife. 2000. Fish passage barrier and surface water diversion screening assessment and prioritization manual. Olympia, Washington.

Weaver, J. L. 2001. *The Transboundary Flathead: A critical landscape for carnivores in the Rocky Mountains*. Wildlife Conservation Society, Bronx, New York.

Weaver, J. L., P. C. Paquet, and L. F. Ruggiero. 1996. Resilience and conservation of large carnivores. *Conservation Biology* 10:964–976.

Weiss, S. 1999. Cars, cows, and checkerspot butterflies: Nitrogen deposition and management of nutrient-poor grasslands for a threatened species. *Conservation Biology* 13:1476–1486.

Wells, P., J. Woods, G. Bridgewater, and H. Morrison. 1999. Wildlife mortalities on railways: monitoring methods and mitigation strategies. Pages 85–88 in G. L. Evink, P. Garrett, and D. Zeigler, eds. *Proceedings of the Third International Conference on Wildlife Ecology and Transportation*. FL-ER-73-99. Florida Department of Transportation, Tallahassee.

Wemple, B. C., and J. A. Jones. 2003. Runoff production on forest roads in a steep, mountain catchment. *Water Resources Research* 39:1–17.

Western Governors' Association. 1999. *Enlibra: A shared environmental doctrine*. Western Governors' Association, Denver.

———. 2008. Wildlife Corridors Initiative: June 2008 report. Western Governor's Association, Jackson, Wyoming.

Whatcom County Public Works. 2006. Whatcom County fish passage barrier inventory. Final report. Bellingham, Washington.

White, R. G., and B. M. Mefford. 2002. Assessment of behavior and swimming ability of Yellowstone River sturgeon for design of fish passage devices. U.S. Bureau of Reclamation. Denver.

White, T. 2007. *Linking conservation and transportation: Using the state wildlife action plans to protect wildlife from road impacts*. Defenders of Wildlife, Washington, DC.

Widenmaier, K., and L. Fahrig. 2006. Inferring white-tailed deer (*Odocoileus virginianus*) population dynamics from wildlife collisions in the city of Ottawa. Pages 589–602 in C. L. Irwin, P. Garrett, and K. P. McDermott, eds. *Proceedings of the 2005 International Conference on Ecology and Transportation*, Center for Transportation and the Environment, North Carolina State University, Raleigh.

Wilbur Smith Associates. 2006. Vermont long-range transportation plan: Survey update draft summary report. Prepared for the Vermont Agency of Transportation.

Wildlife Collision Prevention Program. 2008. Wildlife collision prevention program Available at http://www.wildlifeaccidents.ca/default.htm.

Williams, R. N., ed. 2006. Return to the river: Restoring salmon to the Columbia River. Elsevier Academic, Boston, Massachussetts.

Winston, M. R., C. M. Taylor, and J. Pigg. 1991. Upstream extirpation of four minnow species due to damming of a prairie stream. *Transactions of the American Fisheries Society* 120:98–105.

Witmer, G. W., and D. S. deCalesta. 1985. Effect of forest roads on habitat use by Roosevelt elk. *Northwest Science* 59:122–125.

Wofford, J. E., R. E. Gresswell, and M. A. Banks. 2005. Influence of barriers to movement on within-watershed genetic variation of coastal cutthroat trout. *Ecological Applications* 15:628–637.

Woltz, H. W., J. Gibbs, and P. K. Ducey. 2008. Road crossing structures for amphibians and reptiles: Informing design through behavioural analysis. *Biological Conservation* 141:2745–2750.

Wood, P., and M. L. Wolfe. 1988. Intercept feeding as a means of reducing deer vehicle collisions. *Wildlife Society Bulletin* 16:376–380.

Wooding, J. B., and R. C. Maddrey. 1994. Impacts of roads on black bears. *Eastern Workshop on Black Bear Research and Management* 12:124–129.

Woods, J. G. 1990. Effectiveness of fences and underpasses on the trans-Canada highway and their impact on ungulate populations. Report to Banff National Park Warden Service, Banff, Alberta.

Wright, K. G., G. J. Robertson, and R. I.Goudie. 1998. Evidence of spring staging and migration route of individual breeding harlequin ducks, *Histrionicus histrionicus*, in southern British Columbia. *Canadian Field-Naturalist* 112:518–519.

Wydeven, A. P., D. J. Mlandenoff, T. A. Sickley, B. E. Kohn, R. P. Thiel, and J. L. Hansen. 2001. Road density as a factor in habitat selection by wolves and other carnivores in the great lakes region. *Endangered Species Update* 18:110–114.

Yamada, K., J. Elith, M. McCarthy, and A. Zerger. 2003. Eliciting and integrating expert knowledge for wildlife habitat modeling. *Ecological Modelling* 165:251–264.

Zacks, J. L. 1986. Do white-tail deer avoid red? An evaluation of the premise underlying the design of Swareflex wildlife reflectors. *Transportation Research Record* 1075:35–43.

Zee, F. F. van der, J. Wiertz, C. J .F. ter Braak, R. C. van Apeldoorn, and J. Vink. 1992. Landscape change as a possible cause of the badger *Meles meles* L. decline in the Netherlands. *Biological Conservation* 61:17–22.

INDEX